Do the Math Workbook

Elementary & Intermediate Algebra
Third Edition

Michael Sullivan, III
Joliet Junior College

Katherine Struve
Columbus State Community College

Janet Mazzarella
Southwestern College

PEARSON

Boston Columbus Indianapolis New York San Francisco Upper Saddle River
Amsterdam Cape Town Dubai London Madrid Milan Munich Paris Montreal Toronto
Delhi Mexico City São Paulo Sydney Hong Kong Seoul Singapore Taipei Tokyo

The author and publisher of this book have used their best efforts in preparing this book. These efforts include the development, research, and testing of the theories and programs to determine their effectiveness. The author and publisher make no warranty of any kind, expressed or implied, with regard to these programs or the documentation contained in this book. The author and publisher shall not be liable in any event for incidental or consequential damages in connection with, or arising out of, the furnishing, performance, or use of these programs.

Reproduced by Pearson from electronic files supplied by the author.

Copyright © 2014, 2010, 2006 Pearson Education, Inc.
Publishing as Pearson, 75 Arlington Street, Boston, MA 02116.

All rights reserved. No part of this publication may be reproduced, stored in a retrieval system, or transmitted, in any form or by any means, electronic, mechanical, photocopying, recording, or otherwise, without the prior written permission of the publisher. Printed in the United States of America.

ISBN-13: 978-0-321-88128-1
ISBN-10: 0-321-88128-1

www.pearsonhighered.com

PEARSON

Do the Math workbook

Elementary & Intermediate Algebra, Third Edition

Table of Contents

Chapter 1	**1**		**Chapter 4**	**133**
Section 1.2	1		Section 4.1	133
Section 1.3	7		Section 4.2	139
Section 1.4	13		Section 4.3	145
Section 1.5	19		Section 4.4	151
Section 1.6	25		Section 4.5	157
Section 1.7	31		Section 4.6	163
Section 1.8	37			

Chapter 2	**43**		**Chapter 5**	**169**
Section 2.1	43		Section 5.1	169
Section 2.2	49		Section 5.2	175
Section 2.3	55		Section 5.3	181
Section 2.4	61		Section 5.4	187
Section 2.5	67		Section 5.5	193
Section 2.6	73		Section 5.6	199
Section 2.7	79			
Section 2.8	85			

Chapter 3	**91**		**Chapter 6**	**205**
Section 3.1	91		Section 6.1	205
Section 3.2	97		Section 6.2	211
Section 3.3	103		Section 6.3	217
Section 3.4	109		Section 6.4	225
Section 3.5	115		Section 6.5	231
Section 3.6	121		Section 6.6	237
Section 3.7	127		Section 6.7	243

Chapter 7	**249**		**Chapter 11**	**441**
Section 7.1	249		Section 11.1	441
Section 7.2	255		Section 11.2	447
Section 7.3	261		Section 11.3	453
Section 7.4	267		Section 11.4	459
Section 7.5	273		Section 11.5	465
Section 7.6	279			
Section 7.7	285			
Section 7.8	291			

Chapter 8	**297**		**Chapter 12**	**471**
Section 8.1	297		Section 12.1	471
Section 8.2	303		Section 12.2	477
Section 8.3	309		Section 12.3	483
Section 8.4	315		Section 12.4	489
Section 8.5	321		Section 12.5	495
Section 8.6	327		Section 12.6	501
Section 8.7	333			
Section 8.8	339			

Chapter 9	**345**		**Chapter 13**	**507**
Section 9.1	345		Section 13.1	507
Section 9.2	351		Section 13.2	513
Section 9.3	357		Section 13.3	519
Section 9.4	363		Section 13.4	525
Section 9.5	369			
Section 9.6	375			
Section 9.7	381		**Appendix A**	531
Section 9.8	387		**Appendix B**	537
Section 9.9	393			

Chapter 10	**399**		**Appendix C**	543
Section 10.1	399		Section C.1	543
Section 10.2	405		Section C.2	549
Section 10.3	411		Section C.3	555
Section 10.4	417		Section C.4	561
Section 10.5	423			
Section 10.6	429			
Section 10.7	435		**Answers**	**AN-1**
			Graphing Answers	**GA-1**

Name: Date:
Instructor: Section:

Five-Minute Warm-Up 1.2
Fractions, Decimals, and Percents

Definitions and vocabulary. In your own words, write a brief definition for each of the following.

1. Natural numbers _____

2. Natural numbers that are *prime* _____

3. Natural numbers that are *composite* _____

4. Least common multiple _____

5. In the statement $4 \cdot 9 = 36$,
   (a) list the factor(s).   **(b)** list the product(s).   5a. _____

   5b. _____

6. List three common multiples of 8 and 12.   6. _____

7. List three common factors of 8 and 12.   7. _____

8. List all possible factors of 18.   8. _____

9. Write the prime factorization of 18.   9. _____

Sullivan/Struve/Mazzarella, *Elementary & Intermediate Algebra*, 3e
Copyright © 2014 Pearson Education, Inc.

Name:
Instructor:

Date:
Section:

Guided Practice 1.2
Fractions, Decimals, and Percents

Objective 1: Factor a Number as a Product of Prime Factors

1. We use a *factor tree* to find the prime factorization of a number. The process begins with finding two factors of the given number. Continue to factor until all factors are prime.

Find the prime factorization of each natural number. (*See textbook Example 1*)
    (a) 30   (b) 120   1a. _____

                                                                                                                     1b. _____

Objective 2: Find the Least Common Multiple of Two or More Numbers

2. Find the least common multiple of the numbers 8 and 12: (*See textbook Examples 2 and 3*)

Step 1: Write each number as the product of prime factors, aligning common factors vertically.	Factor 8 as the product of primes:	(a) _____
	Factor 12 as the product of primes:	(b) _____
Step 2: Write down the factor(s) that the numbers share, if any. Then write down the remaining factors the greatest number of times that the factors appear in any number.	List the common factors:	(c) _____
	List the remaining factors:	(d) _____
Step 3: Multiply the factors listed in Step 2. The product is the least common multiple (LCM).	Multiply. LCM =	(e) _____

Objective 3: Write Equivalent Fractions

3. *Equivalent fractions* are fractions that represent the same part of a whole. Sometimes we simplify fractions, using smaller numbers to represent an equivalent part of a whole. Other times we multiply by one, written as a quantity over itself, to find larger numbers to represent the equivalent fraction.

Write the fraction $\frac{5}{8}$ as an equivalent fraction with a denominator of 24. (*See textbook Example 4*)

    (a) Determine the missing numerator and denominator to multiply by 1: $\frac{5}{8} \cdot \frac{?}{?} = \frac{}{24}$ 3a. _____

    (b) Find the equivalent fraction by determining the product: $\frac{5}{8} \cdot 1 = \frac{5}{8} \cdot \frac{\square}{\square} = \frac{?}{24}$ 3b. _____

Sullivan/Struve/Mazzarella, *Elementary & Intermediate Algebra*, 3e
Copyright © 2014 Pearson Education, Inc.

Guided Practice 1.2

4. Write $\frac{7}{36}$ and $\frac{11}{24}$ as equivalent fractions with the least common denominator. *(See textbook Example 5)*

Step 1: Find the least common denominator of the fractions.

Factor 36 as the product of primes: **(a)** _____

Factor 24 as the product of primes: **(b)** _____

Write the factors in the LCD: **(c)** _____

Multiply to determine the LCD: **(d)** _____

Step 2: Rewrite each fraction with the least common denominator.

To write $\frac{7}{36}$ on the LCD, multiply by? **(e)** _____

To write $\frac{11}{24}$ on the LCD, multiply by? **(f)** _____

Multiply to find the equivalent fractions. **(g)** $\frac{7}{36} \cdot \frac{\square}{\square} = \frac{?}{?} \;\; ; \;\; \frac{11}{24} \cdot \frac{\square}{\square} = \frac{?}{?}$

Objective 5: Round Decimals

5. It is important to identify the place value of digits in a number and to be able to round to a specific place.

Use the number 94,204.6375 to round to the required place. *(See textbook Example 7)*

(a) thousands (b) tens

5a. _____

5b. _____

(c) whole number (or ones) (d) tenth

5c. _____

5d. _____

(e) hundreds (f) hundredth

5e. _____

5f. _____

Objective 6: Convert Between Fractions and Decimals

6. Convert $\frac{9}{25}$ to a decimal. *(See textbook Example 8)* 6. _____

7. Convert 0.8 to a fraction and write in lowest terms. *(See textbook Example 9)* 7. _____

Objective 7: Convert Between Percents and Decimals

8. Write 75% as a decimal. *(See textbook Example 10)* 8. _____

9. Write 0.005 in percent form. *(See textbook Example 11)* 9. _____

Name:
Instructor:

Date:
Section:

Do the Math Exercises 1.2
Fractions, Decimals, and Percents

In Problems 1–2, find the prime factorization of each number.

1. 54

2. 63

1. _____

2. _____

In Problems 3–4, find the LCM of each set of numbers.

3. 8 and 70

4. 9, 15, and 20

3. _____

4. _____

In Problems 5–6, write each fraction with the given denominator.

5. Write $\dfrac{4}{5}$ with denominator 15.

6. Write $\dfrac{5}{14}$ with denominator 28.

5. _____

6. _____

In Problems 7–8, write the equivalent fractions with the least common denominator.

7. $\dfrac{1}{12}$ and $\dfrac{5}{18}$

8. $\dfrac{5}{12}$ and $\dfrac{7}{15}$

7. _____

8. _____

In Problems 9–10, write each fraction in lowest terms.

9. $\dfrac{9}{15}$

10. $\dfrac{24}{27}$

9. _____

10. _____

Do the Math Exercises 1.2

In Problems 11–12, tell the place value of the digit in the given number.

11. 9124.786; the 7

12. 539.016; the 9

11. _____

12. _____

In Problems 13–14, round each number to the given place.

13. 7298.0845 to the nearest hundred

14. 37.439 to the nearest tenth

13. _____

14. _____

In Problems 15–16, write each fraction as a terminating or repeating decimal.

15. $\dfrac{2}{9}$

16. $\dfrac{11}{32}$

15. _____

16. _____

In Problems 17–18, write each decimal as a fraction in lowest terms.

17. 0.4

18. 0.358

17. _____

18. _____

19. Write 59% as a decimal.

20. Write 0.349 as a percent.

19. _____

20. _____

21. Talladega Raceway At Talladega, one of the crew chiefs discovered that in a given time interval, Jeff Gordon completed 18 laps, Jimmie Johnson completed 12 laps, and Tony Stewart completed 15 laps. Supposed all three drivers begin at the same time. How many laps would need to be completed so that all three drivers were at the finish line at exactly the same time?

21. _____

Name:
Instructor:

Date:
Section:

Five-Minute Warm-Up 1.3
The Number Systems and the Real Number Line

In Problems 1 – 4, simplify each of the following expressions.

1. $\dfrac{0}{7}$ 2. $\dfrac{12}{0}$ 3. $\dfrac{9}{9}$ 4. $\dfrac{3}{1}$

1. _____

2. _____

3. _____

4. _____

5. In the fraction $\dfrac{3}{7}$, identify **(a)** the numerator and **(b)** the denominator.

5a. _____

5b. _____

In Problems 6 – 9, use the set of numbers $\left\{\dfrac{15}{0}, -100, \dfrac{3}{6}, 0.001, \dfrac{0}{1}, -1, \dfrac{4}{2}\right\}$. *List all the elements that are...*

6. Nonnegative   7. Positive

8. Positive and less than one   9. Neither positive nor negative

6. _____

7. _____

8. _____

9. _____

In Problems 10 and 11, write the fraction as either a repeating decimal or a terminating decimal.

10. $\dfrac{5}{6}$ 11. $\dfrac{4}{5}$

10. _____

11. _____

Sullivan/Struve/Mazzarella, *Elementary & Intermediate Algebra*, 3e
Copyright © 2014 Pearson Education, Inc.

Name:
Instructor:

Date:
Section:

Guided Practice 1.3
The Number Systems and the Real Number Line

1. A *set* is a collection of objects. We use braces { } to indicate the set and enclose the objects, or *elements*, in the set. When the set has no elements, we say that the set is an *empty set* and use the notation ∅ or { } to indicate an empty set. It is incorrect to use {∅} to denote an empty set.

If set A is the set of digits $\{0, 1, 2, 3, 4, 5, 6, 7, 8, 9\}$, *(See textbook Example 1)*

    (a) write a set B which contains elements from set A which are even.   1a. _____

    (b) write a set C which contains elements from set A which are odd.   1b. _____

Objective 1: Classify Numbers

2. Review the definitions of the sets of numbers listed in textbook Section 1.3. These sets are:

    natural numbers   whole numbers   integers
    rational numbers   irrational numbers   real numbers

List the numbers in the set $\left\{\dfrac{-8}{2}, 2.\overline{6}, \dfrac{0}{3}, 14, -10, \dfrac{7}{0}, -1.050050005...\right\}$ that are... *(See textbook Example 2)*

    (a) natural _____   **(b)** whole _____

    (c) integer _____   **(d)** rational _____

    (e) irrational _____   **(f)** real _____

3. List the sets of numbers to which the given number belongs. *(See textbook Example 2)*

    (a) π _____   **(b)** $\dfrac{12}{-6}$ _____

    (c) $1.25252525...$ _____   **(d)** $\dfrac{-100}{-20}$ _____

Objective 2: Plot Points on a Real Number Line

4. On the real number line, label the points with coordinates: $-\dfrac{5}{2}, 3, 1.75, -0.5$ *(See textbook Example 3)*

Sullivan/Struve/Mazzarella, *Elementary & Intermediate Algebra*, 3e
Copyright © 2014 Pearson Education, Inc.

Guided Practice 1.3

Objective 3: Use Inequalities to Order Real Numbers

5. An important property of the real number line follows from the fact that given two numbers (points) a and b, either a is to the left of b ($a < b$), a is equal to b ($a = b$), or a is to the right of b ($a > b$). It is also possible that a is less than or equal to b ($a \leq b$) or a is greater than or equal to b ($a \geq b$).

Replace the question mark by <, >, or =, whichever is correct. *(See textbook Example 4)*

(a) $-10 \,?\, -10.5$ (b) $\dfrac{3}{4} \,?\, 0.75$ (c) $\dfrac{8}{6} \,?\, \dfrac{18}{15}$

5a. _____

5b. _____

5c. _____

6. Determine whether the statement is True or False.

(a) $0 > -2$ (b) $5 \geq \dfrac{-10}{-2}$ (c) $-5 \leq -5.\overline{1}$

6a. _____

6b. _____

6c. _____

7. If x represents any real number, how would you use inequality symbols to write
(a) x is positive?   (b) x is negative?

7a. _____

7b. _____

Objective 4: Compute the Absolute Value of a Real Number

8. The *absolute value* of a number a, written $|a|$, is the distance from 0 to a on the real number line.

Because distance is always positive, the absolute value of a any non-zero number is always _____.

9. Evaluate each of the following: *(See textbook Example 5)*

(a) $\left|\dfrac{0}{-5}\right|$ (b) $|-4.2|$ $|120|$

9a. _____

9b. _____

9c. _____

Name:
Instructor:

Date:
Section:

Do the Math Exercises 1.3
The Number Systems and the Real Number Line

In Problems 1–3, write each set.

1. *B* is the set of *natural* numbers less than 25.

   1. _____

2. *C* is the set of *integers* between –6 and 4, not including –6 or 4.

   2. _____

3. *F* is the set of odd *natural* numbers less than 1.

   3. _____

In Problems 4–6, list the elements in the set $\left\{-4, 3, -\dfrac{13}{2}, 0, 2.303003000...\right\}$ *that are described.*

4. whole numbers   5. rational numbers   4. _____

   5. _____

6. real numbers   6. _____

7. Plot the points in the set $\left\{\dfrac{3}{4}, \dfrac{0}{2}, -\dfrac{5}{4}, -0.5, 1.5\right\}$ on a real number line.

Do the Math Exercises 1.3

In Problems 8–9, determine whether the statement is True or False.

8. $0 < -5$ **9.** $-3 \geq -5$

8. _____

9. _____

In Problems 10–12, replace the ? with the correct symbol: $>, <, =$.

10. $-8 \; ? \; -8.5$ **11.** $\dfrac{5}{12} \; ? \; \dfrac{2}{3}$

10. _____

11. _____

12. $\dfrac{5}{11} \; ? \; 0.\overline{45}$

12. _____

In Problems 13–15, evaluate each expression.

13. $|-8|$ **14.** $|7|$

13. _____

14. _____

15. $\left| -\dfrac{13}{9} \right|$

15. _____

In Problems 16 – 19, determine whether the statement is True or False.

16. Every decimal is a rational number.

17. 0 is a positive number.

18. Every *integer* is also a *real* number.

19. -1 is a nonpositive *integer*.

16. _____

17. _____

18. _____

19. _____

Name:
Instructor:

Date:
Section:

Five-Minute Warm-Up 1.4
Adding, Subtracting, Multiplying, and Dividing Integers

In Problems 1 – 4, what operation is indicated by each of the following words?

1. Quotient

2. Difference

3. Sum

4. Product

1. _____

2. _____

3. _____

4. _____

5. Write an expression for 9 divided by 18 using each of the following symbols:

   (a) $\dfrac{a}{b}$ (b) a/b (c) $a \div b$ (b) $a\overline{)b}$

5a. _____

5b. _____

5c. _____

5d. _____

6. Evaluate each of the following:

   (a) $|-13|$ (b) $\left|\dfrac{15}{4}\right|$ (c) $|0|$

6a. _____

6b. _____

6c. _____

7. Write $\dfrac{90}{36}$ as a fraction in lowest terms.

7. _____

Sullivan/Struve/Mazzarella, *Elementary & Intermediate Algebra*, 3e
Copyright © 2014 Pearson Education, Inc.

Name:
Instructor:
Date:
Section:

Guided Practice 1.4
Adding, Subtracting, Multiplying, and Dividing Integers

Objective 1: Add Integers

1. To add two integers that are the same sign, add their absolute values.

    (a) If both numbers are positive, the sign of the sum is _____.   1a. _____

    (b) If both numbers are negative, the sign of the sum is _____.   1b. _____

2. To add two integers that are different signs, subtract the absolute value of the smaller number from the absolute value of the larger number. The sign of the sum is _____.

3. Find the sum: $-45 + (-22)$ *(See textbook Example 5)*

Step 1: Add the absolute values of the two integers.	Find the absolute value of $\|-45\|$:	(a) _____
	Find the absolute value of $\|-22\|$:	(b) _____
	Add (a) and (b):	(c) _____
Step 2: Attach the common sign, either positive or negative.	The sign of both numbers is negative so the sign of the sum is:	(d) _____
	Write the sum:	(e) _____

4. Find the sum: $73 + (-81)$ *(See textbook Example 6)*

Step 1: Subtract the smaller absolute value from the larger absolute value.	Find the absolute value of $\|73\|$:	(a) _____
	Find the absolute value of $\|-81\|$:	(b) _____
	Subtract (a) from (b):	(c) _____
Step 2: Attach the sign of the integer with the larger absolute value.	Which number has the larger absolute value:	(d) _____
	What is the sign of (d):	(e) _____
	Therefore, the sign of the sum is:	(f) _____
	Write the sum:	(g) _____

Sullivan/Struve/Mazzarella, *Elementary & Intermediate Algebra*, 3e
Copyright © 2014 Pearson Education, Inc.

Guided Practice 1.4

Objective 2: Determine the Additive Inverse of a Number

5. For any real number a other than 0, there is a real number $-a$, called the *additive inverse*, or *opposite*, of a, having the property: $a + (-a) = -a + a = 0$.

Determine the additive inverse of the given real number. *(See textbook Example 7)*

(a) $\dfrac{9}{2}$ (b) -10.3

5a. _____

5b. _____

Objective 3: Subtract Integers

6. Find the difference: $-32 - (-61)$ *(See textbook Example 8)*

Step 1: Change the subtraction problem to an equivalent addition problem.	Write the equivalent addition problem:	(a) _____
Step 2: Find the sum.	Use the rules from Objective 1 to find the sum from (a):	(b) _____

Objective 4: Multiply Integers

7. To multiply two integers, multiply the numbers without regard to the sign.

    (a) If the two factors have the same sign, the sign of the product is _____.   7a. _____

    (b) If the two factors have different signs, the sign of the product is _____.   7b. _____

    (c) _____ is neither positive nor negative.   7c. _____

8. Find the product. *(See textbook Example 11)*

    (a) $6(-12)$ (b) $(-4)(-15)$ (c) $-13 \cdot 9$

8a. _____

8b. _____

8c. _____

Objective 5: Divide Integers

In Problems 9 – 10, identify the (a) dividend, (b) divisor, and (c) the quotient.

9. $\dfrac{36}{12}$ **10.** $(-72) \div 72$

9a. _____ 9b. _____ 9c. _____

10a. _____ 10b. _____ 10c. _____

11. Find the multiplicative inverse (reciprocal) of -3. *(See textbook Example 13)*   11. _____

Name:
Instructor:

Date:
Section:

Do the Math Exercises 1.4
Adding, Subtracting, Multiplying, and Dividing Integers

In Problems 1–4, find the sum.

1. $-4 + 12$

2. $-13 + (-5)$

3. $-145 + (-68)$

4. $(-13) + 37 + (-22)$

1. _____

2. _____

3. _____

4. _____

In Problems 5–6, determine the additive inverse of each real number.

5. -34

6. 7

5. _____

6. _____

In Problems 7–9, find the difference.

7. $12 - 19$

8. $-15 - 9$

9. $46 - (-25)$

7. _____

8. _____

9. _____

In Problems 10–13, find the product.

10. $7 \cdot 9$

11. $(-22)(-5)$

10. _____

11. _____

Sullivan/Struve/Mazzarella, *Elementary & Intermediate Algebra*, 3e
Copyright © 2014 Pearson Education, Inc.

Do the Math Exercises 1.4

12. $(-128)7$ **13.** $-6 \cdot 4 \cdot 8$

12. _____

13. _____

In Problems 14–15, find the multiplicative inverse (or reciprocal) of each number.

14. 10 **15.** -3

14. _____

15. _____

In Problems 16–18, find the quotient.

16. $36 \div 9$ **17.** $\dfrac{-144}{6}$

16. _____

17. _____

18. $\dfrac{-80}{-12}$

18. _____

In Problems 19–20, write each expression using mathematical symbols. Then evaluate the expression.

19. the sum of 32 and -64 **20.** -40 divided by 100

19. _____

20. _____

21. Find two integers whose sum is -8 and whose product is 15.

21. _____

22. Find two integers whose sum is 2 and whose product is -24.

22. _____

Name:
Instructor:

Date:
Section:

Five-Minute Warm-Up 1.5
Adding, Subtracting, Multiplying, and Dividing Rational Numbers

In Problems 1 – 4, perform the indicated operations.

1. $(-15)(-6)$

2. $\dfrac{125}{-5}$

3. $-36 - 42$

4. $27 + (-163)$

1. _____

2. _____

3. _____

4. _____

5. Write $\dfrac{24}{96}$ as a fraction in lowest terms.

5. _____

6. Find the least common denominator of $\dfrac{4}{15}$ and $\dfrac{5}{24}$.

6. _____

7. Rewrite $\dfrac{5}{12}$ as an equivalent fraction with a denominator of 60.

7. _____

Sullivan/Struve/Mazzarella, *Elementary & Intermediate Algebra*, 3e
Copyright © 2014 Pearson Education, Inc.

Name:
Instructor:

Date:
Section:

Guided Practice 1.5
Adding, Subtracting, Multiplying, and Dividing Rational Numbers

1. Write each rational number in lowest terms.

   (a) $\dfrac{-12}{15}$ (b) $-\dfrac{24}{3}$ (c) $\dfrac{-5}{-25}$

   1a. _____

   1b. _____

   1c. _____

Objective 1: Multiply Rational Numbers in Fractional Form

2. Find the product: $-\dfrac{7}{72} \cdot \dfrac{81}{35}$ (See textbook Example 1)

   2. _____

Objective 2: Divide Rational Numbers in Fractional Form

3. Find the quotient: $\dfrac{8}{15} \div \dfrac{4}{45}$ (See textbook Example 2)

Step 1: Write the equivalent multiplication problem.

$\dfrac{8}{15} \div \dfrac{4}{45} =$

(a) _____

Step 2: Write the product in factored form and divide out common factors.

Write the prime factorization of the numerator and denominator:

(b) _____

Divide out common factors: (c) _____

Step 3: Multiply the remaining factors.

$\dfrac{8}{15} \div \dfrac{4}{45} =$ (d) _____

Objective 3: Add and Subtract Rational Numbers in Fractional Form

4. Adding or subtracting fractions with the same denominator use the following properties:

$\dfrac{a}{c} + \dfrac{b}{c} = \dfrac{a+b}{c}, c \neq 0 \qquad \dfrac{a}{c} - \dfrac{b}{c} = \dfrac{a-b}{c}, c \neq 0$

Find the difference and write in lowest terms, if necessary: $\dfrac{7}{12} - \dfrac{11}{12}$ (See textbook Example 4)

   (a) Use the property $\dfrac{a}{c} - \dfrac{b}{c} = \dfrac{a-b}{c}$ to write the difference over a single denominator: 4a. _____

   (b) Rewrite as an addition problem: 4b. _____

   (c) Add the numerators: 4c. _____

   (d) Factor and divide out common factors. State the difference: $\dfrac{7}{12} - \dfrac{11}{12} =$ 4d. _____

Sullivan/Struve/Mazzarella, *Elementary & Intermediate Algebra*, 3e
Copyright © 2014 Pearson Education, Inc.

Guided Practice 1.5

5. Find the sum: $\dfrac{5}{12} + \dfrac{4}{9}$ *(See textbook Example 5)*

Step 1: Find the least common denominator of the denominators.

Write 12 as the product of primes: **(a)** _____

Write 9 as the product of primes: **(b)** _____

Write the factors in the LCD: **(c)** _____

Multiply to determine the LCD: **(d)** _____

Step 2: Write each rational number with the denominator found in Step 1.

To write $\dfrac{5}{12}$ on the LCD, multiply by? **(e)** _____

To write $\dfrac{4}{9}$ on the LCD, multiply by? **(f)** _____

Multiply to find the equivalent fractions.

(g) $\dfrac{5}{12} \cdot \dfrac{\Box}{\Box} = \dfrac{?}{?}$; $\dfrac{4}{9} \cdot \dfrac{\Box}{\Box} = \dfrac{?}{?}$

Step 3: Add the numerators and write the result over the common denominator.

Add the fractions: **(h)** _____

Step 4: Write in lowest terms.

The result is already in lowest terms. State the sum $\dfrac{5}{12} + \dfrac{4}{9}$: **(i)** _____

6. Evaluate and write in lowest terms, if necessary: $-3 + \dfrac{5}{8}$ *(See textbook Example 7)*

(a) Write the integer -3 as a fraction: 6a. _____

(b) Write each fraction as an equivalent fraction over the common denominator: 6b. _____

(c) Add the numerators and write the result over the common denominator: 6c. _____

(d) Write the fractions in lowest terms, if necessary. State the sum of $-3 + \dfrac{5}{8}$:   6d. _____

Objective 4: Add, Subtract, Multiply, and Divide Rational Numbers in Decimal Form

7. To add or subtract decimals, we arrange the numbers in a column with the decimal points aligned. Place the decimal point in the answer directly below the decimal points in the problem. When adding or subtracting decimals, any number that has no decimal point has an implied decimal point located _____.

8. The number of digits to the right of the decimal point in the product is the same as the _____ of the number of digits to the right of the decimal point in the factors.

Name:
Instructor:

Date:
Section:

Do the Math Exercises 1.5
Adding, Subtracting, Multiplying, and Dividing Rational Numbers

In Problems 1–4, find the product, and write in lowest terms.

1. $\dfrac{7}{8} \cdot \dfrac{10}{21}$

2. $-\dfrac{3}{7} \cdot 63$

1. _____

2. _____

3. $-\dfrac{5}{2} \cdot \dfrac{16}{25}$

4. $-\dfrac{60}{75} \cdot \left(-\dfrac{25}{4}\right)$

3. _____

4. _____

In Problems 5–6, find the reciprocal of each number.

5. $\dfrac{9}{4}$

6. -8

5. _____

6. _____

In Problems 7–10, find the quotient, and write in lowest terms.

7. $\dfrac{1}{2} \div \dfrac{3}{6}$

8. $-\dfrac{1}{4} \div 4$

7. _____

8. _____

9. $\dfrac{4}{3} \div \left(-\dfrac{9}{10}\right)$

10. $\dfrac{44}{63} \div \dfrac{88}{21}$

9. _____

10. _____

Sullivan/Struve/Mazzarella, *Elementary & Intermediate Algebra*, 3e
Copyright © 2014 Pearson Education, Inc.

Do the Math Exercises 1.5

In Problems 11–14, find the sum or difference and write in lowest terms.

11. $\dfrac{6}{11} + \dfrac{16}{11}$

12. $-\dfrac{7}{8} + 4$

11. _____

12. _____

13. $-\dfrac{2}{5} + \left(-\dfrac{2}{3}\right)$

14. $-\dfrac{29}{6} - \left(-\dfrac{29}{20}\right)$

13. _____

14. _____

In Problems 15–20, perform the indicated operation.

15. $-(-32.9) + 10.3$

16. $32 - 5.68$

15. _____

16. _____

17. $3.1 \cdot (10.9)$

18. $0.065 \cdot 340$

17. _____

18. _____

19. $\dfrac{332.59}{7.9}$

20. $\dfrac{-48}{0.03}$

19. _____

20. _____

21. **Halloween Candy** Henry decided to make $\dfrac{2}{3}$-oz bags of candy for treats at Halloween. If he bought 16 oz of candy, how many bags will he have to give away?

21. _____

Name:
Instructor:

Date:
Section:

Five-Minute Warm-Up 1.6
Properties of Real Numbers

In Problems 1 – 4, perform the indicated operations.

1. $-5 + 5 + (-2)$

    1. _____

2. $\dfrac{5}{9} \cdot \dfrac{9}{5} \cdot (-12)$

    2. _____

3. $\dfrac{8}{1} \cdot \dfrac{1}{12}$

    3. _____

4. $-33 \cdot \left(-\dfrac{15}{11}\right)$

    4. _____

5. $\dfrac{-75}{-15}$

    5. _____

Sullivan/Struve/Mazzarella, *Elementary & Intermediate Algebra*, 3e
Copyright © 2014 Pearson Education, Inc.

Name: Date:
Instructor: Section:

Guided Practice 1.6
Properties of Real Numbers

Objective 1: Use the Identity Properties of Addition and Multiplication

1. Based on the stated property, write the right side of the equation.

   (a) $\dfrac{4}{7} + 0 =$ _____ ; Identity Property of Addition   1a. _____

   (b) $\dfrac{9}{9} \cdot \left(-\dfrac{1}{4}\right) =$ _____ ; Multiplicative Identity   1b. _____

2. Perform the following conversions. *(See textbook Example 1)*

   (a) 198 inches = ? feet   [1 foot = 12 inches]   2b. _____

   (b) 15 minutes = ? seconds   [1 minute = 60 seconds]   2b. _____

Objective 2: Use the Commutative Properties of Addition and Multiplication

3. The commutative properties of addition and multiplication say that the _____ in which a   3. _____
problem is written will not change the result.

4. The following operations are *not* commutative: _____ and _____

5. Based on the stated property, write the right side of the equation. *(See textbook Example 2)*

   (a) $(-12) + 6 =$ _____ ; Commutative Property of Addition   5a. _____

   (b) $(4 + 20) \cdot \dfrac{1}{2} =$ _____ ; Commutative Property of Multiplication   5b. _____

6. Evaluate the following expressions. *(See textbook Examples 3 and 4)*

   (a) $(-17) + 4 + 17$ 6a. _____

   (b) $42 \cdot 5 \cdot \left(\dfrac{3}{7}\right)$ 6b. _____

Objective 3: Use the Associative Properties of Addition and Multiplication

7. The associative properties of addition and multiplication say that if a problem contains   7. _____
only addition or contains only multiplication, we can keep the order the same but change the
_____ of the numbers in the problem and the result will not change.

Sullivan/Struve/Mazzarella, *Elementary & Intermediate Algebra, 3e*
Copyright © 2014 Pearson Education, Inc.

Guided Practice 1.6

8. Based on the stated property, write the right side of the equation. *(See textbook Example 5)*

   (a) $(-9 + 12) + 2 =$ _____ ; Associative Property of Addition   8a. _____

   (b) $3 \cdot \left(\dfrac{2}{3} \cdot 15\right) =$ _____ ; Associative Property of Multiplication   8b. _____

9. Use an Associative Property to evaluate the following expressions. *(See textbook Examples 6 and 7)*

   (a) $(-7) + 17 + 10$ 9a. _____

   (b) $12 \cdot \left(\dfrac{5}{6} \cdot 5\right)$ 9b. _____

Objective 4: Understand the Multiplication and Division Properties of 0

10. Complete the following. *(See textbook Example 8)*

    (a) $a \cdot 0 = 0 \cdot a =$ _____   (b) $\dfrac{0}{a} =$ _____ ; $a \neq 0$ (c) $\dfrac{a}{0} =$ _____ ; $a \neq 0$

Name:
Instructor:

Date:
Section:

Do the Math Exercises 1.6
Properties of Real Numbers

In Problems 1–4, convert each measurement to the indicated unit of measurement. Use the following conversions:

1 gallon = 4 quarts
16 ounces = 1 pound

3 feet = 1 yard
100 centimeters = 1 meter

1. 130 feet to yards

2. 5900 centimeters to meters

1. _____

2. _____

3. 58 quarts to gallons

4. 120 ounces to pounds

3. _____

4. _____

In Problems 5–8, state the property of real numbers that is being illustrated.

5. $4 \cdot 63 \cdot \frac{1}{4} = 4 \cdot \frac{1}{4} \cdot 63$

6. $(4 \cdot 5) \cdot 7 = 4 \cdot (5 \cdot 7)$

5. _____

6. _____

7. $\frac{-8}{0}$

8. $\frac{5}{12} \cdot \frac{12}{5} = 1$

7. _____

8. _____

In Problems 9–17, evaluate each expression by using the properties of real numbers.

9. $46 + 59 + (-46)$

10. $\frac{4}{9} \cdot \frac{9}{4} \cdot 28$

9. _____

10. _____

Do the Math Exercises 1.6

11. $36 \cdot (-12) \cdot \dfrac{1}{6}$

12. $\dfrac{13}{2} \cdot \dfrac{8}{39} \cdot \dfrac{39}{4}$

11. _____

12. _____

13. $\dfrac{0}{100}$

14. $4000(0.5)(0.001)$

13. _____

14. _____

15. $104 \cdot \dfrac{1}{104}$

16. $30 \cdot \dfrac{4}{4}$

15. _____

16. _____

17. $\dfrac{7}{48} \cdot \left(-\dfrac{21}{4}\right) \cdot \dfrac{12}{7}$

17. _____

18. Insert parentheses to make the statement true: $-6 - 4 + 10 = -20$

18. _____

19. Convert 60 miles per hour to feet per second. *Hint: 1 mile = 5280 feet*

19. _____

Name:
Instructor:

Date:
Section:

Five-Minute Warm-Up 1.7
Exponents and the Order of Operations

1. Find the sum: $-56 + (-25)$

    1. _____

2. Find the difference: $-17 - 31$

    2. _____

3. Find the product: $-36 \cdot \left(\dfrac{8}{9}\right)$

    3. _____

4. Find the quotient: $\dfrac{135}{-3}$

    4. _____

5. Evaluate the expression: $35 - 43 + (-18) + 96$

    5. _____

6. Find the product: $(-3) \cdot (-3) \cdot (-3) \cdot (-3)$

    6. _____

Name:
Instructor:

Date:
Section:

Guided Practice 1.7
Exponents and the Order of Operations

Objective 1: Evaluate Exponential Expressions *(See textbook Examples 1 – 4)*

1. Integer exponents provide a shorthand device for representing repeated multiplications of a real number. In the expression 2^5,

   (a) 2 is called the _____   1a. _____

   (b) 5 is called the _____   1b. _____

   (c) To evaluate the expression, multiply _____ = _____ .   1c. _____

2. Identify the base for each expression.

   (a) -9^2 (b) $(-9)^4$ 2a. _____

   2b. _____

3. Consider the number -6.

   (a) Another word for raising the number -6 to the second power is to say -6 ___.   3a. _____

   (b) This is written _____ .   3b. _____

   (c) To evaluate the expression, multiply _____ = _____ .   3c. _____

4. Consider the number -5.

   (a) Another word for raising the number -5 to the third power is to say -5 ____.   4a. _____

   (b) This is written _____ .   4b. _____

   (c) To evaluate the expression, multiply _____ = _____ .   4c. _____

Objective 2: Apply the Rules for Order of Operations

5. Complete the following table.

		Order of Operations	
(a)	P	_____	This includes all possible grouping symbols.
(b)	E	_____	Work from left to right.
(c)	M/D	_____	Work from left to right.
(d)	A/S	_____	Work from left to right.

Sullivan/Struve/Mazzarella, *Elementary & Intermediate Algebra*, 3e
Copyright © 2014 Pearson Education, Inc.

Guided Practice 1.7

6. We use a variety of grouping symbols. When multiple pairs of grouping symbols exist and are nested inside one another, we evaluate the information in the innermost grouping symbol first and work our way outward.

Besides nested parentheses, what grouping symbols might appear in an expression with multiple operations?

6. _____

7. Evaluate the expression: $5 - 24 \div 6 \cdot 9$ *(See textbook Example 5)*

7. _____

8. Evaluate the expression: $\dfrac{4 \cdot 2^3 + (-16)}{3(4-6)^2}$ *(See textbook Example 9)*

Step 1: Evaluate the expression in parentheses first.

$\dfrac{4 \cdot 2^3 + (-16)}{3(4-6)^2}$ = (a) _____

Step 2: Evaluate the exponents.

= (b) _____

Step 3: Find products.

= (c) _____

Step 4: Add terms in numerator.

= (d) _____

Step 5: Write in lowest terms.

= (e) _____

Sullivan/Struve/Mazzarella, *Elementary & Intermediate Algebra*, 3e
Copyright © 2014 Pearson Education, Inc.

Name:
Instructor:

Date:
Section:

Do the Math Exercises 1.7
Exponents and the Order of Operations

In Problems 1–2, write in exponential form.

1. $4 \cdot 4 \cdot 4 \cdot 4 \cdot 4$

2. $(-8)(-8)(-8)$

1. _____

2. _____

In Problems 3–8, evaluate each exponential expression.

3. 2^5

4. $\left(\dfrac{5}{2}\right)^4$

3. _____

4. _____

5. $(0.04)^2$

6. -5^4

5. _____

6. _____

7. 1^6

8. $\left(-\dfrac{3}{2}\right)^5$

7. _____

8. _____

In Problems 9–19, evaluate each expression.

9. $12 + 8 \cdot 3$

10. $50 \div 5 \cdot 4$

9. _____

10. _____

Sullivan/Struve/Mazzarella, *Elementary & Intermediate Algebra*, 3e
Copyright © 2014 Pearson Education, Inc.

35

Do the Math Exercises 1.7

11. $(7-5) \cdot \dfrac{5}{2}$

12. $\dfrac{5+3}{3+15}$

11. _____

12. _____

13. $12 - [7 + (-6)3]$

14. $(-11.8 - 15.2) \div (-2)$

13. _____

14. _____

15. $-10 - 4^2$

16. $-5^2 + 3^2 \div (3^2 + 9)$

15. _____

16. _____

17. $\left(\dfrac{7-5^2}{8+4\cdot 2}\right)^2$

18. $3 \cdot [6 \cdot (5-2) - 2 \cdot 5]$

17. _____

18. _____

19. $\left(\dfrac{3}{4} + \dfrac{1}{2}\right)\left(\dfrac{2}{3} - \dfrac{1}{2}\right)$

19. _____

20. Express 675 as the product of prime factors. Write the answer in exponential form.

20. _____

21. Insert grouping symbols so that the expression has the desired value: $4 \cdot 7 - 4^2 = -36$

21. _____

Name:
Instructor:

Date:
Section:

Five-Minute Warm-Up 1.8
Simplifying Algebraic Expressions

1. Find the sum: $-14 + 37$

    1. _____

2. Find the difference: $-26 - (-16)$

    2. _____

3. Find the product: $72 \cdot \left(-\dfrac{5}{8}\right)$

    3. _____

4. Find the quotient: $\dfrac{-105}{-15}$

    4. _____

5. Evaluate each of the following:
   (a) $(-4)^2$ (b) -4^2

    5a. _____

    5b. _____

6. Simplify: $(-13 + (-2))^2$

    6. _____

Name: Date:
Instructor: Section:

Guided Practice 1.8
Simplifying Algebraic Expressions

Objective 1: Evaluate Algebraic Expressions

1. A letter that represents any number from a set of numbers is called a(n) _____. 1._____

2. A fixed number, such as 3, or a letter or symbol that represents a fixed number is a(n) _____. 2._____

3. Any combination of variables, constants, grouping symbols, and mathematical operations is called a(n) _____ _____. 3._____

4. When we substitute a numerical value for each variable in an expression and then simplify the result, we are _____ an algebraic expression. 4._____

5. Evaluate the algebraic expression $2x^2 - 5x + 3$ for $x = -2$. *(See textbook Example 1)* 5._____

Objective 2: Identify Like Terms and Unlike Terms

6. A constant or the product of a constant and one or more variables raised to a power is called a _____. 6._____

7. The numerical factor of a term is called a(n) _____. 7._____

8. When terms have the same variable(s) and the same exponent(s) on the variables, we say that the expressions are _____ _____. 8._____

9. Identify the terms in the algebraic expression $x^2 - 2xy - y^2$. *(See textbook Example 3)* 9._____

10. Determine the coefficient of each term. *(See textbook Example 4)*

    (a) $\dfrac{x}{2}$ (b) $-y$ (c) 14   (d) $-\dfrac{2z}{5}$ 10a._____

    10b._____

    10c._____

    10d._____

Sullivan/Struve/Mazzarella, *Elementary & Intermediate Algebra*, 3e
Copyright © 2014 Pearson Education, Inc.

Guided Practice 1.8

11. Determine if the following pairs of terms are *like* or *unlike*. *(See textbook Example 5)*

    (a) $4xy^2$ and $2x^2y$ (b) $\frac{3}{4}n^3$ and $5n^3$ 11a. _____

    11b. _____

Objective 3: Use the Distributive Property

12. Use the Distributive Property to remove the parentheses. *(See textbook Example 6)*

    (a) $2(3x-1)$ (b) $-\frac{1}{5}(10y+5)$ 12a. _____

    12b. _____

Objective 4: Simplify Algebraic Expressions by Combining Like Terms

13. Combine like terms. *(See textbook Examples 7 and 8)*

    (a) $x+3x$ (b) $4x^2 - x + 2 - 4x^2 + 7x + 1$ 13a. _____

    13b. _____

14. Simplify the algebraic expression: $5 + 2(6x+1) - (3x-5)$ *(See textbook Example 9)*   14. _____

Name:
Instructor:

Date:
Section:

Do the Math Exercises 1.8
Simplifying Algebraic Expressions

In Problems 1–2, evaluate each expression using the given values of the variables.

1. $n^2 - 4n + 3$ for $n = 2$

2. $-2p^2 + 5p + 1$ for $p = -3$

1. _____

2. _____

In Problems 3–4, for each expression, identify the terms and then name the coefficient of each term.

3. $3m^4 - m^3n^2 + 4n - 1$

4. $t^3 - \dfrac{t}{4}$

3. _____

4. _____

In Problems 5–6, determine if the terms are like or unlike.

5. -13 and 38

6. x^2y^3 and y^2x^3

5. _____

6. _____

In Problems 7–8, use the Distributive Property to remove the parentheses.

7. $3(4s + 2)$

8. $-5(k - n)$

7. _____

8. _____

Sullivan/Struve/Mazzarella, *Elementary & Intermediate Algebra*, 3e
Copyright © 2014 Pearson Education, Inc.

Do the Math Exercises 1.8

In Problems 9–14, simplify each expression by using the Distributive Property to remove parentheses and combining like terms.

9. $x + 2y + 5x + 7y$

10. $-7p^5 + 2p^5$

9. _____

10. _____

11. $-(-6m + 9n - 8p)$

12. $18m - (6 + 9m)$

11. _____

12. _____

13. $\dfrac{3}{5}y + \dfrac{7}{10}y$

14. $\dfrac{1}{5}(60 - 15x) + \dfrac{3}{4}(12 - 24x)$

13. _____

14. _____

15. **Ticket Sales** A community college theatre group sold tickets to a recent production. Student tickets cost $5 and nonstudent tickets cost $8. The algebraic expression $5s + 8n$ represents the total revenue from selling s student tickets and n nonstudent tickets. Evaluate $5s + 8n$ for $s = 76$ and $n = 63$.

15. _____

Name:
Instructor:

Date:
Section:

Five-Minute Warm-Up 2.1
Linear Equations: The Addition and Multiplication Properties of Equality

1. Determine the additive inverse of $-\dfrac{2}{3}$.

    1. _____

2. Determine the multiplicative inverse of 14.

    2. _____

3. Evaluate: $-\dfrac{5}{2}\left(-\dfrac{2}{5}\right)$

    3. _____

4. Use the Distributive Property to simplify: $-3(6x + 5)$

    4. _____

5. Simplify: $9 - \dfrac{1}{2}(12x - 4)$

    5. _____

6. Evaluate $\dfrac{7}{4}x + 2$ for $x = -16$.

    6. _____

Name: Date:
Instructor: Section:

Guided Practice 2.1
Linear Equations: The Addition and Multiplication Properties of Equality

Objective 1: Determine Whether a Number Is a Solution of an Equation

1. How do you determine if a value of a variable is a solution to an equation?

2. Determine if the given value of the variable is a solution of the equation $-3x + 5 = -10$.
(See textbook Example 1)

(a) $x = 5$ (b) $x = -5$

2a. _____

2b. _____

Objective 2: Use the Addition Property of Equality to Solve linear Equations

3. In your own words, state the Addition Property of Equality.

4. Will the Addition Property of Equality allow you to subtract the same number from both sides of an equation to form an equivalent equation? Explain your reasoning.

5. The goal in solving a linear equation is to get the variable by itself with a coefficient of 1. We call this process _____ _____ _____.

6. Solve the linear equation: $x + 13 = -11$ *(See textbook Example 2)*

Step 1: Isolate the variable *x* on the left side of the equation.		$x + 13 = -11$
	Subtract 13 from both sides:	(a) $x + 13 - \_\_\_\_ = -11 - \_\_\_\_$
Step 2: Simplify the left and right sides of the equation.	Apply the Additive Inverse Property, $a + (-a) = 0$:	(b) _____
	Apply the Additive Identity Property, $a + 0 = a$	(c) _____
Step 3: Check Verify your value for *x* in the original equation.	Replace *x* in the original equation to see if a true statement results.	$x + 13 = -11$
	State the solution set:	(d) _____

Guided Practice 2.1

Objective 3: Use the Multiplication Property of Equality to Solve Linear Equations

7. In your own words, state the Multiplication Property of Equality. Do you believe this property will hold true for division as well? Why or why not?

8. Solve the linear equation: $-9x = 324$ *(See textbook Example 5)*

Step 1: Get the coefficient of the variable *x* to be 1.		$-9x = 324$
	Multiply both sides by the reciprocal of the coefficient, $-\frac{1}{9}$:	(a) _____ $\cdot (-9x) =$ _____ $\cdot (324)$
Step 2: Simplify the left and right sides of the equation.	Regroup factors using the Associative Property of Multiplication:	(b) _____
	Apply the Multiplicative Inverse Property, $a \cdot \frac{1}{a} = 1$:	(c) _____
	Apply the Multiplicative Identity, $1 \cdot a = a$:	(d) _____
Step 3: Check Verify your value for *x* in the original equation.	Replace *x* in the original equation to see if a true statement results.	$-9x = 324$
	State the solution set:	(e) _____

9. State the first step necessary to isolate the variable. *(See textbook Examples 6 – 9)*

(a) $3p = -45$ 9a. _____

(b) $\dfrac{x}{15} = -15$ 9b. _____

(c) $-18 = -\dfrac{2}{3}y$ 9c. _____

(d) $\dfrac{8}{3} = \dfrac{4}{9}n$ 9d. _____

Name:
Instructor:

Date:
Section:

Do the Math Exercises 2.1
Linear Equations: The Addition and Multiplication Properties of Equality

In Problems 1–4, determine if the given value is a solution to the equation. Answer Yes or No.

1. $4t + 2 = 16;\ t = 3$

2. $3(x + 1) - x = 5x - 9;\ x = -3$

1. _____

2. _____

3. $-15 = 3x - 16;\ x = \dfrac{1}{3}$

4. $3s - 6 = 6s - 3.4;\ s = -1.2$

3. _____

4. _____

In Problems 5–8, solve the equation using the Addition Property of Equality. Be sure to check your solution.

5. $13 = u - 6$

6. $-2 = y + 13$

5. _____

6. _____

7. $x - \dfrac{1}{8} = \dfrac{3}{8}$

8. $\dfrac{3}{8} = y - \dfrac{1}{6}$

7. _____

8. _____

Do the Math Exercises 2.1

In Problems 9–14, solve the equation using the Multiplication Property of Equality. Be sure to check your solution.

9. $4z = 30$

10. $-8p = 20$

9. _____

10. _____

11. $\dfrac{4}{3}b = 16$

12. $-\dfrac{6}{5}n = -36$

11. _____

12. _____

13. $\dfrac{1}{4}w = \dfrac{7}{2}$

14. $\dfrac{3}{10}q = -\dfrac{1}{6}$

13. _____

14. _____

15. **New Kayak** The total cost for a new kayak is $862.92, including sales tax of $63.92. To find the cost of the kayak without tax, solve the equation $k + 63.92 = 862.92$, where k represents the cost of the kayak.

15. _____

16. Explain the difference between an algebraic expression and an algebraic equation. Write an equation involving the algebraic expression $x + 10$. Then solve the equation.

Name:
Instructor:

Date:
Section:

Five-Minute Warm-Up 2.2
Linear Equations: Using the Properties Together

1. Simplify by combining like terms: $3x - 2(5x + 4) - 4x$

   1. _____

2. Evaluate the expression $-7(4x - 3) - 12$ for $x = -3$.

   2. _____

3. Simplify: $\dfrac{4}{3} \cdot \left(\dfrac{3}{4} x \right)$

   3. _____

4. Simplify by combining like terms: $8x + 3 - 12 - 6x$

   4. _____

5. Simplify: $\dfrac{-12}{-28}$

   5. _____

6. Divide: $\dfrac{106}{0.8}$

   6. _____

Sullivan/Struve/Mazzarella, *Elementary & Intermediate Algebra*, 3e
Copyright © 2014 Pearson Education, Inc.

Name: Date:
Instructor: Section:

Guided Practice 2.2
Linear Equations: Using the Properties Together

Objective 1: Apply the Addition and the Multiplication Properties of Equality to Solve Linear Equations

1. Solve the equation: $6x - 9 = -27$ *(See textbook Example 1)*

Step 1: Isolate the term containing the variable.

Apply the Addition Property of Equality and add ___ to both sides of the equation:

$6x - 9 = -27$

(a) _____

(b) $6x - 9 + \_\_\_ = -27 + \_\_\_$

Simplify:

(c) _____

Step 2: Get the coefficient of the variable to be 1.

Apply the Multiplication Property of Equality and divide both sides of the equation by ___ (this is the same as multiplying both sides by ___):

(d) _____ ; _____

(e) $\dfrac{6x}{\boxed{?}} = \dfrac{-18}{\boxed{?}}$

Simplify:

(f) _____

Step 3: Check Verify your value for x in the original equation.

Replace x in the original equation to see if a true statement results.

$6x - 9 = -27$

State the solution set:

(g) _____

2. Solve the equation: $\dfrac{5}{4}n + 3 = -12$ *(See textbook Example 2)*

$\dfrac{5}{4}n + 3 = -12$

(a) Subtract 3 from both sides of the equation:

(a) _____

(b) Simplify:

(b) _____

(c) Multiply both sides of the equation by $\dfrac{4}{5}$:

(c) _____

(d) Simplify:

(d) _____

(e) State the solution set:

(e) _____

Objective 2: Combine Like Terms and Use the Distributive Property to Solve Linear Equations

3. Before using the Addition or Multiplication Property of Equality to solve an equation, we must begin by simplifying each side of the equation. If a side of the equation has two or more like terms, we begin by

Guided Practice 2.2

4. When an equation contains parentheses, we use the _____ Property to remove the parentheses before we use the Addition or Multiplication Property of Equality to solve the equation.

5. In the following equations, combine like terms and/or use the Distributive Property to remove the parentheses. What equation is left to solve? *DO NOT SOLVE THE EQUATION.*
(See textbook Examples 3 and 4)

(a) $5x + 3 - 7x + 2 = -4$

(b) $4 + 8(3n + 4) = -2(n - 1) - 3$

5a. _____

5b. _____

Objective 3: Solve a Linear Equation with the Variable on Both Sides of the Equation

6. Our goal is to get the terms that contain the variable on one side of the equation and the constant terms on the other side. In the equation $6 - 5x = 4 + 3x$, what steps would accomplish this goal? **Hint:** There is more than one option. *(See textbook Example 5)*

7. Solve the equation: $3(z + 5) - 16z = 2 - (z + 3)$ *(See textbook Example 6)*

$3(z + 5) - 16z = 2 - (z + 3)$

Step 1: Remove any parentheses using the Distributive Property.

Use the Distributive Property: (a) _____

Step 2: Combine like terms on each side of the equation.

Combine like terms: (b) _____

Step 3: Use the Addition Property of Equality to get the terms with the variable on one side of the equation and the constants on the other side.

Add z to both sides of the equation: (c) _____

Simplify: (d) _____

Subtract 15 from both sides: (e) _____

Simplify: (f) _____

Step 4: Use the Multiplication Property of Equality to get the coefficient of the variable to be 1.

Divide both sides by -12: (g) _____

Simplify: (h) _____

Step 5: Check the solution to verify that is satisfies the original equation.

We leave the check to you.

State the solution set: (i) _____

Name:
Instructor:

Date:
Section:

Do the Math Exercises 2.2
Linear Equations: Using the Properties Together

In Problems 1–4, solve the equation. Check your solution.

1. $-4x + 3 = 15$

2. $6z - 7 = 3$

1. _____

2. _____

3. $\dfrac{5}{4}a + 3 = 13$

4. $\dfrac{1}{5}p - 3 = 2$

3. _____

4. _____

In Problems 5–8, solve the equation. Check your solution.

5. $5r + 2 - 3r = -14$

6. $2b + 5 - 8b = 23$

5. _____

6. _____

7. $3(t - 4) = -18$

8. $-5(6 + z) = -20$

7. _____

8. _____

In Problems 9–12, solve the equation. Check your solution.

9. $3 + 8x = 21 - x$

10. $6 - 12m = -3m + 3$

9. _____

10. _____

Sullivan/Struve/Mazzarella, *Elementary & Intermediate Algebra*, 3e
Copyright © 2014 Pearson Education, Inc.

Do the Math Exercises 2.2

11. $-4(10 - 7x) = 3x + 10$ **12.** $-8 + 4(p + 6) = 10p$

11. _____

12. _____

13. Wendy's A Wendy's Mandarin Chicken salad contains 10 grams of fat less than a Taco Bell Zesty Chicken Border Bowl. If there are 60 grams of fat in the two salads, find the number of grams of fat in the Taco Bell Zesty Chicken Border Bowl by solving the equation $x + (x - 10) = 60$, where x represents the number of grams of fat in a Taco Bell Border Bowl and $x - 10$ represents the number of grams of fat in the Wendy's Mandarin Chicken salad.

13.

14. Overtime Pay Juan worked a total of 50 hours last week and earned $855. He earned 1.5 times his regular hourly rate for 6 hours, and double his hourly rate for 4 holiday hours. Solve the equation $40x + 6(1.5x) + 4(2x) = \$855$, where x is Juan's regular hourly rate.

14. _____

15. Perimeter of a Triangle The perimeter of a triangle is 210 inches. If the sides are made up of 3 consecutive even integers, find the lengths of each of the 3 sides by solving the equation $x + (x + 2) + (x + 4) = 210$, where x represents the length of the shortest side.

15. _____

16. Determine the value of d to make the statement true.
In the equation $5d - 2x = -2$, the solution is -6.

16.

Sullivan/Struve/Mazzarella, *Elementary & Intermediate Algebra*, 3e
Copyright © 2014 Pearson Education, Inc.

Name:
Instructor:

Date:
Section:

Five-Minute Warm-Up 2.3
Solving Linear Equations Involving Fractions and Decimals; Classifying Equations

1. Find the LCD of $\dfrac{6}{7}$ and $\dfrac{2}{5}$.
   1. _____

2. Find the LCD of $-\dfrac{4}{25}$ and $\dfrac{7}{45}$.
   2. _____

3. Use the Distributive Property to simplify: $-4\left(\dfrac{3}{4}x - 1\right)$
   3. _____

4. Multiply: $100(0.43x - 2.7)$
   4. _____

5. Simplify by combining like terms: $3 - 0.5(10x - 50) + 6x$
   5. _____

6. Simplify by combining like terms: $\dfrac{4}{3}\left[9 + 3(x - 5)\right]$
   6. _____

7. Simplify by combining like terms: $6\left(\dfrac{5x + 1}{3} - 2\right)$
   7. _____

Sullivan/Struve/Mazzarella, *Elementary & Intermediate Algebra*, 3e
Copyright © 2014 Pearson Education, Inc.

Name: Date:
Instructor: Section:

Guided Practice 2.3
Solving Linear Equations Involving Fractions and Decimals; Classifying Equations

Objective 1: Use the Least Common Denominator to Solve a Linear Equation Containing Fractions

1. Solve the linear equation: $\frac{3}{2}x - \frac{4}{3}x = -\frac{9}{4}$ *(See textbook Example 1)*

Step 1: Apply the Distributive Property to remove parentheses.

Determine the LCD: (a) _____

Use the Multiplication Property of Equality: (b) ___ $\cdot \left(\frac{3}{2}x - \frac{4}{3}x\right) = $ ___ $\cdot \left(-\frac{9}{4}\right)$

Use the Distributive Property: (c) ___ $\cdot \left(\frac{3}{2}x\right) - $ ___ $\cdot \left(\frac{4}{3}x\right) = $ ___ $\cdot \left(-\frac{9}{4}\right)$

Simplify (c): (d) _____

Step 2: Combine like terms.

Combine like terms: (e) _____

Step 3: Use the Addition Property of Equality to get the terms with the variable on one side of the equation and the constants on the other side.

This step is not necessary as the variable term is on the left side and the constant term on right.

Step 4: Get the coefficient of the variable to be 1.

Divide both sides by 2: (f) _____

Simplify: (g) _____

Step 5: Check Verify your value for x in the original equation.

Replace x in the original equation to see if a true statement results.

State the solution set: (h) _____

2. Solve the equation $\frac{4x+3}{9} + 1 = \frac{2x+1}{2}$. *(See textbook Example 2)*

We will multiply both sides of the equation by the LCD to remove the fractions from the equation. *DO NOT FINISH SOLVING THE EQUATION.*

   (a) Identify the LCD of the denominators in the equation: **(a)** _____

   (b) Apply the Multiplication Property of Equality. **(b)** _____

   (c) Distribute the LCD to each terms and divide out common factors: **(c)** _____

   (d) Use the Distributive Property to remove parentheses: **(d)** _____

Sullivan/Struve/Mazzarella, *Elementary & Intermediate Algebra*, 3e
Copyright © 2014 Pearson Education, Inc.

Guided Practice 2.3

(e) Combine like terms on each side of the equation: (e) _____

Objective 2: Solve a Linear Equation Containing Decimals

3. When encountering an equation that contains decimals, an option is to multiply both sides of the equation by some power of 10 that will clear the decimals. Try solving an equation that contains decimals using this technique. Then, try solving the equation again but use decimal operations at each step of the solution. Which method do you prefer?

Solve: $0.06x + 0.075(7500 - x) = 517.5$ *(See textbook Example 5)*   3. _____

Objective 3: Classify a Linear Equation as an Identity, a Conditional Equation, or a Contradiction

4. Solve the linear equation $-2(3x - 1) = 4 - 6(x + 3)$. State whether the equation is an identity, contradiction or conditional equation. *(See textbook Example 6)*

(a) Is the last statement of the solution true or false? _____

(b) Therefore the solution set is _____

(c) Is this equation an identity, contradiction or a conditional equation? _____

Objective 4: Use Linear Equations to Solve Problems

5. Martina saves dimes and quarters in a piggy bank. She opened the bank and discovered that she had $24.85 from a total of 127 coins. Martina wants to know how many dimes and how many quarters are in the bank….apparently, she doesn't want to just count them. To find the number of quarters, q, she solves the equation: $0.10(127 - q) + 0.25q = 24.85$. *(See textbook Example 9)*

(a) Solve the equation. What value did you find for the value of q? _____

(b) Check. Number of dimes plus number of quarters = 127? _____ Total value of $24.85? _____

(c) Answer the question: How many dimes were there? ____ How many quarters were there? ____

Sullivan/Struve/Mazzarella, *Elementary & Intermediate Algebra*, 3e
Copyright © 2014 Pearson Education, Inc.

Name:
Instructor:

Date:
Section:

Do the Math Exercises 2.3
Solving Linear Equations Involving Fractions and Decimals; Classifying Equations

In Problems 1–4, solve the equation. Check your solution.

1. $\dfrac{3}{2}n - \dfrac{4}{11} = \dfrac{91}{22}$

2. $\dfrac{3m}{8} - 1 = \dfrac{5}{6}$

1. _____

2. _____

3. $\dfrac{3}{2}b - \dfrac{4}{5}b = \dfrac{28}{5}$

4. $\dfrac{2}{3}(6 - x) = \dfrac{5x}{6}$

3. _____

4. _____

In Problems 5–8, solve the equation. Check your solution.

5. $0.3z = 6$

6. $p + 0.04p = 260$

5. _____

6. _____

7. $0.7y - 4.6 = 0.4y - 2.2$

8. $5 - 0.2(m - 2) = 3.6m + 1.6$

7. _____

8. _____

In Problems 9–12, solve the equation. State whether the equation is a contradiction, an identity, or a conditional equation.

9. $4(y - 2) = 5y - (y + 1)$

10. $-3x + 2 + 5x = 2(x + 1)$

9. _____

10. _____

Sullivan/Struve/Mazzarella, *Elementary & Intermediate Algebra*, 3e
Copyright © 2014 Pearson Education, Inc.

Do the Math Exercises 2.3

11. $7b + 2(b - 4) = 8b - (3b + 2)$

12. $\dfrac{2m+1}{4} - \dfrac{m}{6} = \dfrac{m}{3} - 1$

11. _____

12. _____

13. Team Sweatshirt Your favorite college has logo sweatshirts on sale for $33.60. The sweatshirt has been marked down by 30%. To find the original price of the sweatshirt, x, solve the equation $x - 0.30x = 33.60$.

13. _____

14. Clean Car Pablo cleaned out his car and found nickels and quarters in the car seats. He found $4.25 in change and noticed that the number of quarters was 5 less than twice the number of nickels. Solve the equation $0.05n + 0.25(2n - 5) = 4.25$ to find n, the number of nickels Pablo found.

14. _____

15. Paying Your Taxes You are married and just determined that you paid $13,310 in federal income taxes in 2012. The solution to the equation $13,310 = 0.25(x - 70,700) + 9735$ represents the adjusted gross income of you and your spouse in 2012. Determine the adjusted gross income of you and your spouse in 2012.

15. _____

Five-Minute Warm-Up 2.4
Evaluating Formulas and Solving Formulas for a Variable

1. Evaluate the expression $2L + 2W$ for $L = 10.5$ and $W = 3.5$

   1. _____

2. Round the expression 12.3762 to the hundredths place.

   2. _____

3. Write the percent as a decimal: 7.25%

   3. _____

4. Evaluate and round your answer to the nearest tenth: 25π

   4. _____

5. Solve: $-4x + 5 = -11$

   5. _____

Name:
Instructor:

Date:
Section:

Guided Practice 2.4
Evaluating Formulas and Solving Formulas for a Variable

Objective 1: Evaluate a Formula

1. A formula is an equation that describes how two or more variables are related. If you travel in Europe and the temperature is 20° Celsius, use the formula $F = \frac{9}{5}C + 32$ to find the equivalent temperature in degrees Fahrenheit. *(See textbook Example 1)*

   1. _____

2. Many times we solve problems that involve simple interest. Simple interest is money that is paid for the use of money, without any effect of compounding (receiving interest on the interest). For each of the variables in the simple interest formula, identify what the variable represents.

$I = Prt$			
I		r	
P		t	

3. List the formulas for each of the following geometric figures. *(Refer to textbook Section 2.4, as necessary.)*

(a) **Square**	(b) **Rectangle**
Area:_____ Perimeter:_____	Area:_____ Perimeter:_____
(c) **Triangle**	(d) **Trapezoid**
Area:_____ Perimeter:_____	Area:_____ Perimeter:_____
(e) **Parallelogram**	(f) **Circle**
Area:_____ Perimeter:_____	Area:_____ Circumference:_____
(g) **Cube**	(h) **Rectangular Solid**
Volume:_____ Surface Area:_____	Volume:_____ Surface Area:_____
(i) **Sphere**	(j) **Right Circular Cylinder**
Volume:_____ Surface Area:_____	Volume:_____ Surface Area:_____
(k) **Cone**	
Volume:_____	

4. Find the number of meters of fencing that must be purchased to enclose a dog run that is 8.5 meters long and 6.75 meters wide. If fencing cost $15 per meter, how much will it cost to install the fence?
   (a) What formula will you need to solve this problem? *(See textbook Examples 4 and 5)* 4a. _____

   (b) Substitute the values in the problem into the formula to find how much fencing is required.

   4b. _____

   (c) Multiply to find the cost. 4c. _____

Guided Practice 2.4

Objective 2: Solve a formula for a Variable

5. Solve $6x - 12y = -24$ for y. (See textbook Example 9)

		$6x - 12y = -24$
Step 1: Isolate the term containing the variable, y.	Subtract $6x$ from both sides of the equation to isolate the term $-12y$:	(a) _____
	Simplify:	(b) _____
Step 2: Get the coefficient of the variable to be 1.	Divide both sides of the equation by -12:	(c) _____
	Simplify using $\dfrac{a-b}{c} = \dfrac{a}{c} - \dfrac{b}{c}$:	(d) $y =$ _____

6. The amount of profit P, earned by a manufacturer is given by the formula $P = R - C$, where R represents the manufacturer's revenue and C represents the manufacturer's costs.

    (a) Solve the equation for C, the manufacturer's costs.   6a. _____

    (b) Find the amount of costs incurred if the manufacturer has a $5200 profit and $7300 in revenue.   6b. _____

Name:
Instructor:

Date:
Section:

Do the Math Exercises 2.4
Evaluating Formulas and Solving Formulas for a Variable

In Problems 1–5, substitute the given values into the formula and then evaluate to find the unknown quantity. Label units in the answer. If the answer is not exact, round your answer to the nearest hundredth.

1. **Length of a Bridge** The formula $m = 0.3048f$ converts length in feet f to length in meters m. The George Washington Bridge in New York City is 3500 feet long. How long is the George Washington Bridge in meters?

    1. _____

2. **Salesperson's Earnings** The formula $E = 750 + 0.07S$ is a formula for the earnings, E, of a salesperson who receives $750 per week plus 7% commission on all sales, S. Find the earnings of a salesperson who had weekly sales of $1200.

    2. _____

3. **Investing an Inheritance** Christopher invested his $5000 inheritance from his grandmother in a 9-month Certificate of Deposit that earns 4% simple interest per annum. Use the formula $I = Prt$ to find the amount of interest Christopher's investment will earn.

    3. _____

4. Consider the rectangle:

    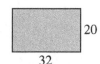

    a) Find the perimeter of the rectangle.   4a. _____

    (b) Find the area of the rectangle.   4b. _____

5.

    (a) Find the circumference of the circle. Use $\pi \approx 3.14$.   5a. _____

    (b) Find the area of the circle.   5b. _____

Do the Math Exercises 2.4

In Problems 6–9, solve each formula for the stated variable.

6. $F = mv^2$; solve for m

7. $V = \dfrac{1}{3}Bh$; solve for B

6. _____

7. _____

8. $S = a + b + c$; solve for b

9. $P = 2l + 2w$; solve for l

8. _____

9. _____

In Problems 10–13, solve each formula for y.

10. $-2x + y = 18$

11. $12x - 6y = 18$

10. _____

11. _____

12. $5x + 6y = 18$

13. $\dfrac{2}{3}x - \dfrac{5}{2}y = 5$

12. _____

13. _____

In Problems 14–15, (a) solve for the indicated variable, and then (b) find the value of the unknown quantity. When given, label units in the answer.

14. **Profit = Revenue − Cost:** $P = R - C$

(a) Solve for R.

(b) Find R when P = $4525 and C = $1475.

14a. _____

14b. _____

15. **Simple Interest:** $I = Prt$

(a) Solve for t.

(b) Find t when I = $42, P = $525, and r = 4%.

15a. _____

15b. _____

Name:
Instructor:

Date:
Section:

Five-Minute Warm-Up 2.5
Problem Solving: Direct Translation

1. Solve the equation: $x - 17.38 = -72.1$

   1. _____

2. Solve the equation: $x + 0.05x = 33.6$

   2. _____

3. Each of the following words or phrases can be replaced with one of the symbols: $+ \,/\, - \,/\, \times \,/\, \div$. Select the correct symbol for each word or phrase.

   (a) twice   **(b)** less than   **(c)** of   **(d)** sum

   (e) per   **(f)** increased by   **(g)** difference   **(h)** quotient

   (i) fewer   **(j)** greater than   **(k)** ratio   **(l)** decreased by

   (m) product   **(n)** exceeds by   **(o)** more than   **(p)** double

   (q) less   **(r)** times   **(s)** half   **(t)** altogether

   3a. _____
   3b. _____
   3c. _____
   3d. _____
   3e. _____
   3f. _____
   3g. _____
   3h. _____
   3i. _____
   3j. _____
   3k. _____
   3l. _____
   3m. _____
   3n. _____
   3o. _____
   3p. _____
   3q. _____
   3r. _____
   3s. _____
   3t. _____

Sullivan/Struve/Mazzarella, *Elementary & Intermediate Algebra*, 3e
Copyright © 2014 Pearson Education, Inc.

Name:
Instructor:

Date:
Section:

Guided Practice 2.5
Problem Solving: Direct Translation

Objective 1: Translate English Phrases into Algebraic Expressions

1. Express each English phrase as an algebraic expression. *(See textbook Example 1)*
   (a) The sum of -5 and $3x$
   (b) 5 times the sum of r and 25

   (c) The sum of 3 times a number x and 12
   (d) The ratio of twice a number p and 7

   (e) 18 less than half of a number z
   (f) The difference of w and 900

   1a. _____
   1b. _____
   1c. _____
   1d. _____
   1e. _____
   1f. _____

2. Order matters when subtracting and dividing. Translate the English phrase as an algebraic expression.
   (a) a less than b
   (b) a less b
   (c) a subtracted from b

   (d) a divided by b
   (e) a divided into b

   2a. _____
   2b. _____
   2c. _____
   2d. _____
   2e. _____

Objective 2: Translate English Sentences into Equations

3. Translate each sentence into an equation. Do not solve the equation. *(See textbook Example 4)*
   (a) The sum of 5 and x is 19.
   (b) 14 less than twice y is 30.

   (c) The difference of z and 9 is the same as 6 more than two-fifths of z.

   3a. _____
   3b. _____
   3c. _____

4. List six steps for solving problems with mathematical models.

   _____   _____
   _____   _____
   _____   _____

Objective 3: Build Models for Solving Direct Translation Problems

5. If n represents the first of three unknown consecutive integers, how would you represent the next two integers? (a) _____ (b) _____

Guided Practice 2.5

6. If x represents the first of four unknown consecutive even integers, how would you represent the next three even integers? (a) _____ (b) _____ (c) _____

7. If p represents the first of four unknown consecutive odd integers, how would you represent the next three odd integers? (a) _____ (b) _____ (c) _____

8. The sum of three consecutive odd integers is 453. Find the integers. *(See textbook Example 6)*

    Step 1: Identify We are looking for three consecutive odd integers whose sum is 453.

    (a) **Step 2: Name** Let n represent the first integer. What expression represents the next odd integer? _____ the third odd integer? _____

    (b) **Step 3: Translate** Write an equation for the sum of the three odd integers. _____

    (c) **Step 4: Solve** your equation from Step 3. What value did you find for n? _____

    Step 5: Check Find the other integers. Do they meet the criteria in the problem?

    (d) **Step 6: Answer the Question** _____

9. Lucky you! A long-lost uncle left you $20,000 in his will. You have decided to invest the money in two different accounts, a conservative bond fund and a more aggressive stock account. If you will invest $5000 less in the bond fund than the stock account, how much of the $20,000 will you invest in each account? *(See textbook Example 8)*

    Step 1: Identify We want to know how much to invest in each account.

    (a) **Step 2: Name** Let x represent the amount in the stock account. What algebraic expression represents $5000 less going in the bond fund? _____

    (b) **Step 3: Translate** Write an equation for the sum of the two accounts: _____

    (c) **Step 4: Solve** your equation from Step 3. What value did you find for x? _____

    Step 5: Check Use your value for the stock account and Step 2 to find how much is in the bond account. Do these answers seem reasonable? Do they meet all of the criteria?

    (d) **Step 6: Answer the Question** _____

Name:
Instructor:

Date:
Section:

Do the Math Exercises 2.5
Problem Solving: Direct Translation

In Problems 1–4, translate each phrase to an algebraic expression. Let x represent the unknown number.

1. double a number

2. 8 less than a number

3. the quotient of –14 and a number

4. 21 more than 4 times a number

1. _____

2. _____

3. _____

4. _____

In Problems 5–7, choose a variable to represent one quantity. State what that quantity represents and then express the second quantity in terms of the first.

5. The Toronto Blue Jays scored 3 fewer runs than the Cleveland Indians.

5. _____

6. Beryl has $0.25 more than 3 times the amount Ralph has.

6. _____

7. There were 12,765 fans at a recent NBA game. Some held paid admission tickets and some held special promotion tickets.

7. _____

In Problems 8–11, translate each statement into an equation. Let x represent the unknown number. DO NOT SOLVE.

8. The sum of 43 and a number is –72.

9. 49 is 3 less than twice a number.

8. _____

9. _____

Sullivan/Struve/Mazzarella, *Elementary & Intermediate Algebra*, 3e
Copyright © 2014 Pearson Education, Inc.

Do the Math Exercises 2.5

10. The quotient of a number and –6, decreased by 15, is 30.

11. Twice the sum of a number and 5 is the same as 7 more than the number.

10. _____

11. _____

In problems 12–15, translate the given information into an equation and solve.

12. **Consecutive Integers** The sum of three consecutive odd integers is 81. Find the numbers.

12. _____

13. **Towers** The tallest buildings in the world (those having the most stories) are Burj Khalifa in Dubai, UAE, and the Abraj Al-Bait Tower in Mecca, Saudi Arabia. The Abraj Al-Bait Tower has 43 fewer floors than Burj Khalifa. The two buildings together have 283 floors. Find the number of floors in each building.

13. _____

14. **Investments** Jack and Diane have $40,000 to invest. Their financial advisor has recommended that they diversify by placing some of the money in stocks and some in bonds. Based upon current market conditions, he has recommended that the amount in bonds should equal two-thirds of the amount invested in stocks. How much should be invested in stocks? How much should be invested in bonds?

14. _____

15. **Cellular Telephones** You need a new cell phone for emergencies only. Company A charges $12 per month plus $0.10 per minute, while Company B charges $0.15 per minute with no monthly service charge. For how many minutes will the monthly cost be the same?

15. _____

Name:
Instructor:

Date:
Section:

Five-Minute Warm-Up 2.6
Problem Solving: Problems Involving Percent

1. Write as a decimal.

    (a) 62%   (b) 1.75%

    1a. _____

    1b. _____

2. Write as a percent.

    a) 0.055   (b) 1.5

    2a. _____

    2b. _____

3. Combine like terms: $p + 0.75p$

    3. _____

4. Solve: $x - 0.20x = 10.60$

    4. _____

Name:
Instructor:
Date:
Section:

Guided Practice 2.6
Problem Solving: Problems Involving Percent

Objective 1: Solve Problems Involving Percent

1. Percent means "divided by 100". For instance, 20% means $\frac{20}{100}$ or $\frac{1}{5}$. Write each of these percents as fractions. Simplify, if possible.

    (a) 75%   (b) 100%   (c) $5\frac{1}{4}$%

    1a. _____

    1b. _____

    1c. _____

2. To divide by 100, shift the decimal point 2 decimal places to the left. Write each of these percents as decimals.

    (a) 68%   (b) 2.5%   (c) 150%

    2a. _____

    2b. _____

    2c. _____

3. If a question says, "what percent" and the solution to our equation is either a fraction or a decimal, we convert the result to percent by reversing the processes above. Write each of the following as a percent.

    (a) 0.08   (b) $\frac{7}{10}$ (c) 0.3

    3a. _____

    3b. _____

    3c. _____

4. A number is 72% of 30. Find the number. *(See textbook Example 1)*
    (a) Translate the statement into an equation.   4a. _____

    (b) Solve the equation and answer the question.   4b. _____

5. The number 150 is what percent of 120? *(See textbook Example 2)*
    (a) Translate the statement into an equation.   5a. _____

    (b) Solve the equation and answer the question.   5b. _____

Guided Practice 2.6

Objective 2: Solve Business Problems That Involve Percent

6. You just purchased a new computer. The price of the computer, including 7.5% sales tax, was $1343.75. How much did the computer cost before sales tax? *(See textbook Example 5)*

   Step 1: Identify We want to know the price of the computer before sales tax.

   Step 2: Name Let p represent the price of the computer before sales tax.

   (a) **Step 3: Translate** If p is the price of the computer, what algebraic expression calculates the amount of sales tax on the computer?

   _____

   (b) Write the equation: the original price plus the sales tax equals the total price: _____

   (c) **Step 4: Solve** Solve your equation from Step 3. _____

   Step 5: Check Does your answer seem reasonable? Have you satisfied the criteria in the question?

   (d) **Step 6: Answer** Always be careful to answer the question that is being asked. _____

7. Suppose you just learned that the local sporting goods store is having an end-of-the-season sale on ski equipment. Everything has been marked down by 60%. The sale price on a pair of Black Diamond skis is $224. What was the original price? *(See textbook Example 6)*

   Step 1: Identify We are looking for the original price of the skis, which have been marked down by 60%, and we know that the sale price is $224.

   Step 2: Name Let p represent the original price of the skis.

   (a) **Step 3: Translate** If p is the original price of the skis, what algebraic expression calculates the amount of discount?

   _____

   (b) Write the equation: the original price less the discount equals the sale price:

   _____

   (c) **Step 4: Solve** Solve your equation from Step 3. _____

   Step 5: Check Does your answer seem reasonable? Have you satisfied the criteria in the question?

   (d) **Step 6: Answer** Always be careful to answer the question that is being asked. _____

Name:
Instructor:

Date:
Section:

Do the Math Exercises 2.6
Problem Solving: Problems Involving Percent

In Problems 1–8, find the unknown in each percent question.

1. What is 80% of 50?

2. 75% of 20 is what number?

1. _____

2. _____

3. What number is 150% of 9?

4. 40% of what number is 122?

3. _____

4. _____

5. 45% of what number is 900?

6. 11 is 5.5% of what number?

5. _____

6. _____

7. 4 is what percent of 25?

8. What percent of 16 is 12?

7. _____

8. _____

9. **Sales Tax** The sales tax in Franklin County, Ohio, is 5.75%. The total cost of purchasing a used Honda Civic, including sales tax, is $8460. Find the cost of the car before sales tax.

9. _____

10. **Pay Raise** MaryBeth works from home as a graphic designer. Recently she raised her hourly rate by 5% to cover increased costs. Her new hourly rate is $29.40. Find MaryBeth's previous hourly rate.

10. _____

Sullivan/Struve/Mazzarella, *Elementary & Intermediate Algebra*, 3e
Copyright © 2014 Pearson Education, Inc.

Do the Math Exercises 2.6

11. Good Investment Perry just learned that his house increased in value by 4% over the past year. The value of the house is now $208,000. What was the value of the home one year ago?

11. _____

12. Hailstorm Toyota Town had a 15%-off sale on cars that had been damaged in a hailstorm. The original price of a truck was $28,000. What is the sale price?

12. _____

13. Business: Marking up the Price of Books A college bookstore marks up the price that it pays the publisher for a book by 30%. If the selling price of a book is $117, how much did the bookstore pay for the book?

13. _____

14. Vacation Package The Liberty Travel Agency advertised a 5-night vacation package in Jamaica for 30% off the regular price. The sale price of the package is $1139. To the nearest dollar, how much was the vacation package before the 30% off sale?

14. _____

15. Census Data Based on data obtained from the U.S. Census Bureau, 24% of the 118 million females aged 18 years or older have never married. How many females aged 18 years or older have never married?

15. _____

Name:
Instructor:

Date:
Section:

Five-Minute Warm-Up 2.7
Problem Solving: Geometry and Uniform Motion

1. Solve: $a + 2(a - 30) = 90$

   1. _____

2. Solve: $2(x + 5) + 2(3x - 8) = 22$

   2. _____

3. If x represents the unknown quantity, translate each of the following English phrases into an algebraic expression.

   (a) 12 more than twice an unknown number

   3a. _____

   (b) 25 less than half of the width

   3b. _____

   (c) double the difference of 45 and the measure of an angle

   3c. _____

   (d) the sum of 15 and twice the cost

   3d. _____

   (e) twice the sum of the speed and 30

   3e. _____

4. If an 8-foot board is cut into two pieces and one piece has length p, express the length of the other piece in terms of p.

   4. _____

Name: Date:
Instructor: Section:

Guided Practice 2.7
Problem Solving: Geometry and Uniform Motion

Objective 1: Set Up and Solve Complementary and Supplementary Angle Problems

1. Two angles are *complementary* if the sum of their measures is _____. Each angle is called the *complement* of the other.

2. Two angles are *supplementary* if the sum of their measures is _____. Each angle is called the *supplement* of the other.

3. Find the measures of two supplementary angles such that the measure of the smaller angle is 20° less than the measure of the larger angle. *(See textbook Example 1)*

   Step 1: Identify Since the angles are supplementary, we know the sum of their measures is 180°.

   Step 2: Name Let x represent the measure of the larger angle.

   (a) **Step 3: Translate** If x is the measure of the larger angle, what algebraic expression calculates the measure of the smaller angle?

   _____

   (b) Write the equation: the sum of the measures is equal to 180: _____

   (c) **Step 4: Solve** your equation from Step 3. _____

   Step 5: Check Does your answer seem reasonable? Have you satisfied the criteria in the question?

   (d) **Step 6: Answer** Always be careful to answer the question that is being asked. _____

Objective 2: Set Up and Solve "Angles of a Triangle" Problems

4. The sum of the measures of the three angles of a triangle is always _____.

5. The measure of the third angle of a triangle is 30° more than measure of the second angle of a triangle and the measure of the second angle is twice the measure the first angle. Find the measure of each angle of the triangle. *(See textbook Example 2)*

   Step 1: Identify Since these are angles of a triangle, we know the sum of their measures is 180°.

   Step 2: Name Let x represent the measure of the first angle.

   (a) **Step 3: Translate** If x is the measure of the first angle, what algebraic expression calculates the measure of the second angle?

   _____

Sullivan/Struve/Mazzarella, *Elementary & Intermediate Algebra*, 3e
Copyright © 2014 Pearson Education, Inc.

Guided Practice 2.7

5. **(b)** What algebraic expression calculates the measure of the third angle? _____

   (c) Write the equation: the sum of the measures is equal to 180: _____

   (d) Step 4: Solve your equation from Step 3. _____

   Step 5: Check Does your answer seem reasonable? Have you satisfied the criteria in the question?

   (e) Step 6: Answer Always be careful to answer the question that is being asked. _____

Objective 3: Use Geometry Formulas to Solve Problems

6. **Tablecloth** Mary is making a rectangular tablecloth that is 20 inches longer than it is wide. If she has 200 inches of trim that will just fit around the edge of the tablecloth, find the dimensions of the tablecloth. *(See textbook Example 3)*

   (a) If x is width of the tablecloth, what algebraic expression represents the length? _____

   (b) What formula from geometry is needed to solve this problem? _____

   (c) Write an equation that models the information in the problem: _____

   (d) Solve your equation from (c) and answer the question: _____

Objective 4: Set Up and Solve Uniform Motion Problems

7. **Boats** Two boats are traveling toward the same port from opposite directions. They are 50 miles apart and one boat is traveling 4 mph faster than the other. If the boats both reach the port in 2 hours and 30 minutes, find the speed for each boat. *(See textbook Examples 5 and 6)*

   (a) What do you know about the distances traveled by the two boats? _____

   (b) Complete the following table.

	Rate, mph	Time, hours	Distance, miles
Slower boat			
Faster boat			
Total			

   (c) Write and solve an equation that models the distance traveled by the boats.

   (d) Answer the question. _____

Name:
Instructor:

Date:
Section:

Do the Math Exercises 2.7
Problem Solving: Geometry and Uniform Motion

Identify the measure of each of the angles.

1. Find two complementary angles if the measure of the first angle is 25° less than the measure of the second.

   1. _____

2. In a triangle, the second angle measures four times the first. The measure of the third angle is 18° more than the second. Find the measures of the three angles.

   2. _____

Use a formula from geometry to solve for the unknown quantity.

3. The width of a rectangle is 10 meters less than the length. If the perimeter is 56 meters, find the length of each side of the rectangle.

   3. _____

Solve the uniform motion problem. Set up the uniform motion problem following steps 4a–4d.

4. Two trains leave Albuquerque, traveling the same direction on parallel tracks. One train is traveling at 72 mph, and the other is traveling at 66 mph. How long before they are 45 miles apart?

   (a) Write an algebraic expression for the distance traveled by the faster train.

   4a. _____

   (b) Write an algebraic expression for the distance traveled by the slower train.

   4b. _____

   (c) Write an algebraic expression for the difference in distance between the two.

   4c. _____

   (d) Write an equation to answer the question.

   4d. _____

Sullivan/Struve/Mazzarella, *Elementary & Intermediate Algebra*, 3e
Copyright © 2014 Pearson Education, Inc.

Do the Math Exercises 2.7

Fill in the table from the information given. Then write the equation that will solve the problem. DO NOT SOLVE.

5. A 580-mile trip in a small plane took a total of 5 hours. The first two hours were flown at one rate and then the plane encountered a headwind and was slowed by 10 mph.

5. _____

	Rate	·	Time	=	Distance
Beginning of trip					
Rest of the trip					
Total					

Write an equation that will solve each problem. Then solve.

6. **Garden** The length of a rectangular garden is 9 feet. If 26 feet of fencing are required to fence the garden, find the width of the garden.

6. _____

7. **Buying Fertilizer** Melinda has to buy fertilizer for a flower garden in the shape of a right triangle. If the area of the garden is 54 square feet and the base of the garden measures 9 feet, find the height of the triangular garden.

7. _____

8. **Table** The Jacksons are having a custom rectangular table made for a small dining area. The length of the table is 18 inches more than the width, and the perimeter is 180 inches. Find the length and width of the table.

8. _____

9. **Cyclists** Two cyclists leave a city at the same time, one going east and the other going west. The west-bound cyclist bikes at 4 mph faster than the east-bound cyclist. After 5 hours they are 200 miles apart. How fast is the east-bound cyclist riding?

9. _____

10. **River Trip** Max lives on a river, 30 miles from town. Max travels downstream (with the current) at 20 mph. Returning upstream (against the current) his rate is 12 mph. If the total trip to town and back took 4 hours, how long did he spend returning from town?

10. _____

Name:
Instructor:

Date:
Section:

Five-Minute Warm-Up 2.8
Solving Linear Inequalities in One Variable

In Problems 1 – 6, replace the question mark by <, >, or = to make the statement true.

1. −0.2 ? 0

2. −5 ? −4.5

3. $\dfrac{7}{12}$? $\dfrac{9}{16}$

4. 0.625 ? $\dfrac{5}{8}$

5. $\dfrac{-13}{8}$? $\dfrac{-11}{4}$

6. 0.003 ? 0.0031

1. _____

2. _____

3. _____

4. _____

5. _____

6. _____

7. Plot the set of points $\left\{-3.5, \dfrac{3}{2}, -\dfrac{2}{3}, 3\right\}$ on the number line.

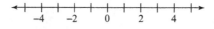

Name:
Instructor:

Date:
Section:

Guided Practice 2.8
Solving Linear Inequalities in One Variable

Objective 1: Graph Inequalities on a Real Number Line

1. When representing an inequality on a number line, we graph the inequality using a(n) _____ if the endpoint in included and a(n) _____ if the endpoint is not included.

2. Graph $x > -2$ on a real number line. *(See textbook Example 1)*

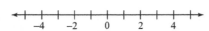

Objective 2: Use Interval Notation

3. Write the inequality in interval notation. *(See textbook Example 2)*
    (a) $x < -5$ (b) $x \geq 10$

3a. _____

3b. _____

Objective 3: Solve Linear Inequalities Using Properties of Inequalities

4. *True or False* The Addition Property of Inequality states that the direction of the inequality does not change regardless of the quantity that is added to each side of the inequality.

4. _____

5. *True or False* When multiplying both sides of an inequality by a negative number, we reverse the direction of the inequality.

5. _____

6. *True or False* When multiplying both sides of an inequality by a positive number, we reverse the direction of the inequality.

6. _____

7. Solve the inequality $3x + 10 > 1$. Graph the solution set. *(See textbook Example 3)*

Step 1: Get the expression containing the variable on the left side of the inequality.	This is already done.	$3x + 10 > 1$
Step 2: Isolate the variable x on the left side.	Subtract 10 from both sides:	(a) $3x + 10$ _____ > 1 _____
	Simplify:	(b) _____
	Divide both sides by 3 and simplify:	(c) _____
	Write the answer in set-builder notation:	(d) _____
	Write the answer using interval notation:	(e) _____
	Graph the solution set:	(f) number line from −4 to 4

Guided Practice 2.8

8. When multiplying or dividing both sides of an inequality by a negative real number, remember to reverse the direction of the inequality symbol. Solve the inequality and state the solution using interval notation. *(See textbook Example 6)*

   (a) $-6x < -72$

   (b) $-\dfrac{2}{5}x \geq 10$

   8a. _____

   8b. _____

9. Solve the inequality $\dfrac{1}{6}(4x - 5) \leq \dfrac{1}{2}(2x + 3)$. *(See textbook Examples 7 and 8)*

Step 1: To rewrite the inequality as an equivalent inequality without fractions, we multiply both sides of the inequality by the LCD. Then remove parentheses.	Multiply both sides by the LCD:	$\dfrac{1}{6}(4x-5) \leq \dfrac{1}{2}(2x+3)$ (a) ___ $\cdot \dfrac{1}{6}(4x-5) \leq$ ___ $\cdot \dfrac{1}{2}(2x+3)$
	Divide out common factors:	(b) _____
	Use the Distributive Property:	(c) _____
Step 2: Combine like terms on each side of the inequality.		This is already done.
Step 3: Get the variable expressions on the left side of the inequality and the constants on the right side.	Subtract $6x$ from both sides:	(d) _____
	Add 5 to both sides:	(e) _____
Step 4: Get the coefficient of the variable to be one.	Divide both sides by -2: Remember to reverse the inequality symbol!	(f) _____
	Write the answer in set-builder notation:	(g) _____
	Write the answer using interval notation:	(h) _____

Objective 4: Model Inequality Problems

10. Write the appropriate inequality symbol for each phrase.

Phrase	Inequality Symbol	Phrase	Inequality Symbol
(a) At least	(a) _____	(b) No less than	(b) _____
(c) More than	(c) _____	(d) Greater than	(d) _____
(e) No more than	(e) _____	(f) At most	(f) _____
(g) Fewer than	(g) _____	(h) Less than	(h) _____

Name:
Instructor:

Date:
Section:

Do the Math Exercises 2.8
Solving Linear Inequalities in One Variable

In Problems 1–4, graph each inequality on a number line, and write each inequality in interval notation.

1. $n > 5$

2. $x \leq 6$

1. _____

2. _____

3. $x \geq -2$

4. $y < -3$

3. _____

4. _____

In Problems 5–6, use interval notation to express the inequality shown in each graph.

5.

6.

5. _____

6. _____

In Problems 7–8, fill in the blank with the correct symbol. State which property of inequality is being utilized.

7. If $3x - 2 > 7$, then $3x$ ____ 9

8. If $\dfrac{5}{3}x \geq -10$, then x ____ -6

7. _____

8. _____

In Problems 9–17, solve the inequality and express the solution set in set-builder notation and interval notation. Graph the solution set on a real number line.

9. $x - 2 < 1$

10. $4x > 12$

9. _____

10. _____

Sullivan/Struve/Mazzarella, *Elementary & Intermediate Algebra*, 3e
Copyright © 2014 Pearson Education, Inc.

89

Do the Math Exercises 2.8

11. $-7x \geq 28$

12. $2x+5 > 1$

11. _____

12. _____

13. $2x-2 \geq 3+x$

14. $2-3x \leq 5$

13. _____

14. _____

15. $-3(1-x) > x+8$

16. $2y-5+y < 3(y-2)$

15. _____

16. _____

17. $3(p+1)-p \geq 2(p+1)$

17. _____

In Problems 18–19, write the given statement using inequality symbols. Let x represent the unknown quantity.

18. The cost of a new lawnmower is at least $250.

18. _____

19. There are fewer than 25 students in your math class on any given day.

19. _____

20. Truck Rental A truck can be rented from Acme Truck Rental for $80 per week plus $0.28 per mile. How many miles can be driven if you have at most $100 to spend on truck rental?

20. _____

Five-Minute Warm-Up 3.1
The Rectangular Coordinate System and Equations in Two Variables

1. Plot the following points on the real number line: $-1, \dfrac{3}{2}, 3, -2.5$

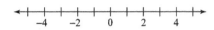

2. Evaluate: $-2x - 3$
   (a) for $x = -5$ (b) for $x = 0$

   2a. _____

   2b. _____

3. Evaluate: $7x + 3y$
   (a) for $x = -1, y = -6$ (b) for $x = 3, y = -4$

   3a. _____

   3b. _____

4. Solve: $-2x + 9 = -1$

   4. _____

5. Solve: $4(2x - 1) - 6x = 2 - 5(x - 3)$

   5. _____

Name: Date:
Instructor: Section:

Guided Practice 3.1
The Rectangular Coordinate System and Equations in Two Variables

Objective 1: Plot Points in the Rectangular Coordinate System

1. Label each quadrant and axis in the rectangular or Cartesian coordinate system.

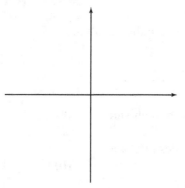

2. What name do we give the ordered pair $(0, 0)$? _____

In Problem 3, circle one answer for each underlined choice.

3. To plot the ordered pair $(-3, 2)$, you would move 3 units <u>up, down, left or right</u> from the origin?

4. Identify the coordinates of each point labeled below. *(See textbook Example 2)*

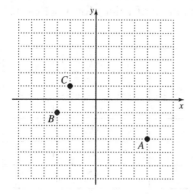

4.
A _____
B _____
C _____

Objective 2: Determine Whether an Ordered Pair Satisfies an Equation

5. Determine which, if any, of the following ordered pairs satisfy the equation $2x - y = -5$.
(See textbook Example 3)

(a) $(1, -3)$ (b) $(-5, -5)$

5a. _____

5b. _____

Sullivan/Struve/Mazzarella, *Elementary & Intermediate Algebra*, 3e
Copyright © 2014 Pearson Education, Inc.

Guided Practice 3.1

6. Find an ordered pair that satisfies the equation $4x + 2y = 7$. *(See textbook Example 4)*

Step 1: Choose any value for one of the variables in the equation.	You may choose any value of x or y that you wish. In this example, we will let $x = 3$.	
Step 2: Substitute the value of the variable chosen in Step 1 into the equation and then use the techniques learned in Chapter 2 to solve for the remaining variable.	Substitute 3 for x in the equation $4x + 2y = 7$. Simplify and solve for y.	$4x + 2y = 7$ $4(3) + 2y = 7$ $12 + 2y = 7$
	Subtract 12 from both sides:	(a) _____
	Divide both sides by 2 and simplify:	(b) _____
	Write the complete solution as an ordered pair:	(c) $(x, y) =$ _____

Objective 3: Create a Table of Values That Satisfy an Equation

7. To create a table of values that satisfy an equation, choose any value for one of the variables in the equation. Substitute the value of the variable into the equation and then use the techniques learned in Chapter 2 to solve for the remaining variable. Complete the table of values for $y = 3x - 1$.
(See textbook Example 5)

	x	y	(x, y)
(a)	−2	3(___) − 1 = ___	(−2, ___)
(b)	−1	3(___) − 1 = ___	(−1, ___)
(c)	0	3(___) − 1 = ___	(0, ___)
(d)	1	3(___) − 1 = ___	(1, ___)
(e)	2	3(___) − 1 = ___	(2, ___)

Name: Date:
Instructor: Section:

Do the Math Exercises 3.1
The Rectangular Coordinate System and Equations in Two Variables

In Problems 1–2, plot the following ordered pairs in the rectangular coordinate system. Tell which quadrant each point lies in or state that the point lies on the x-axis or y-axis.

1. $P(-3, -2); Q(2, -4); R(4, 3); S(-1, 4);$
   $T(-2, -4); U(3, -3)$

2. $P\left(\dfrac{3}{2}, -2\right); Q\left(0, \dfrac{5}{2}\right); R\left(-\dfrac{9}{2}, 0\right); S(0, 0);$
   $T\left(-\dfrac{3}{2}, -\dfrac{9}{2}\right); U\left(3, \dfrac{1}{2}\right) V\left(\dfrac{5}{2}, -\dfrac{7}{2}\right)$

1. _____

2. _____

Identify the coordinates of each point labeled in the figure. Name the quadrant in which each point lies or state that the point lies on the x- or y-axis.

3.

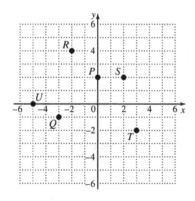

3. _____

In Problems 4–5, determine whether or not the ordered pair satisfies the equation.

4. $y = 2x - 3$ $A(-1, -5)$ $B(4, -5)$ $C(-2, -7)$

5. $5x - y = 12$ $A(2, 0)$ $B(0, 12)$ $C(-2, -22)$

4. _____

5. _____

Sullivan/Struve/Mazzarella, *Elementary & Intermediate Algebra*, 3e
Copyright © 2014 Pearson Education, Inc.

Do the Math Exercises 3.1

6. Find an ordered pair that satisfies the equation $x + y = 7$ by letting $x = 2$.

7. Find an ordered pair that satisfies the equation $5x - 3y = 11$ by letting $y = 3$.

6. _____

7. _____

In Problems 8–9, use the equation to complete the table. Use the table to list some of the ordered pairs that satisfy the equation.

8. $y = 4x - 5$

x	y	(x, y)
-3		
1		
2		

9. $3x + 4y = 2$

x	y	(x, y)
-2		
2		
4		

In Problems 10–11, for each equation find the missing value in the ordered pair.

10. $y = 5x - 4$ $A(-1, \_\_\_)$ $B(\_\_\_, 31)$ $C\left(-\dfrac{2}{5}, \_\_\_\right)$

11. $\dfrac{1}{3}x + 2y = -1$ $A(-4, \_\_\_)$ $B\left(\_\_\_, -\dfrac{3}{4}\right)$ $C(0, \_\_\_)$

10. _____

11. _____

12. **Taxi Ride** The cost to take a taxi is $1.70 plus $2.00 per mile for each mile driven. The total cost, C, is given by the equation $C = 1.7 + 2m$ where m represents the total miles driven.

(a) How much will it cost to take a taxi 5 miles?

12a. _____

(b) How much will it cost to take a taxi 20 miles?

12b. _____

(c) If you spent $32.70 on cab fare, how far was your trip?

12c. _____

(d) If (m, C) represents any ordered pair that satisfies $C = 1.7 + 2m$, interpret the meaning of (14, 29.7) in the context of this problem.

12d. _____

Five-Minute Warm-Up 3.2
Graphing Equations in Two Variables

1. Solve: $-5x = 45$

   1. _____

2. Solve: $12y = -40$

   2. _____

3. Solve: $5y + 12 = 2$

   3. _____

4. Evaluate: $-2x + 8y$
   (a) for $x = -3$, $y = 2$ (b) for $x = 10$, $y = 4$

   4a. _____

   4b. _____

5. Solve using the given information: $y = \dfrac{3}{2}x + 5$
   (a) If $x = -4$, find y.   (b) If $y = -1$, find x.

   5a. _____

   5b. _____

6. Plot the ordered pairs: $(0, -3)$; $(2, 0)$; $(-3, 2)$; $(1, -4)$; $(-2, -1)$

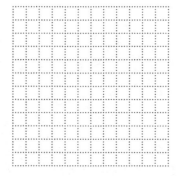

Name:
Instructor:

Date:
Section:

Guided Practice 3.2
Graphing Equations in Two Variables

Objective 1: Graph a Line by Plotting Points

1. Point-plotting is one method we use to graph a line. Find several ordered pairs that satisfy the equation. Plot the points in a rectangular coordinate system and then connect the points in a smooth curve or line.

Graph the equation $y = -2x - 3$ using the point-plotting method. *(See textbook Example 1)*

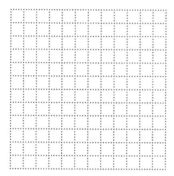

2. When a linear equation in two variables is written in the form $Ax + By = C$ where A, B, and C are real numbers and both A and B cannot be 0, we say the linear equation is written in _____ form.

Objective 2: Graph a Line Using Intercepts

3. To find the *x*-intercept(s), if any, of the graph of an equation, let _____ in the equation and solve for *x*.

4. To find the *y*-intercept(s), if any, of the graph of an equation, let _____ in the equation and solve for *y*.

5. Graph the linear equation $3x - 2y = -6$ by finding its intercepts. *(See textbook Example 6)*

   (a) Identify the *x*-intercept: _____

   (b) Identify the *y*-intercept: _____

   (c) Draw the axes. Plot the intercepts and connect the points in a straight line to obtain the graph.

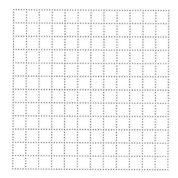

Sullivan/Struve/Mazzarella, *Elementary & Intermediate Algebra*, 3e
Copyright © 2014 Pearson Education, Inc.

Guided Practice 3.2

Objective 3: Graph Vertical and Horizontal Lines

6. A vertical line is given by an equation of the form $x = a$ where a is the x-intercept.

Graph the equation $x = -2$. *(See textbook Example 9)*

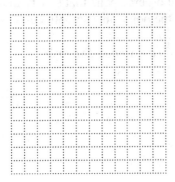

7. A horizontal line is given by an equation of the form $y = b$ where b is the y-intercept.

Graph the equation $y = 3$. *(See textbook Example 10)*

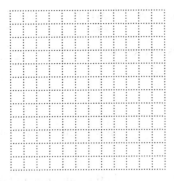

Do the Math Exercises 3.2
Graphing Equations in Two Variables

In Problems 1–2, determine whether or not the equation is a linear equation in two variables.

1. $y^2 = 2x + 3$

2. $y - 2x = 9$

1. _____

2. _____

In Problems 3–4, graph each linear equation using the point-plotting method.

3. $y = -3x - 1$

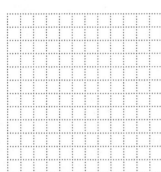

4. $y = x - 6$

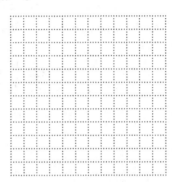

In Problems 5–8, find the intercepts of each graph.

5.

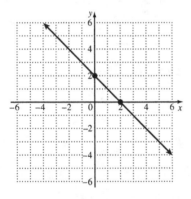

6.

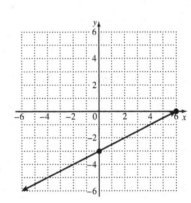

5. _____

6. _____

7.

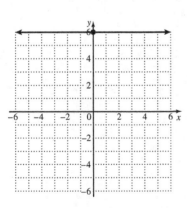

8.

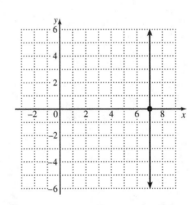

7. _____

8. _____

Do the Math Exercises 3.2

In Problems 9–10, find the intercepts of each equation.

9. $3x - 5y = 30$

10. $\dfrac{x}{2} - \dfrac{y}{8} = 1$

9. _____

10. _____

In Problems 11–12, graph each linear equation by finding its intercepts.

11. $3x - 5y = 15$

12. $\dfrac{1}{3}x - y = 0$

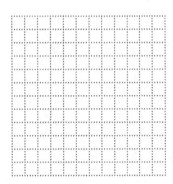
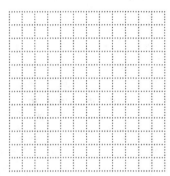

In Problems 13–14, graph each horizontal or vertical line.

13. $x = -7$

14. $y = 2$

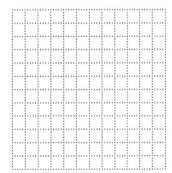
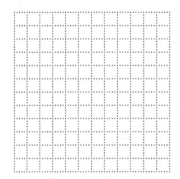

Name:
Instructor:

Date:
Section:

Five-Minute Warm-Up 3.3
Slope

In Problems 1 – 7, simplify each expression.

1. $\dfrac{9-3}{15-3}$

2. $\dfrac{6-15}{2-5}$

3. $\dfrac{-2-8}{4-(-2)}$

4. $\dfrac{-5-(-9)}{3-5}$

5. $\dfrac{5-12}{-6-(-6)}$

6. $\dfrac{-18-(-18)}{4-(-4)}$

7. $\dfrac{\dfrac{1}{2}-\left(-\dfrac{3}{4}\right)}{-\dfrac{2}{3}-\dfrac{5}{6}}$

1. _____

2. _____

3. _____

4. _____

5. _____

6. _____

7. _____

Sullivan/Struve/Mazzarella, *Elementary & Intermediate Algebra*, 3e
Copyright © 2014 Pearson Education, Inc.

Name:
Instructor:
Date:
Section:

Guided Practice 3.3
Slope

Objective 1: Find the Slope of a Line Given Two Points

1. The ratio of the rise (vertical change) to the run (horizontal change) is called the _____ of the line.

2. The slope of a line which passes through the ordered pairs (x_1, y_1) and (x_2, y_2) is given by the formula:

$$m = \frac{\rule{2cm}{0.4pt}}{\rule{2cm}{0.4pt}}$$

3. Find the slope of the line containing the points $(-1, 7)$ and $(3, -13)$. *(See textbook Example 1)*

3. _____

Objective 2: Find the Slope of Vertical and Horizontal Lines

4. Calculate the slope of the line through the points $(-8, 3)$ and $(-8, -1)$. *(See textbook Example 3)*

4. _____

5. Is the graph of the line in #4 vertical or horizontal?

5. _____

6. Calculate the slope of the line through the points $(-1, 5)$ and $(4, 5)$. *(See textbook Example 4)*

6. _____

7. Is the graph of the line in #6 vertical or horizontal?

7. _____

8. *Properties of Slope* Describe the graph of the line with the following slope:

   (a) Positive _____

   (b) Negative _____

   (c) Zero _____

   (d) Undefined _____

Guided Practice 3.3

Objective 3: Graph a Line Using Its Slope and a Point on the Line

9. Draw a graph of the line that contains the point $(-1, 3)$ and has slope $-\frac{3}{4}$. *(See textbook Example 5)*

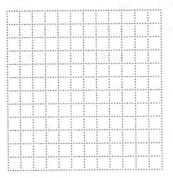

Objective 4: Work with Applications of Slope

10. The slope of a line is also called the _____ _____ of _____ of y with respect to x.

11. A road rises 2 feet for every 40 feet of horizontal distance covered.

    (a) Write a fraction that represents the average rate of change for this road.   11a. _____

    (b) Find the grade of the road by writing your answer from part (a) as a percent.   11b. _____

Name:
Instructor:

Date:
Section:

Do the Math Exercises 3.3
Slope

In Problems 1–2, find the slope of the line whose graph is given.

1.
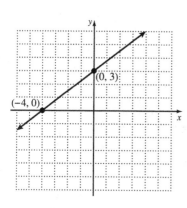

2.
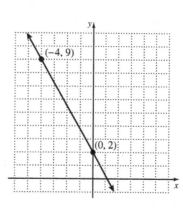

1. _____

2. _____

In Problems 3–4, (a) plot the points in a rectangular coordinate system, (b) draw a line through the points, (c) and find and interpret the slope of the line.

3. $(2, 6)$ and $(-2, -4)$

4. $(4, -5)$ and $(-2, -4)$

3c. _____

4c. _____

In Problems 5–6, find the slope of the line whose graph is given.

5.
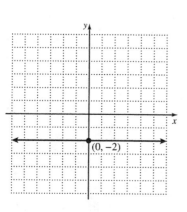

6.
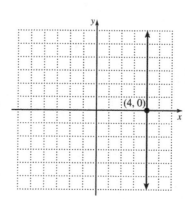

5. _____

6. _____

Sullivan/Struve/Mazzarella, *Elementary & Intermediate Algebra*, 3e
Copyright © 2014 Pearson Education, Inc.

Do the Math Exercises 3.3

In Problems 7–8, draw a graph of the line that contains the given point and has the given slope.

7. $(3, -1)$; $m = -1$

8. $(-2, -3)$; $m = \dfrac{5}{2}$

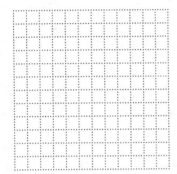
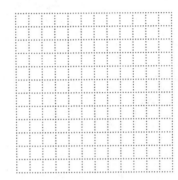

9. **Roof Pitch** A canopy is set up on the football field. On the 45-yard line, the height of the canopy is 68 inches. The peak of the canopy is at the 50-yard marker where the height is 84 inches. What is the pitch of the roof of the canopy?

9. _____

10. **Earning Potential** On average, a person who graduates from high school can expect to have lifetime earnings of 1.2 million dollars. It takes four years to earn a bachelor's degree, but the lifetime earnings will increase to 2.1 million dollars. Use the ordered pairs (0, 1.2 million) and (4, 2.1 million) to find and interpret the slope of the line representing the increase in earnings due to finishing college.

10. _____

Name:
Instructor:

Date:
Section:

Five-Minute Warm-Up 3.4
Slope-Intercept Form of a Line

1. Solve $7x + 5y = -35$ for y.

    1. _____

2. Solve $3x - 8y = 12$ for y.

    2. _____

3. Solve $15 = 3 - 9x$ for x.

    3. _____

4. Identify the following information using: $y = -\dfrac{5}{3}x - \dfrac{1}{2}$

    4a. _____

    (a) the coefficient of x (b) the constant

    4b. _____

Name:
Instructor:
Date:
Section:

Guided Practice 3.4
Slope-Intercept Form of a Line

Objective 1: Use the Slope-Intercept Form to Identify the Slope and y-Intercept of a Line

1. The *Slope-Intercept Form* of a line is an equation written in the form $y = mx + b$, where the slope of the line is ____, the coefficient of *x*. The *y*-intercept is ____, the constant.

1. _____

   _____

2. Find the slope and *y*-intercept of the line whose equation is $y = -\frac{1}{2}x + 6$.
(See textbook Example 1)

2. slope = _____

   y-intercept = _____

3. Find the slope and *y*-intercept of the line whose equation is $6x + 3y = 15$.
(See textbook Example 2)

3. slope = _____

   y-intercept = _____

Objective 2: Graph a Line Whose Equation Is in Slope-Intercept Form

4. Graph the line $y = \frac{2}{5}x - 1$ using the slope and *y*-intercept. *(See textbook Example 4)*

   (a) What is the slope? _____

   (b) What is the *y*-intercept? _____

   (c) Graph the line.

Objective 3: Graph a Line Whose Equation Is in the Form Ax + By = C

5. Graph the line $2x + 3y = 6$ using the slope and *y*-intercept. *(See textbook Example 5)*

   (a) Write the equation in slope-intercept form.

   _____

   (b) What is the slope? _____

   (c) What is the *y*-intercept? _____

   (d) Graph the line.

Guided Practice 3.4

Objective 4: Find the Equation of a Line Given Its Slope and y-Intercept

6. Find the equation of a line whose slope is $-\dfrac{1}{4}$ and whose *y*-intercept is 3.   6. _____

(See textbook Example 6)

Objective 5: Work with Linear Models in Slope-Intercept Form

7. The cost of renting a particular car for a day is $22 plus $0.07 per mile. Find the following. *(See textbook Example 8)*

   (a) Write a linear equation that relates the daily cost of renting the car, *y*, to the number of miles driven in a day, *x*.   7a. _____

   (b) What is the cost of driving this car 100 miles?   7b. _____

Do the Math Exercises 3.4
Slope-Intercept Form of a Line

In Problems 1–4, find the slope and y-intercept of the line whose equation is given.

1. $y = -6x + 2$
2. $y = -x - 12$

1. _____
2. _____

3. $6x - 8y = -24$
4. $10x + 6y = 24$

3. _____
4. _____

In Problems 5–6, use the slope and y-intercept to graph each line whose equation is given.

5. $y = -4x - 1$
6. $y = \dfrac{4}{3}x - 3$

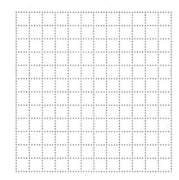
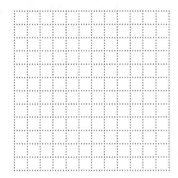

In Problems 7–8, graph each line using the slope and y-intercept.

7. $x - 2y = -4$
8. $4x + 3y = -6$

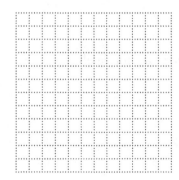
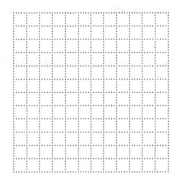

Do the Math Exercises 3.4

In Problems 9–14, find the equation of the line with the given slope and intercept.

9. slope is 1; y-intercept is 10

10. slope is $\frac{4}{7}$; y-intercept is –9

11. slope is $\frac{1}{4}$; y-intercept is $\frac{3}{8}$

12. slope is 0; y-intercept is –2

13. slope is undefined; x-intercept is 4

14. slope is –3; y-intercept is 0

9. _____

10. _____

11. _____

12. _____

13. _____

14. _____

15. **Counting Calories** According to a 1989 National Academy of Sciences Report, the recommended daily intake of calories for males between the ages of 7 and 15 can be calculated by the equation $y = 125x + 1125$, where x represents the boy's age and y represents the recommended calorie intake.

(a) What is the recommended caloric intake for a 12-year-old boy?

15a. _____

(b) What is the age of a boy whose recommended caloric intake is 2250 calories?

15b. _____

(c) Interpret the slope of $y = 125x + 1125$.

15c. _____

(d) Why would this equation not be accurate for a 3-year-old male?

15d. _____

(e) Graph the equation in a rectangular coordinate system. Label the axes appropriately.

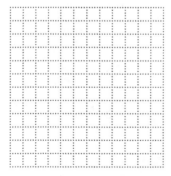

Five-Minute Warm-Up 3.5
Point-Slope Form of a Line

1. Solve $y - (-2) = -4(x - 3)$ for y.

    1. _____

2. Solve $9x - 3y = -12$ for y.

    2. _____

3. Evaluate: $\dfrac{8 - 5}{2 - 11}$

    3. _____

4. Evaluate: $\dfrac{0.4 - (-0.2)}{9.9 - 10.2}$

    4. _____

5. Use the Distributive Property to simplify: $-\dfrac{7}{4}(4x - 12)$

    5. _____

6. Find the slope of the line that contains $(-7, 10)$ and $(-3, 6)$.

    6. _____

Name:
Instructor:
Date:
Section:

Guided Practice 3.5
Point-Slope Form of a Line

Objective 1: Find the Equation of a Line Given a Point and a Slope

1. What equation do we use to write the *Point-Slope Form* of a line with slope m and containing (x_1, y_1)?

   _____

2. Find the equation of a line whose slope is 9 and that contains the point $(-2, -1)$.
   (See textbook Example 1)

   (a) Identify values: $m = $ _____; $x_1 = $ _____; $y_1 = $ _____

   (b) Point-slope form of a line: _____

   (c) Substitute the values from (a) into the equation from (b): _____

   (d) Simplify: _____

   (e) Use the Distributive Property: _____

   (f) Simplify and write the equation in slope-intercept form: _____

3. Find the equation of a line whose slope is $-\dfrac{5}{6}$ and that contains the point $(0, 12)$.
   (See textbook Example 2)

   (a) Identify values: $m = $ _____; $x_1 = $ _____; $y_1 = $ _____

   (b) Point-slope form of a line: _____

   (c) Substitute the values from (a) into the equation from (b): _____

   (d) Simplify and use the Distributive Property: _____

4. Find the equation a horizontal line that contains the point $(5, -11)$. *(See textbook Example 3)*

   (a) Identify values: $m = $ _____; $x_1 = $ _____; $y_1 = $ _____

   (b) Point-slope form of a line: _____

   (c) Substitute the values from (a) into the equation from (b): _____

   (d) Simplify: _____

Guided Practice 3.5

Objective 2: Find the Equation of a Line Given Two Points

5. To find the equation of a line given two points, the first step is to find _____.

6. Find the equation of the line through the points $(1, 5)$ and $(-1, -1)$. If possible, write the equation in slope-intercept form. *(See textbook Example 4)*

Step 1: Find the slope of the line containing the points.

Formula for the slope of a line through two points: **(a)** _____

Identify values: **(b)** $x_1 =$ _____ ; $y_1 =$ _____

   $x_2 =$ _____ ; $y_2 =$ _____

Substitute values into (a) and simplify: **(c)** _____

Step 2: Substitute the slope found in Step 1 and either point into the point-slope form of a line to find the equation.

Point-slope form of a line: **(d)** _____

Identify values: **(e)** $m =$ _____ ; $x_1 =$ _____ ; $y_1 =$ _____

Substitute values into (e):   **(f)** _____

Step 3: Solve the equation for y.

Distribute the 3:   **(g)** _____

Add _____ to both sides: **(h)** _____

Is the equation in slope-intercept form?   **(i)** _____

Try using (x_2, y_2) in (d). Does this change the equation of the line found in (i)?   **(j)** _____

Identify the slope: **(k)** _____

Identify the y-intercept: **(l)** _____

7. Find the equation of a line through the points $(-5, -1)$ and $(-5, 3)$. *(See textbook Example 5)*

   (a) Find the slope of the line: _____

   (b) What do we know about the graph of this line? _____

   (c) No matter what value of y we choose, the x-coordinate of the point on the line will be _____

   (d) Write the equation of the line: _____

Name:
Instructor:

Date:
Section:

Do the Math Exercises 3.5
Point-Slope Form of a Line

In Problems 1–4, find the equation of the line that contains the given point and with the given slope. Write the equation in slope-intercept form and graph the line.

1. (4, 1); slope = 6

2. (6, –3); slope = –5

1. _____

2. _____

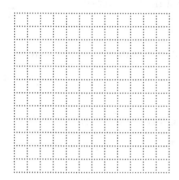

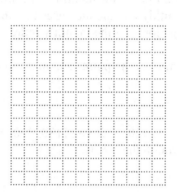

3. (–8, 2); slope = $-\dfrac{1}{2}$

4. (–7, –1); slope = 0

3. _____

4. _____

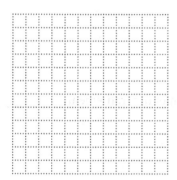

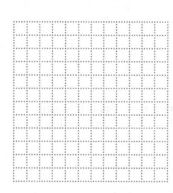

In Problems 5–6, find the equation of the line that contains the given point and satisfies the given information. Write the equation in slope-intercept form, if possible.

5. Horizontal line that contains (–6, –1)

6. Vertical line that contains (4, –3)

5. _____

6. _____

In Problems 7–10, find the equation of the line that contains the given points. Write the equation in slope-intercept form, if possible.

7. (–2, 4) and (2, –2)

8. (4, 18) and (–1, 3)

7. _____

8. _____

Do the Math Exercises 3.5

9. (−6, 5) and (7, 5)   10. (−3, 8) and (−3, 1)   9. _____

10.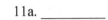

11. **Retirement Plans** Based on the retirement plan available by his employer, Kei knows that if he retires after 20 years, his monthly retirement income will be $3150. If he retires after 30 years, his monthly income increases to $3600. Let x represent the number of years of service and y represent the monthly retirement income.

(a) Interpret the meaning of the point (30, 3600) in the context of this problem.   11a. _____

(b) Plot the ordered pairs (20, 3150) and (30, 3600) in a rectangular coordinate system and graph the line through the points.

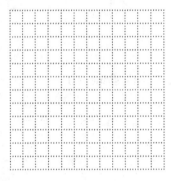

(c) Find the linear equation, in slope-intercept form, that relates the monthly retirement income, y, to the number of years of service, x.   11c. _____

(d) Use the equation found in part (c) to find the monthly income for 15 years of service.   11d. _____

(e) Interpret the slope.   11e. _____

Five-Minute Warm-Up 3.6
Parallel and Perpendicular Lines

1. Determine the reciprocal of -5.

    1. _____

2. Determine the reciprocal of $\frac{1}{3}$.

    2. _____

3. Determine the reciprocal of $-\frac{9}{4}$.

    3. _____

4. Solve $-4x + y = -12$ for y.

    4. _____

5. Identify the following information using: $-x + 4y = -2$
   (a) the slope of the line   (b) the y-intercept of the line

    5a. _____

    5b. _____

6. Find the equation of the line that has slope $-\frac{3}{2}$ and contains the point $(5, -2)$.

    6. _____

Name:
Instructor:
Date:
Section:

Guided Practice 3.6
Parallel and Perpendicular Lines

Objective 1: Determine Whether Two Lines Are Parallel

1. In your own words, write a definition of parallel lines.

2. Two lines are parallel if and only if they have the same _____ and different _____.

3. If $L_1 : y = m_1 x + b_1$ and $L_2 : y = m_2 x + b_2$ and L_1 is parallel to L_2, then m_1 _____ m_2 and b_1 _____ b_2.

4. Determine whether the line $4x - y = 1$ is parallel to $8x - 2y = -10$. *(See textbook Example 2)*

    To find the slope and *y*-intercept, we solve each equation for *y* so that each is in slope-intercept form.

    (a) Equation 1, $4x - y = 1$, written in slope-intercept form: _____

    (b) Determine the slope and *y*-intercept of the line in equation 1:   slope: _____ ; *y*-intercept: _____

    (c) Equation 2, $8x - 2y = -10$, written in slope-intercept form: _____

    (d) Determine the slope and *y*-intercept of the line in equation 2:   slope: _____ ; *y*-intercept: _____

    (e) Conclusion: _____

Objective 2: Find the Equation of a Line Parallel to a Given Line

5. Find an equation for the line that is parallel to $5x - 2y = -1$ and contains the point $(-3, -4)$. *(See textbook Example 3)*

Step 1: Find the slope of the given line by putting the equation in slope-intercept form.   (a) _____

Step 2: Use the point-slope form of a line with the given point and the slope found in Step 1 to find the equation of the parallel line.

    Slope of the line found in Step 1:   (b) _____

    Slope of the parallel line: (c) _____

    Identify values: (d) $m =$ _____ ; $x_1 =$ _____ ; $y_1 =$ _____

    Point-slope form of a line: (e) _____

    Substitute values into (e):   (f) _____

Step 3: Put the equation in slope-intercept form by solving for *y*.

    Solve for *y*:   (g) _____

Sullivan/Struve/Mazzarella, *Elementary & Intermediate Algebra, 3e*
Copyright © 2014 Pearson Education, Inc.

Guided Practice 3.6

Objective 3: Determine Whether Two Lines are Perpendicular

6. In your own words, write a definition of perpendicular lines.

7. Two lines are perpendicular if and only if the product of their slopes is _____.

8. If $L_1 : y = m_1 x + b_1$ and $L_2 : y = m_2 x + b_2$ and L_1 is perpendicular to L_2, then $m_1 \cdot m_2 =$ ___ or $m_1 =$ ___.

9. Determine whether the line $y = \dfrac{3}{4}x + 6$ is perpendicular to $y = \dfrac{4}{3}x - 2$. *(See textbook Example 6)*

   (a) Determine the slope and y-intercept of the line in equation 1:   slope: _____; y-intercept: _____

   (b) Determine the slope and y-intercept of the line in equation 2:   slope: _____; y-intercept: _____

   (c) What is the product of the slopes of the two lines?   _____

   (d) Conclusion: _____

10. Find an equation of the line that is perpendicular to the line $-2x + y = 7$ and contains the point $(3, -12)$. Write the equation in slope-intercept form. *(See textbook Example 8)*

Step 1: Find the slope of the given line by putting the equation in slope-intercept form.		(a) _____
Step 2: Find the slope of the perpendicular line.	Slope of the line found in Step 1:	(b) _____
	Slope of the perpendicular line:	(c) _____
	Identify values:	(d) $m =$ ___; $x_1 =$ ___; $y_1 =$ ___
Step 3: Use the point-slope form of a line with the given point and the slope found in Step 2 to find the equation of the perpendicular line.	Point-slope form of a line:	(e) _____
	Substitute values into (e):	(f) _____
Step 4: Put the equation in slope-intercept form by solving for y.	Solve for y:	(g) _____

Name:
Instructor:

Date:
Section:

Do the Math Exercises 3.6
Parallel and Perpendicular Lines

In Problems 1–3, fill in the chart with the missing slopes.

	Slope of the Given Line	Slope of a Line Parallel to the Given Line	Slope of a Line Perpendicular to the Given Line
1.	$m = 4$		
2.	$m = -\dfrac{1}{8}$		
3.	$m = \dfrac{5}{2}$		

In Problems 4–7, determine if the lines are parallel, perpendicular, or neither.

4. $L_1 : y = -4x + 3$
   $L_2 : y = 4x - 1$

5. $L_1 : y = 3x - 1$
   $L_2 : y = 6 - \dfrac{x}{3}$

4. _____

5. _____

6. $L_1 : x - 4y = 24$
   $L_2 : 2x - 8y = -8$

7. $L_1 : x - 2y = -8$
   $L_2 : x + 2y = 2$

6. _____

7. _____

In Problems 8–11, find the equation of the line that contains the given point and is parallel to the given line. Write the equation in slope-intercept form, if possible.

8. $(7, -5); \ y = 2x + 6$

9. $(-4, 5); \ x = -3$

8. _____

9. _____

Sullivan/Struve/Mazzarella, *Elementary & Intermediate Algebra*, 3e
Copyright © 2014 Pearson Education, Inc.

Do the Math Exercises 3.6

10. $(4, -8)$; $y = -1$

11. $(6, 7)$; $2x + 3y = 9$

10. _____

11. _____

In Problems 12–15, find the equation of the line that contains the given point and is perpendicular to the given line. Write the equation in slope-intercept form, if possible.

12. $(4, 7)$; $y = \dfrac{1}{3}x - 3$

13. $(-2, -5)$; $y = -2x + 5$

12. _____

13. _____

14. $(3, -6)$; y-axis

15. $(11, -6)$; x-axis

14. _____

15. _____

Five-Minute Warm-Up 3.7
Linear Inequalities in Two Variables

1. Solve: $x + 5 < -3$

   1. _____

2. Solve: $5x - 2 \geq -17$

   2. _____

3. Solve: $-3 + 7(x - 2) \geq -2(1 - 5x)$

   3. _____

4. Graph: $y = 2x - 1$

   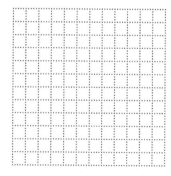

Name:
Instructor:

Date:
Section:

Guided Practice 3.7
Linear Inequalities in Two Variables

Objective 1: Determine Whether an Ordered Pair Is a Solution to a Linear Inequality

1. *True or False* To determine if an ordered pair is a solution to the linear inequality, substitute the values for the variables into the inequality. If a true statement results, then the ordered pair is a solution to the inequality.

1. _____

2. Determine if $(1, 3)$ is a solution to $y \leq 2x + 1$. *(See textbook Example 1)*

2. _____

Objective 2: Graph Linear Inequalities

3. To graph any linear inequality in two variables, you must first graph the corresponding equation. If the inequality is strict ($<$ or $>$), you should use a _____ line. If the inequality is nonstrict ($\leq$ or $\geq$), you should use a _____ line.

4. The graph of a line separates the *xy*-plane into two _____.

5. If a test point satisfies the inequality, then every ordered pair that lies in that half plane also satisfies the inequality. To represent this solution set, we _____ the half plane containing the test point.

6. Graph the linear inequality $y \geq x + 2$. *(See textbook Example 2)*

Step 1: We replace the inequality symbol with an equal sign and graph the corresponding line.	Write the equation:	**(a)** _____
	Identify the slope:	**(b)** _____
	Identify the *y*-intercept:	**(c)** _____
	Graph the line on the grid provided in (h). Is the line connecting these points solid or dashed?	**(d)** _____
Step 2: We select any test point that is not on the line and determine whether the test point satisfies the inequality. When the line does not contain the origin, it is usually easiest to choose the origin, (0, 0), as the test point.	Select a test point:	**(e)** _____
	Substitute the values for the variables into the inequality. Is the statement true or false?	**(f)** _____
	Which half plane should be shaded; the half plane containing the test point or the opposite half plane?	**(g)** _____
	Graph the solution set by using the slope and *y*-intercept, graphing the line, and shading the appropriate half plane.	**(h)**

$y \geq x + 2$

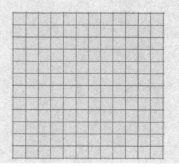

Sullivan/Struve/Mazzarella, *Elementary & Intermediate Algebra*, 3e
Copyright © 2014 Pearson Education, Inc.

Guided Practice 3.7

7. Graph the linear inequality $4x - y > 0$. *(See textbook Example 4)*

Replace the inequality symbol with an equal sign to obtain the graph of $4x - y > 0$. Is the line solid or dashed?

(a) _____

Select a test point. Can you use (0, 0)?

(b) _____

Determine if the inequality is satisfied by your test point? If the statement is true, shade the half plane containing the test point. If the statement is false, shade the opposite half plane.

(c)

Objective 3: Solve Problems Involving Linear Inequalities

8. Wood Flooring Mary Jo owns a discount flooring store and sells two different grades of wood flooring. There is a domestic cherry wood that sells for $4.32 per square foot and a reclaimed Douglas fir that sells for $2.50 per square foot. How many square feet of wood flooring does she need to sell to have an income of at least $5000? *(See textbook Example 6)*

(a) Write a linear inequality that describes Mary Jo's options for selling wood flooring.

Step 1: **Identify** We want to determine how many square feet of each she should sell to earn at least $5000. This requires an inequality in two variables.

Step 2: **Name the Unknowns** Let x represent the number of square feet of cherry flooring sold and let y represent the number of square feet of Douglas fir sold.

Step 3: **Translate** If cherry wood sells for $4.32 per square foot and Douglas fir sells for $2.50 per square foot, write an inequality the represents her total income greater than or equal to $5000.

(a) _____

(b) Will Mary Jo make her sales goal if she sells 600 square feet of cherry wood and 850 square feet of Douglas fir?

(b) _____

Name:
Instructor:

Date:
Section:

Do the Math Exercises 3.7
Linear Inequalities in Two Variables

In Problems 1–4, determine which of the following ordered pairs, if any, are solutions to the given linear inequality in two variables.

1. $y \leq x - 5$ $A(-4, -6)$ $B(0, -8)$ $C(3, 10)$
2. $y > -2x + 1$ $A(0, 0)$ $B(-2, 3)$ $C(2, -1)$

1. _____

2. _____

3. $3x + 5y \geq 4$ $A(2, 0)$ $B(-3, 4)$ $C(6, -2)$
4. $x > 10$ $A(3, 12)$ $B(11, -6)$ $C(10, 16)$

3. _____

4. _____

In Problems 5–12, graph each linear inequality.

5. $y \leq 4x + 2$

6. $y > x - 3$

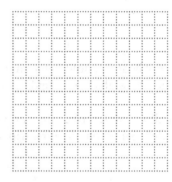

7. $y < 4$

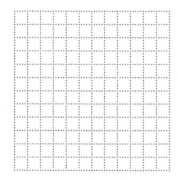

8. $y \geq \dfrac{3}{2}x - 2$

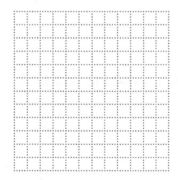

Do the Math Exercises 3.7

9. $2x + 6y < -6$

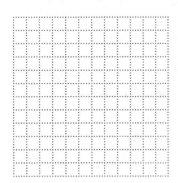

10. $6x - 8y \geq 24$

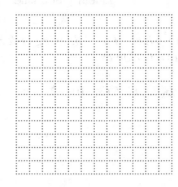

11. $x - y > 0$

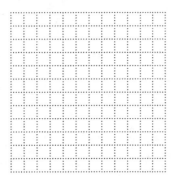

12. $x \geq 2$

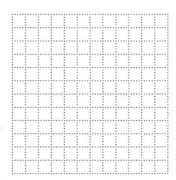

13. **Fishing Trip** Patrick's rowboat can hold a maximum of 500 pounds. In Patrick's circle of friends, the average adult weighs 160 pounds and the average child weighs 75 pounds.

(a) Write a linear inequality that describes the various combinations of the number of adults a and children c that can go on a fishing trip in the rowboat.

13a. _____

(b) Will the boat sink with 2 adults and 4 children?

13b. _____

(c) Will the boat sink with 1 adult and 5 children?

13c. _____

Five-Minute Warm-Up 4.1
Solving Systems of Linear Equations by Graphing

1. Graph each of the following linear equations.

   (a) $6x - 3y = -12$

   (b) $y = -\dfrac{3}{2}x - 1$

   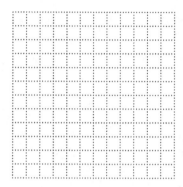
   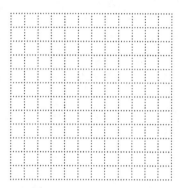

2. Determine whether $2x + 3y = -3$ is parallel to $-4x - 6y = 18$.

   2.

3. Determine whether $3x + 2y = 6$ is parallel to $y = -\dfrac{3}{2}x + 3$.

   3.

Name:
Instructor:

Date:
Section:

Guided Practice 4.1
Solving Systems of Linear Equations by Graphing

Objective 1: Determine Whether an Ordered Pair Is a Solution of a System of Linear Equations

1. Determine whether the given ordered pairs are solutions of the system of equations. *(See textbook Example 2)*

$$\begin{cases} x + y = 0 \\ 2x - 3y = 5 \end{cases}$$

(a) $(-2, 2)$ (b) $(1, -1)$

1a. _____

1b. _____

Objective 2: Solve a System of Linear Equations by Graphing

2. Solve the following system by graphing: $\begin{cases} x - y = -5 \\ 2x + y = -1 \end{cases}$ *(See textbook Example 3)*

Step 1: Graph the first equation in the system.	The equation $x - y = -5$ can be graphed using the intercepts or by writing the equation in slope-intercept form. We will use the intercepts. Name the x and y-intercept and then graph the line in the coordinate plane below.	(a) x-intercept _____ (b) y-intercept _____
Step 2: Graph the second equation in the system.	The equation $2x + y = -1$ can also be graphed by either method. This time we will use slope-intercept form. Write the equation in slope-intercept form. Name the slope and y-intercept and then graph the line in the coordinate plane below.	(c) equation _____ (d) slope _____ (e) y-intercept _____
Step 3: Determine the point of intersection of the two lines, if any.	Name the point where the lines appear to intersect.	(f) _____
Step 4: Verify that the point of intersection determined in Step 3 is a solution of the system.	Check that the point satisfies both of the equations in the system.	(g)

State the solution to the system: (h) _____

Sullivan/Struve/Mazzarella, *Elementary & Intermediate Algebra*, 3e
Copyright © 2014 Pearson Education, Inc.

Guided Practice 4.1

3. Solve the system by graphing: $\begin{cases} 2x + y = 1 \\ -4x - 2y = 6 \end{cases}$ (See textbook Example 5)   3. _____

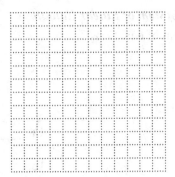

Objective 3: Classify Systems of Linear Equations as Consistent or Inconsistent

4. Complete the following chart which describes the solutions to a system of linear equations in two variables.

Number of Solutions	Classification of the System	Appearance of the Graph of the Two Lines
(a) no solution		
(b) infinitely many solutions		
(c) exactly one solution		

5. (a) Without graphing, determine the number of solutions of the system:
$\begin{cases} 4x - 3y = -6 \\ 8x - 6y = -12 \end{cases}$ (See textbook Example 7)   5a. _____

(b) State whether the system is consistent or inconsistent.   5b. _____

(c) If the system is consistent, state whether the equations are dependent or independent. 5c. _____

Name:
Instructor:

Date:
Section:

Do the Math Exercises 4.1
Solving Systems of Linear Equations by Graphing

In Problems 1–2, determine whether the ordered pair is a solution of the system of equations.

1. $\begin{cases} 2x + 5y = 0 \\ x - 3y = 11 \end{cases}$

   (a) $(5, -2)$
   (b) $(-5, 2)$
   (c) $(2, -3)$

2. $\begin{cases} x - 4y = 2 \\ 5x - 8y = -6 \end{cases}$

   (a) $(-8, -2)$
   (b) $(2, 2)$
   (c) $\left(\dfrac{-10}{3}, -\dfrac{4}{3}\right)$

1a. _____
1b. _____
1c. _____

2a. _____
2b. _____
2c. _____

In Problems 3–6, solve each system of equations by graphing.

3. $\begin{cases} y = -4x + 6 \\ y = -2x \end{cases}$

4. $\begin{cases} -2x + 5y = -20 \\ 4x - 10y = 10 \end{cases}$

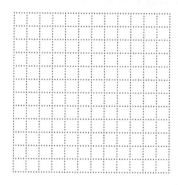
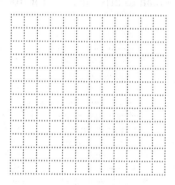

5. $\begin{cases} 2x + 5y = 2 \\ -x + 2y = -1 \end{cases}$

6. $\begin{cases} 3y = -4x - 4 \\ 8x + 2 = -6y - 6 \end{cases}$

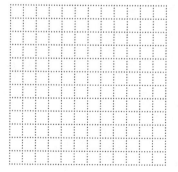
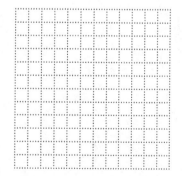

Sullivan/Struve/Mazzarella, *Elementary & Intermediate Algebra*, 3e
Copyright © 2014 Pearson Education, Inc.

Do the Math Exercises 4.1

In Problems 7–10, without graphing, determine the number of solutions of each system of equations. State whether the system is consistent or inconsistent. For those systems that are consistent, state whether the equations are dependent or independent.

7. $\begin{cases} 2x + y = -3 \\ -2x - y = 6 \end{cases}$

8. $\begin{cases} -x + y = 5 \\ x - y = -5 \end{cases}$

7. _____

8. _____

9. $\begin{cases} y - 3 = 0 \\ x - 3 = 0 \end{cases}$

10. $\begin{cases} 5x + y = -1 \\ 10x + 2y = -2 \end{cases}$

9. _____

10. _____

Set up a system of linear equations in two variables that models the problem. Then solve the system of linear equations by graphing.

11. **Car Rental** The Speedy Car Rental agency charges $60 per day plus $0.14 per mile while the Slow-but-Cheap Car Rental agency charges $30 per day plus $0.20 per mile. Determine the number of miles for which the cost of the car rental will be the same for both companies. If you plan to drive the car for 400 miles, which company should you use?

11. _____

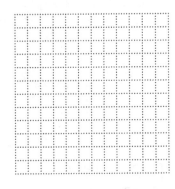

Five-Minute Warm-Up 4.2
Solving Systems of Linear Equations Using Substitution

1. Solve $5x - 3y = -18$ for y.

    1. _____

2. Solve $3x + 8y = -12$ for x.

    2. _____

3. Solve: $\dfrac{3}{2}x - \dfrac{5}{2}(x + 6) = -8$

    3. _____

4. Solve: $-\left(\dfrac{7}{3}y + 6\right) - \dfrac{2}{3}y = -15$

    4. _____

Guided Practice 4.2
Solving Systems of Linear Equations Using Substitution

Objective 1: Solve a System of Linear Equations Using the Substitution Method

1. Solve the following system by substitution: $\begin{cases} 4x + 3y = -23 & (1) \\ y = -x - 7 & (2) \end{cases}$ *(See textbook Example 1)*

Step 1: Solve one of the equations for one of the unknowns. Equation (2) is already solve for y: $y = -x - 7$

Step 2: Substitute the expression for y into equation (1).

Equation (1): **(a)** _____

Substitute into equation (1): **(b)** _____

Step 3: Solve the equation for x.

Distribute the 3: **(c)** _____

Combine like terms: **(d)** _____

Add 21 to both sides: **(e)** $x =$ _____

Step 4: Substitute your value for x into equation 2 and solve for y.

Equation (2): **(f)** _____

Substitute for x: **(g)** _____

Solve for y: **(h)** _____

Step 5: Verify your solution.

Check that your ordered pair satisfies both of the equations.

State the solution: **(i)** _____

2. Solve the following system by substitution: $\begin{cases} 3x - 2y = -8 \\ x + 3y = 34 \end{cases}$ *(See textbook Example 3)*   2. _____

Guided Practice 4.2

3. Solve the following system by substitution: $\begin{cases} 5x + 8y = 4 \\ y = -\dfrac{5}{8}x + \dfrac{1}{2} \end{cases}$ 3. _____

(See textbook Example 5)

Objective 2: Solve Applied Problems Involving Systems of Linear Equations

4. **Unknown Integers** The sum of two numbers is −11 and their difference is −101. Find the numbers.
(See textbook Example 6)

   Step 1: Identify This is a direct translation problem involving unknown integers.

   Step 2: Name Let the two unknown numbers be identified as x and y.

   Step 3: Translate
   (a) Write an equation showing the sum of the two numbers is −11: 4a. _____

   (b) Write an equation showing the difference of the two numbers is −101: 4b. _____

   Step 4: Solve
   (c) Write the system to be solved: 4c. _____

   (d) Solve one of the equations for one of the variables. Substitute this expression into the opposite equation and solve for the unknown. What is the solution to the system? 4d. _____

   Step 5: Check Verify that the values you found satisfy the system and make sense in the original problem.

   Step 6: Answer the question: 4e. _____

Name:
Instructor:

Date:
Section:

Do the Math Exercises 4.2
Solving Systems of Linear Equations Using Substitution

In Problems 1–4, solve each system of equations using substitution.

1. $\begin{cases} 5x + 2y = -5 \\ 3x - y = -14 \end{cases}$

2. $\begin{cases} y = \dfrac{2}{3}x + 1 \\ y = -\dfrac{3}{2}x + 40 \end{cases}$

1. _____

2. _____

3. $\begin{cases} y = 5x - 3 \\ y = 2x - \dfrac{21}{5} \end{cases}$

4. $\begin{cases} y = 4x \\ 2x - 3y = 5 \end{cases}$

3. _____

4. _____

In Problems 5–8, solve each system of equations using substitution. State whether the system is inconsistent, or consistent and dependent.

5. $\begin{cases} 2x - y = 3 \\ y - 2x = 3 \end{cases}$

6. $\begin{cases} y = -x + 1 \\ x + y = 1 \end{cases}$

5. _____

6. _____

7. $\begin{cases} 4 + y = 3x \\ 6x - 2y = -2 \end{cases}$

8. $\begin{cases} x + y = 2 \\ 2x + 2y = 2 \end{cases}$

7. _____

8. _____

Sullivan/Struve/Mazzarella, *Elementary & Intermediate Algebra*, 3e

Do the Math Exercises 4.2

In Problems 9–14, solve each system of equations using substitution.

9. $\begin{cases} y = \dfrac{3}{4}x \\ x = 4(y-1) \end{cases}$

10. $\begin{cases} x = 3y + 6 \\ y - \dfrac{1}{3}x = -2 \end{cases}$

9. _____

10. _____

11. $\begin{cases} 2x + 3y = -1 \\ 2x - 9y = -9 \end{cases}$

12. $\begin{cases} -4x + y = 3 \\ 8x - 2y = 1 \end{cases}$

11. _____

12. _____

13. $\begin{cases} 18x - 6y = -7 \\ x + 2y = 0 \end{cases}$

14. $\begin{cases} \dfrac{2}{3}x + \dfrac{1}{3}y = 1 \\ \dfrac{1}{2}x + \dfrac{1}{4}y = \dfrac{3}{4} \end{cases}$

13. _____

14. _____

15. **Investment** Elaine wants to invest part of her $24,000 Virginia Lottery winnings in an international stock fund that yields 4% annual interest and the remainder in a domestic growth fund that yields 6.5% annually. Use the system of equations to determine the amount Elaine should invest in each account to earn $1360 interest at the end of one year, where x represents the amount invested in the international stock fund and y represents the amount invested in the domestic growth fund.

15. _____

$$\begin{cases} x + y = 24{,}000 \\ 0.04x + 0.065y = 1360 \end{cases}$$

Name:
Instructor:

Date:
Section:

Five-Minute Warm-Up 4.3
Solving Systems of Linear Equations Using Elimination

1. What is the additive inverse of $\dfrac{5}{4}$?

   1. _____

2. What is the additive inverse of -6?

   2. _____

3. Distribute: $-\dfrac{7}{4}(-12x + 8y)$

   3. _____

4. Solve: $-6x + 7 = 3(-2x + 1)$

   4. _____

5. Simplify: $8x - 7y + 3 + -8x - 2y - 12$

   5. _____

6. (a) Find the LCM of 8 and 12.

   6a. _____

   (b) By what factor must 8 be multiplied so that the product of the factor and 8 equals the LCM?

   6b. _____

   (c) By what factor must 12 be multiplied so that the product of the factor and 12 equals the LCM?

   6c. _____

Sullivan/Struve/Mazzarella, *Elementary & Intermediate Algebra*, 3e
Copyright © 2014 Pearson Education, Inc.

Name:
Instructor:
Date:
Section:

Guided Practice 4.3
Solving Systems of Linear Equations Using Elimination

Objective 1: Solve a System of Linear Equations Using the Elimination Method

1. This method is called "elimination" because one of the variables is eliminated through the process of addition. The elimination method is also sometimes referred to as the *addition method*.

Solve the following system by elimination: $\begin{cases} -x + y = 12 \\ x - 3y = -26 \end{cases}$ *(See textbook Example 1)*   1. _____

2. Solve the following system by elimination: $\begin{cases} 2x + y = -4 \quad (1) \\ 3x + 5y = 29 \quad (2) \end{cases}$ *(See textbook Example 2)*

Step 1: Write each equation in standard form, $Ax + By = C$.

Both equations are already in standard form.

Step 2: Our first goal is to get the coefficients on one of the variables to be additive inverses. In looking at this system, we can make the coefficients of y be additive inverses by multiplying equation (1) by -5.

Multiply both sides of (1) by -5, use the Distributive Property, and then write the equivalent system of equations.

$\begin{cases} 2x + y = -4 \\ 3x + 5y = 29 \end{cases}$

(a) $\begin{cases} \underline{\qquad\qquad} \quad (1) \\ \underline{\qquad\qquad} \quad (2) \end{cases}$

Step 3: We now add equations (1) and (2) to eliminate the variable y and then solve for x.

Add (1) and (2):   (b) _____

Divide both sides by -7:   (c) $x =$ _____

Step 4: Substitute your value for x into either equation (1) or equation (2). We will use equation (1) as it looks like less work.

Equation (1): $2x + y = -4$

Substitute your value for x and solve for y.

(d) $y =$ _____

Step 5: Check your answer in both of the original equations. If both equations yield a true statement, you have the correct answer.

Write the ordered pair that is the solution to the system.   (e) _____

Guided Practice 4.3

3. Solve the following system by elimination:
$\begin{cases} \frac{1}{2}x - \frac{1}{5}y = \frac{2}{5} & (1) \\ -\frac{3}{8}x + \frac{5}{4}y = -\frac{5}{2} & (2) \end{cases}$

(See textbook Example 3)

(a) Multiply both sides of equation (1) by the LCD of 2 and 5:   3a. _____

(b) Multiply both sides of equation (2) by the LCD of 2, 4, and 8:   3b. _____

(c) Solve this equivalent system by elimination and state the solution: 3c. _____

4. When solving a system of linear equations by elimination, if both variables were eliminated and the statement $0 = 6$ was left, what is the solution to this system?   4. _____
(See textbook Example 4)

5. When solving a system of linear equations by elimination, if both variables were eliminated and the statement $0 = 0$ was left, what is the solution to this system?   5. _____
(See textbook Example 5)

Objective 2: Solve Applied Problems Involving Systems of Linear Equations

6. **Ticket Prices** Admission to the movie theatre costs $4.50 per child and $7.50 per adult. If the theatre took in $3337.50 and had 525 patrons, how many of each ticket were sold? *(See textbook Example 6)*

(a) If a represents the number of adult tickets sold and c represents the number of children's tickets sold, write an equation that shows that the total number of tickets sold is 525.   6a. _____

(b) Use the rate for a child's ticket and number of children's tickets sold plus the rate for an adult ticket and the number of adult tickets sold to write an equation showing the total revenue for this performance is $3337.50.   6b. _____

(c) Solve the system.   6c. _____

(d) Answer the question in the problem.   6d. _____

Do the Math Exercises 4.3
Solving Systems of Linear Equations Using Elimination

In Problems 1–2, solve each system of equations using elimination.

1. $\begin{cases} 2x + 6y = 6 \\ -2x + y = 8 \end{cases}$

2. $\begin{cases} 3x + 2y = 7 \\ -3x + 4y = 2 \end{cases}$

1. _____

2. _____

In Problems 3–4, solve each system of equations using elimination. State whether the system is inconsistent or consistent and dependent.

3. $\begin{cases} x - y = -4 \\ -2x + 2y = -8 \end{cases}$

4. $\begin{cases} 2x + 5y = 15 \\ -6x - 15y = -45 \end{cases}$

3. _____

4. _____

In Problems 5–10, solve each system of equations using elimination.

5. $\begin{cases} 5x - y = 3 \\ -10x + 2y = 2 \end{cases}$

6. $\begin{cases} 2x + 3y = 2 \\ 5x + 7y = 0 \end{cases}$

5. _____

6. _____

7. $\begin{cases} x + 3y = 6 \\ 9y = -3x + 18 \end{cases}$

8. $\begin{cases} 5x + 7y = 6 \\ 2x - 3y = 11 \end{cases}$

7. _____

8. _____

Do the Math Exercises 4.3

9. $\begin{cases} 12x + 15y = 55 \\ \dfrac{1}{2}x + 3y = \dfrac{3}{2} \end{cases}$

10. $\begin{cases} -2.4x - 0.4y = 0.32 \\ 4.2x + 0.6y = -0.54 \end{cases}$

9. _____

10. _____

In Problems 11–14, solve by any method: graphing, substitution, or elimination.

11. $\begin{cases} 2x - 3y = -10 \\ -3x + y = 1 \end{cases}$

12. $\begin{cases} 0.25x + 0.10y = 3.70 \\ x + y = 25 \end{cases}$

11. _____

12. _____

13. **Carbs** Yvette and José go to McDonald's for breakfast. Yvette orders two sausage biscuits and one 16-ounce orange juice. The entire meal had 98 grams of carbohydrates. José orders three sausage biscuits and two 16-ounce orange juices and his meal had 168 grams of carbohydrates. How many grams of carbohydrates are in a sausage biscuit? How many grams of carbohydrates are in a 16-ounce orange juice? Solve the system

13. _____

$$\begin{cases} 2b + u = 98 \\ 3b + 2u = 168 \end{cases}$$

where b represents the number of grams of carbohydrates in a sausage biscuit and u represents the number of grams of carbohydrates in an orange juice to find the answers.

Name:
Instructor:

Date:
Section:

Five-Minute Warm-Up 4.4
Solving Direct Translation, Geometry, and Uniform Motion Problems Using Systems of Linear Equations

1. If you are traveling at an average speed of 65 miles per hour (mph),
   (a) how far will you travel in 5 hours?
   (b) If you went 195 miles, how long were you traveling?

   1a. _____

   1b. _____

2. Suppose that you have $7000 in a Certificate of Deposit (CD) that pays 4.5% annual simple interest. What is the amount of interest paid
   (a) after 3 years?
   (b) after two months?

   2a. _____

   2b. _____

3. Translate each of the following sentences into an equation. *DO NOT SOLVE.*

   (a) Seven less than twice a number is 45.   3a. _____

   (b) 5 times the sum of 3 and some number is the same as half the number.   3b. _____

   (c) The quotient of a number and 3 is equal to the number subtracted from 20.   3c. _____

Name:
Instructor:

Date:
Section:

Guided Practice 4.4
Solving Direct Translation, Geometry, and Uniform Motion Problems Using Systems of Linear Equations

Objective 1: Model and Solve Direct Translation Problems

1. Fun with Numbers The sum of four times a first number and a second number is 68. If the first number is decreased by twice the second number the result is −1. Find the numbers. *(See textbook Example 1)*

Step 1: Identify We are looking for the two unknown numbers.

Step 2: Name Let x represent the first number and y represent the second number.

Step 3: Translate

(a) Write the equation for the sum of four times a first number and a second number is 68:

1a. _____

(b) Write the equation for the first number decreased by twice the second number results in −1:

1b. _____

(c) **Step 4: Solve** Use either substitution or elimination to find the solution to your system of equations. What is the solution to the system?

1c. _____

Step 5: Check Verify that the values you found satisfy the system and make sense in the original problem.

(d) **Step 6: Answer** the question in the problem:

1d. _____

Objective 2: Model and Solve Geometry Problems

2. Two angles are supplementary if the sum of the measures of the angles is:

2. _____

3. If one angle is 30° more than its supplement, find the measures of the two angles. *(See textbook Example 3)*

(a) If one angle has x degrees and the supplement has y degrees, write an equation showing the sum of the measures of the two angles:

3a. _____

(b) Write an equation showing one angle is 30° more than its supplement:

3b. _____

(c) Solve your system of two linear equations.

3c. _____

(d) Answer the question in the problem.

3d. _____

Sullivan/Struve/Mazzarella, *Elementary & Intermediate Algebra*, 3e
Copyright © 2014 Pearson Education, Inc.

Guided Practice 4.4

Objective 3: Model and Solve Uniform Motion Problems

4. Airplane Speed A plane can fly 2400 miles east, with the wind, in 6 hours. The return trip west to the same point, against the wind, takes 8 hours. Find the airspeed of the plane and the effect wind resistance has on the plane. *(See textbook Example 4)*

Step 1: Identify This is a uniform motion problem. We want to determine the airspeed of the plane and the effect of wind resistance.

Step 2: Name We will let a represent the airspeed of the plane and w represent the effect of wind resistance.

Step 3: Translate
(a) Flying east, with the wind, the ground speed of the plane is: 4a. _____

(b) Flying west, against the wind, the ground speed of the plane is: 4b. _____

Complete the table from the information given, using a for speed of the plane and w for speed of the wind.

	Distance (miles)	Rate (mph)	Time (hours)
(c) With the wind			
(d) Against the wind			

(e) Write a system of equations using the information in your table: 4e. _____

(f) **Step 4: Solve** Use either substitution or elimination to find the solution to your system of equations. 4f. _____

Step 5: Check Does your solution satisfy each of the equations in the system? Does your answer seem reasonable?

(g) **Step 6: Answer** the question in the problem: 4g. _____

Do the Math Exercises 4.4
Solving Direct Translation, Geometry, and Uniform Motion Problems Using Systems of Linear Equations

Complete the system of equations. Do not solve the system.

1. The sum of two numbers is 90. If 20 is added to 3 times the smaller number, the result exceeds twice the larger number by 50. Let a represent the smaller number and let b represent the larger number.

$$\begin{cases} a+b = 90 \\ \underline{} = \underline{} \end{cases}$$

Complete the system of equations. Do not solve the system.

2. The perimeter of a rectangle is 212 centimeters. The length is 8 centimeters less than three times the width. Let l represent the length of the rectangle and let w represent the width of the rectangle.

$$\begin{cases} 2w + 2l = 212 \\ \underline{} = \underline{} \end{cases}$$

Complete the system of equations. Do not solve the system.

3. A plane flew with the wind for 3 hours, covering 1200 miles. It then returned over the same route to the airport against the wind in 4 hours. Let a represent the airspeed of the plane and w represent the effect of wind resistance.

$$\begin{cases} 3(a+w) = 1200 \\ \underline{} = \underline{} \end{cases}$$

4. **Fun with Numbers** Find two numbers whose sum is 55 and whose difference is 17.   4. _____

5. **Fun with Numbers** The sum of two numbers is 32. Twice the larger subtracted from   5. _____
the smaller is –22. Find the numbers.

Do the Math Exercises 4.4

6. **Perimeter of a Parking Lot** A rectangular parking lot has a perimeter of 125 feet. The length of the parking lot is 10 feet more than the width. What is the length of the parking lot? What is the width?

   6. _____

7. **Working with Complements** The measure of one angle is 10° less than the measure of three times its complement. Find the measures of the two angles.

   7. _____

8. **Finding Supplements** The measure of one angle is 20° more than two-thirds the measure of its supplement. Find the measures of the two angles.

   8. _____

9. **Southwest Airlines Plane** A Southwest Airlines plane can fly 455 mph against the wind and 515 mph when it flies with the wind. Find the effect of the wind and the groundspeed of the airplane.

   9. _____

10. **Rowing** On Monday afternoon, Andrew rowed his boat with the current for 4.5 hours and covered 27 miles, stopping in the evening at a campground. On Tuesday morning he returned to his starting point against the current in 6.75 hours. Find the speed of the current and the rate at which Andrew rowed in still water.

    10. _____

11. **Horseback Riding** Monica and Gabriella enjoy riding horses at a dude ranch in Colorado. They decide to go down different trails, agreeing to meet back at the ranch later in the day. Monica's horse is going 4 mph faster than the Gabriella's, and after 2.5 hours, they are 20 miles apart. Find the speed of each horse.

    11. _____

Five-Minute Warm-Up 4.5
Solving Mixture Problems Using Systems of Linear Equations

1. Suppose that Sherry has a credit card balance of $2000. Each month, the credit card charges 18% annual simple interest on any outstanding balances.

    (a) What is the interest that Sherry will be charged on this loan after one month?   1a. _____

    (b) What is the balance on Sherry's credit card after one month?   1b. _____

2. Write the percent as a decimal.   2a. _____
    (a) 40%   (b) 0.15%
    2b. _____

3. Solve by any appropriate method: $\begin{cases} 2x + 5y = -11 \\ 3x + 2y = 11 \end{cases}$ 3. _____

Name:
Instructor:
Date:
Section:

Guided Practice 4.5
Solving Mixture Problems Using Systems of Linear Equations

Objective 1: Draw Up a Plan for Solving Mixture Problems

1. The cost of an adult ticket to the art museum is $7.50 and student tickets cost $4.00. When a group of 40 adults and students entered the museum, the total receipts were $202. How many adults and how many students were in the group that entered the museum? *(See textbook Example 1)*

Fill in the chart that summarizes the information in the problem.

	Number	Cost per Person in Dollars	Amount
(a) Adults			
(b) Students			
(c) Total			

(d) Write a system of equations to solve this problem. Do not solve the system. 1d. _____

Objective 2: Set Up and Solve Money Mixture Problems

2. Johnny has $6.75 in dimes and quarters. If he has 42 coins, how many quarters does Johnny have? *(See textbook Example 2)*

Fill in the chart that summarizes the information in the problem.

	Number of Coins	Value per Coin in Dollars	Total Value
(a) Quarters			
(b) Dimes			
(c) Total			

(d) Write a system of equations to solve this problem. 2d. _____

(e) Solve the system and answer the question in the problem. 2e. _____

Guided Practice 4.5

Objective 3: Set Up and Solve Dry Mixture and Percent Mixture Problems

3. A candy store sells chocolate-covered almonds for $6.50 per pound and chocolate-covered peanuts for $4.00 per pound. The manager decides to make a bridge mix that combines the almonds with the peanuts. She wants the bridge mix to sell for $6.00 per pound, and there should be no loss in revenue from selling the bridge mix versus the almonds and peanuts alone. How many pounds of chocolate-covered almonds and chocolate-covered peanuts are required to create 50 pounds of bridge mix? *(See textbook Example 4)*

Complete the chart that summarizes the information in the problem.

	Price $/Pound	Number of Pounds	Revenue
(a) Almonds			
(b) Peanuts			
(c) Mix			

(d) Write a system of equations to solve this problem. Do not solve the system.   3d. _____

4. The Chemistry stockroom has a jar labeled 25% hydrochloric acid and another labeled 40% hydrochloric acid. If 90 ml of 30% solution are needed for today's lab experiment, how much of each should the stockroom assistant mix together? *(See textbook Example 5)*

Complete the chart that summarizes the information in the problem.

	Number of ml	Concentration	Amount of Pure HCl
(a) 25% HCl Solution			
(b) 40% HCl Solution			
(c) 30% HCl Solution			

(d) Write a system of equations to solve this problem. Do not solve the system.   4d. _____

Name:
Instructor:

Date:
Section:

Do the Math Exercises 4.5
Solving Mixture Problems Using Systems of Linear Equations

In Problems 1–3, fill in the table from the information given. Then write the system that models the problem. DO NOT SOLVE.

1. John Murphy sells jewelry at art shows. He sells bracelets for $10 and necklaces for $15. At the end of one day John found that he had sold 69 pieces of jewelry and had receipts of $895.

   1. _____

	Number ·	Cost per Item =	Total Value
Bracelets			
Necklaces			
Total			

2. Sherry has a savings account that earns 2.75% simple interest per year and a certificate of deposit (CD) that earns 2% simple interest annually. At the end of one year, Sherry received $37.75 in interest on a total investment of $1700 in the two accounts.

   2. _____

	Principal ·	Rate =	Interest
Savings Account			
Certificate of Deposit			
Total			

3. A merchant wishes to mix peanuts worth $5 per pound and trail mix worth $2 per pound to yield 40 pounds of a nutty mixture that will sell for $3 per pound.

   3. _____

	Number of Pounds ·	Price per Pound =	Total Value
Peanuts			
Trail Mix			
Total			

In Problems 4–6, complete the system of linear equations to solve the problem. Do not solve.

4. On a school field trip, 22 people attended a dress rehearsal of the Broadway play, "Avenue Q." They paid $274 for the tickets, which cost $15 for each adult and $7 for each child. How many adults and how many children attended "Avenue Q"? Let a represent the number of adult tickets purchased and let c represent the number of children's tickets purchased.

$$\begin{cases} a + c = 22 \\ \underline{} + \underline{} = 274 \end{cases}$$

Sullivan/Struve/Mazzarella, *Elementary & Intermediate Algebra*, 3e
Copyright © 2014 Pearson Education, Inc.

Do the Math 4.5

5. You have a total of $2650 to invest. Account A pays 5% annual interest and account B pays 6.5% annual interest. How much should you invest in each account if you would like the investment to earn $155 at the end of one year? Let A represent the amount of money invested in the account that earns 5% annual interest and let B represent the amount of money invested in the account that earns 6.5% annual interest.

$$\begin{cases} A + B = \underline{\qquad} \\ \underline{\qquad} + \underline{\qquad} = 155 \end{cases}$$

6. The Latte Shoppe sells Bold Breakfast coffee for $8.60 per pound and Wake-Up coffee for $5.75 per pound. One day, the amount of Bold Breakfast coffee sold was 2 pounds less than twice the amount of Wake-Up coffee and the revenue received from selling both types of coffee was $143.45. How many pounds of each type of coffee were sold that day? Let B represent the number of pounds of Bold Breakfast coffee sold and let W represent the number of pounds of Wake-Up sold. Complete the system of equations:

$$\begin{cases} 8.60B + 5.75W = \underline{\qquad} \\ B = \underline{\qquad} - \underline{\qquad} \end{cases}$$

In Problems 7–9, write a system of equations and then solve.

7. **Ticket Pricing** A ticket on the roller coaster is priced differently for adults and children. One day there were 5 adults and 8 children in a group and the cost of their tickets was $48.50. Another group of 4 adults and 12 children paid $57. What is the price of each type of ticket?

7. _____

8. **Investments** Ann has $5000 to invest. She invests in two different accounts, one expected to return 4.5% and the other expected to return 9%. In order to earn $382.50 for the year, how much should she invest at each rate?

8. _____

9. **Alcohol** A laboratory assistant is asked to mix a 30% alcohol solution with 21 liters of an 80% alcohol solution to make a 60% alcohol solution. How many liters of the 30% alcohol solution should be used?

9. _____

Name:
Instructor:

Date:
Section:

Five-Minute Warm-Up 4.6
Systems of Linear Inequalities

1. Solve $18x + 30 \leq -6$ and write your answer in interval notation.

   1. _____

2. Solve $\frac{5}{2}(4x - 12) < 15x - 25$ and write your answer in interval notation.

   2. _____

3. Graph each of the following linear inequalities.
   (a) $y \geq -x + 2$
   (b) $8x - 4y < -16$

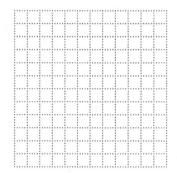
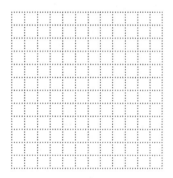

Name:
Instructor:
Date:
Section:

Guided Practice 4.6
Systems of Linear Inequalities

Objective 1: Determine Whether an Ordered Pair Is a Solution of a System of Linear Inequalities

1. Determine which of the following points, if any, satisfy the system of linear inequalities.
(See textbook Example 1)

$$\begin{cases} 2x - y \geq -3 \\ x + 5y < 4 \end{cases}$$

    (a) $(-1, 1)$ (b) $(-4, -3)$ (c) $(1, -1)$ 1. _____

Objective 2: Graph a System of Linear Inequalities

2. The only way to show the solution of a system of linear inequalities is by _____.

3. To graph the inequality $x - 8y > 4$, we use a _____ line as the boundry. *Refer to textbook Section 3.7.*

4. To graph the inequality $3x + 2y \geq -6$, we use a _____ line as the boundry.

5. To graph a linear inequality, we first graph the equation to determine the boundary line. This line divides the plane into two half-planes. To decide which half-plane to shade, we use a test point such as (0, 0), provided that this point does not lie on the line. If the test point satisfies the inequality, we shade the half-plane that contains the point. If the test point does not satisfy the inequality, we shade

_____.

6. When graphing a system of linear inequalities, we are looking for the ordered pairs that satisfy both inequalties simultaneously. Therefore, the solution of the system is the _____ of the graphs of the linear inequalties.

7. Graph the system: $\begin{cases} 2x + 3y > -3 \\ -3x + y \leq 2 \end{cases}$ *(See textbook Examples 2 and 3)*

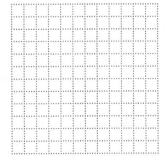

Sullivan/Struve/Mazzarella, *Elementary & Intermediate Algebra*, 3e
Copyright © 2014 Pearson Education, Inc.

165

Guided Practice 4.6

Objective 3: Solve Applied Problems Involving Systems of Linear Inequalities

8. Party Planning Alexis and Sarah are planning a barbeque for their friends. They plan to serve grilled fish and carne asada and want to spend at most $40 on the meat. The fish sells for $8 per pound and the carne asada is $5 per pound. Since most of their friends do not eat red meat, they plan to buy at least twice as much fish as carne asada. Let x represent the amount of carne asada purchased and y represent the amount of fish purchased. *(See textbook Example 5)*

(a) Write an inequality that describes how much Alexis and Sarah will spend on meat.   8a. _____

(b) Write an inequality that describes that amount of carne asada that will be purchased relative to the amount of fish that will be purchased.   8b. _____

(c) Since Alexis and Sarah will purchase a positive quantity of meat, write the two inequalities that describe this constraint.   8c. _____

_____

(d) Graph the system of inequalities.

(e) Can Alexis and Sarah purchase 5 pounds of carne asada and 12 pounds of fish and stay within their budget?   8e. _____

Do the Math Exercises 4.6
Systems of Linear Inequalities

In Problems 1–2, determine which of the following point(s), if any, is a solution of the system of linear inequalities.

1. $\begin{cases} 2x - 3y < 3 \\ 2x + y < -5 \end{cases}$
   (a) $(-4, 1)$
   (b) $\left(-\dfrac{3}{2}, -2\right)$
   (c) $(-1, -2)$

2. $\begin{cases} x - 2y > 2 \\ -3x - 2y \le 6 \end{cases}$
   (a) $\left(-1, -\dfrac{3}{2}\right)$
   (b) $(0, -4)$
   (c) $(4, -1)$

1a. _____

1b. _____

1c. _____

2a. _____

2b. _____

2c. _____

In Problems 3–6, graph each system of linear inequalities.

3. $\begin{cases} y > 2 \\ y > \dfrac{2}{3}x - 1 \end{cases}$

4. $\begin{cases} x + y < 2 \\ 3x - 5y \ge 0 \end{cases}$

Do the Math Exercises 4.6

5. $\begin{cases} -y < \dfrac{2}{3}x+1 \\ -3x+y \leq 2 \end{cases}$

6. $\begin{cases} x - y \leq 3 \\ 2x+3y \leq -9 \end{cases}$

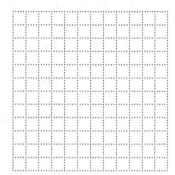
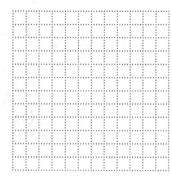

7. **Party Food** Steven and Christopher are planning a party. They plan to buy bratwurst for $4.00 per pound and hamburger patties that cost $3.00 per pound. They can spend at most $70 and think they should have no more than 20 pounds of bratwurst and hamburger patties. A system of inequalities that models the situation is given by

$$\begin{cases} 4b+3h \leq 70 \\ b+h \leq 20 \\ b \geq 0 \\ h \geq 0 \end{cases}$$

where b represents the number of pounds of bratwurst and h represents the number of pounds of hamburger patties.

(a) Graph the system of linear inequalities.

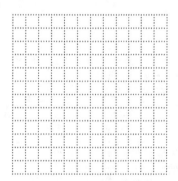

(b) Is it possible to purchase 10 pounds of bratwurst and 10 pounds of hamburger patties? 7b. _____

(c) Is it possible to purchase 15 pounds of bratwurst and 5 pounds of hamburger patties? 7c. _____

Name:
Instructor:

Date:
Section:

Five-Minute Warm-Up 5.1
Adding and Subtracting Polynomials

1. What is the coefficient of $-a^3$?

   1. _____

2. Combine like terms: $-3x^2 + x - 5 + x^2 - x + 9$

   2. _____

3. Simplify the expression: $5 - 2(x + 3y) - 4(2x - 1)$

   3. _____

4. Use the Distributive Property to simplify: $\dfrac{5}{3}(-9x + 6)$

   4. _____

5. Simplify: $x^2y + 6xy^2 - (3xy^2 + 2x^2y)$

   5. _____

6. Evaluate $-\dfrac{3}{4}x^2 - 5x$ for $x = -2$.

   6. _____

Name:
Instructor:

Date:
Section:

Guided Practice 5.1
Adding and Subtracting Polynomials

Objective 1: Define Monomial and Determine the Degree of a Monomial

1. For our study of polynomials, we begin with some definitions regarding monomials. *(See textbook Examples 1 – 3)*

(a) In your own words, what is a *monomial*? _____

(b) What is the coefficient of a monomial? _____

(c) How do you determine the degree of a monomial in one variable? _____

(d) How do you determine the degree of a monomial in more than one variable?
_____

Objective 2: Define Polynomial and Determine the Degree of a Polynomial

2. Next, we move on to some definitions used with polynomials. *(See textbook Example 4)*

(a) In your own words, define *polynomial*. _____

(b) When is a polynomial in *standard form*? _____

(c) How do you determine the degree of a polynomial? _____

3. Some polynomials have special names. Always simplify the polynomial first, if possible, before determining if the polynomial has one of the following specific names:

(a) a polynomial with exactly one term is a _____

(b) a polynomial with two different monomials is a _____

(c) a polynomial with exactly three different monomials is a _____

(d) a polynomial with more than three terms is called a _____

Objective 3: Simplify Polynomials by Combining Like Terms

4. In your own words, define *like terms*. _____

5. To add two polynomials, we need to combine the like terms of the polynomials. Parentheses are included to indicate the first polynomial added to a second polynomial.

Find the sum using horizontal addition: $(6x^3 + 3x^2 - 5x + 2) + (x^3 - 4x^2 + 3)$ *(See textbook Example 6)*

5. _____

Sullivan/Struve/Mazzarella, *Elementary & Intermediate Algebra*, 3e
Copyright © 2014 Pearson Education, Inc.

Guided Practice 5.1

6. Find the sum using vertical addition: $(x^2y - 5x^2y^2 - 3xy^2) + (-2x^2y - 9x^2y^2 + 4xy^2)$
(See textbook Examples 6 and 7)

6. _____

7. Find the difference: $(6z^3 + 2z^2 - 5) - (-3z^3 + 9z^2 - z + 1)$ (See textbook Example 8)

7. _____

8. Perform the indicated operations: $(6n^3 - 2n^2 + n - 1) - (4n^2 - 1) + (-6n^3 + 5n^2 + n)$
(See textbook Example 10)

8. _____

Objective 4: Evaluate Polynomials
9. Evaluate the polynomial $2x^3 - x^2 - 3x + 5$ for (See textbook Example 11)
   (a) $x = -1$ **(b)** $x = 3$

9a. _____

9b. _____

10. Evaluate the polynomial $-a^2b^2 + 2ab - b^3$ for $a = 2$ and $b = -1$.
(See textbook Example 12)

10. _____

Name:
Instructor:

Date:
Section:

Do the Math Exercises 5.1
Adding and Subtracting Polynomials

In Problems 1–2, determine whether the given expression is a monomial (Yes or No). For those that are monomials, state the coefficient and degree.

1. $-x^6 y$

2. y^{-1}

1. _____

2. _____

In Problems 3–4, determine whether the algebraic expression is a polynomial (Yes or No). If it is a polynomial, write the polynomial in standard form, determine its degree, and state if it is a monomial, binomial, or trinomial. If it is a polynomial with more than three terms, identify the expression as a polynomial.

3. $4y^{-2} + 6y - 1$

4. $p^5 - 3p^4 + 7p + 8$

3. _____

4. _____

In Problems 5–8, add the polynomials. Express your answer in standard form.

5. $(x^2 - 2) + (6x^2 - x - 1)$

6. $(3 - 12w^2) + (2w^2 - 5 + 6w)$

5. _____

6. _____

7. $\left(\dfrac{3}{8}b^2 - \dfrac{3}{5}b + 1\right) + \left(\dfrac{5}{6}b^2 + \dfrac{2}{15}b - 1\right)$

8. $(4a^2 + ab - 9b^2) + (-6a^2 - 4ab + b^2)$

7. _____

8. _____

Sullivan/Struve/Mazzarella, *Elementary & Intermediate Algebra*, 3e

Do the Math Exercises 5.1

In Problems 9–12, subtract the polynomials. Express your answer in standard form.

9. $(3x^2 + x - 3) - (x^2 - 2x + 4)$

10. $(m^4 - 3m^2 + 5) - (3m^4 - 5m^2 - 2)$

9. _____

10. _____

11. $\left(\dfrac{7}{4}x^2 - \dfrac{5}{8}x - 1\right) - \left(\dfrac{7}{6}x^2 + \dfrac{5}{12}x + 5\right)$

12. $(4m^2n - 2mn - 4) - (10m^2n - 6mn - 3)$

11. _____

12. _____

In Problems 13–16, evaluate the polynomials for each of the given value(s).

13. $-x^2 + 10$

   (a) $x = 0$
   (b) $x = -1$
   (c) $x = 1$

14. $2 + \dfrac{1}{2}n^2$

   (a) $n = 4$
   (b) $n = 0.5$
   (c) $n = -\dfrac{1}{4}$

13a. _____
13b. _____
13c. _____
14a. _____
14b. _____
14c. _____

15. $-2ab^2 - 2a^2b - b^3$
   for $a = 1$ and $b = -2$

16. $m^2n^2 - mn^2 + 3m^2 - 2$
   for $m = \dfrac{1}{2}$ and $n = -1$

15. _____

16. _____

Name:
Instructor:

Date:
Section:

Five-Minute Warm-Up 5.2
Multiplying Monomials: The Product and Power Rules

1. Write each expression in exponential form.

   (a) $\left(\dfrac{3}{4}\right) \cdot \left(\dfrac{3}{4}\right) \cdot \left(\dfrac{3}{4}\right)$

   (b) $(-5)(-5)(-5)(-5)$

   1a. _____

   1b. _____

2. Evaluate each exponential expression.

   (a) -3^2

   (b) $\left(\dfrac{3}{2}\right)^3$

   (c) $(-4)^2$

   2a. _____

   2b. _____

   2c. _____

   (d) $\left(-\dfrac{4}{3}\right)^3$

   (e) $(-2)^5$

   (f) $-(-6)^2$

   2d. _____

   2e. _____

   2f. _____

3. Evaluate: $-\dfrac{5}{3} \cdot (-18) \cdot \left(-\dfrac{2}{5}\right)$

   3. _____

4. Use the Distributive Property to simplify: $5(4-2x)$

   4. _____

Sullivan/Struve/Mazzarella, *Elementary & Intermediate Algebra*, 3e
Copyright © 2014 Pearson Education, Inc.

Name: Date:
Instructor: Section:

Guided Practice 5.2
Multiplying Monomials: The Product and Power Rules

Objective 1: Simplify Exponential Expressions Using the Product Rule

1. To multiply exponential expressions which have the same base, the product will have the common base and an exponent which is the ____ of the exponents of the factors.   1. _____

2. Simplify each expression. *(See textbook Examples 2 and 3)*

   (a) $r^3 \cdot r^5$ (b) $m^2 \cdot n \cdot n^4 \cdot m$ 2a. _____

   2b. _____

Objective 2: Simplify Exponential Expressions Using the Power Rule

3. To raise exponential expressions containing a power to a power, keep the base and ____ the powers.   3. _____

4. Simplify each expression. Write the answer in exponential form. *(See textbook Examples 4 and 5)*

   (a) $\left(p^6\right)^3$ (b) $\left[(-x)^3\right]^4$ 4a. _____

   4b. _____

5. Raising a negative number to an even exponent results in a _____ number.   5. _____

6. Raising a negative number to an odd exponent results in a _____ number.   6. _____

Objective 3: Simplify Exponential Expressions Containing Products

7. When we raise a product to a power, each factor is _____.

8. Simplify each expression. *(See textbook Example 6)*

   (a) $\left(5xy^2\right)^3$ (b) $\left(-m^2n^3\right)^4$ 8a. _____

   8b. _____

Sullivan/Struve/Mazzarella, *Elementary & Intermediate Algebra*, 3e
Copyright © 2014 Pearson Education, Inc.

Guided Practice 5.2

Objective 4: Multiply a Monomial by a Monomial

9. Multiply and simplify each expression. *(See textbook Example 7)*

    (a) $(-x^4)(12x^7)$ (b) $\left(\frac{7}{4}q^2\right)\left(\frac{12}{21}q^5\right)$

    9a. _____

    9b. _____

10. Multiply and simplify each expression. *(See textbook Example 8)*

    (a) $(-5m^4n^4)(-2mn^3)$ (b) $(8pq^4)(-2p^2)(-3pq^6)$

    10a. _____

    10b. _____

Do the Math Exercises 5.2
Multiplying Monomials: The Product and Power Rules

In Problems 1–4, simplify each expression.

1. $3 \cdot 3^3$

2. $a^3 \cdot a^7$

3. $b^5 \cdot b \cdot b^7$

4. $(-z)(-z)^5$

1. _____

2. _____

3. _____

4. _____

In Problems 5–8, simplify each expression.

5. $(3^2)^2$

6. $[(-n)^7]^3$

7. $(k^8)^3$

8. $[(-a)^6]^3$

5. _____

6. _____

7. _____

8. _____

Do the Math Exercises 5.2

In Problems 9–12, simplify each expression.

9. $(4y^3)^2$

10. $\left(\dfrac{3}{4}n^2\right)^3$

9. _____

10. _____

11. $(-4ab^2)^3$

12. $(-3a^6bc^4)^4$

11. _____

12. _____

In Problems 13–16, multiply the monomials.

13. $(7b^5)(-2b^4)$

14. $\left(\dfrac{3}{8}y^5\right)\left(\dfrac{4}{9}y^6\right)$

13. _____

14. _____

15. $(6a^2b^3)(2a^5b)$

16. $\left(\dfrac{1}{2}b\right)(-20a^2b)\left(-\dfrac{2}{3}a\right)$

15. _____

16. _____

17. Given the rectangle below, find (a) the perimeter and (b) the area.

17a. _____

17b. _____

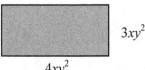

$3xy^2$

$4xy^2$

Five-Minute Warm-Up 5.3
Multiplying Polynomials

In Problems 1 – 6, simplify each expression.

1. $x^6 \cdot x$

2. $(5y^2)(4y^6)$

1. _____

2. _____

3. $(2z)^6$

4. $(-3a^4)^2$

3. _____

4. _____

5. $\left(\dfrac{5}{2}x\right)^2$

6. $(-4xy^2)^2(-x^3y^4)^3$

5. _____

6. _____

7. Use the Distributive Property to simplify: $11(4a - 3b)$

7. _____

8. Simplify: $\dfrac{2}{3}\left(\dfrac{6x}{5} + \dfrac{9}{8}\right)$

8. _____

Name:
Instructor:

Date:
Section:

Guided Practice 5.3
Multiplying Polynomials

Objective 1: Multiply a Polynomial by a Monomial

1. To multiply a monomial and a polynomial we use _____.

2. Multiply and simplify. *(See textbook Examples 2 and 3)*

   (a) $-3ab^2\left(5a^2 - 2ab + 3b^2\right)$ (b) $\left(\dfrac{7}{3}x^2 - \dfrac{1}{6}x - \dfrac{4}{9}\right)18x^5$

   2a. _____

   2b. _____

Objective 2: Multiply Two Binomials Using the Distributive Property

3. Find the product using the Distributive Property. *(See textbook Examples 4 and 5)*

   (a) $(x-4)(x-2)$ (b) $(3x-5)(2x+1)$

   3a. _____

   3b. _____

Objective 3: Multiply Two Binomials Using the FOIL Method

4. (a) What process does FOIL tell how to do? _____

   (b) What does each of the letters stand for?

   F _____ O _____ I _____ L _____

   (c) Can the FOIL method be extended to multiply other types of polynomials? _____

5. Use the FOIL Method to find the product. *(See textbook Examples 6 and 7)*

   (a) $(u-5)(u+3)$ (b) $(7x-2y)(x-3y)$

   5a. _____

   5b. _____

Objective 4: Multiply the Sum and Difference of Two Terms

6. Find the product: $(A-B)(A+B) =$ 6. _____

Guided Practice 5.3

7. Find each product. *(See textbook Examples 8 and 9)*
(a) $(x-9)(x+9)$ (b) $(2a+5b)(2a-5b)$ 7a. _____

7b. _____

Objective 5: Square a Binomial

8. Find the product:

(a) $(A+B)^2 =$ _____   (b) $(A-B)^2 =$ _____

9. Find the product: $(x-3)^2$ *(See textbook Example 10)*

Step 1: Use the $(A-B)^2$ pattern.

Write out the pattern for $(A-B)^2$:   (a) _____

Identify the components. What is the expression for A?   (b) $A =$ _____

What is the expression for B?   (c) $B =$ _____

Step 2: Simplify.

Find each of the components in the pattern.   (d) $A^2 =$ _____

(e) $2AB =$ _____

(f) $B^2 =$ _____

Use the pattern to write the product: $(x-3)^2 =$

(g) _____

10. Find the product using the special products patterns. *(See textbook Example 11)*

(a) $(12-x)^2$ (b) $(4x+5y)^2$ 10a. _____

10b. _____

Objective 6: Multiply a Polynomial by a Polynomial

11. When we find the product of two polynomials, we make repeated use of the Extended Form of the Distributive Property. These multiplication problems can be formatted either horizontally, like textbook Example 12a, or vertically, like textbook Example 12b. Try each method and see which you prefer.

Find the product. *(See textbook Example 13)*
(a) $(3x-1)(x^2+2x-4)$ (b) $(4x+3)(2x^2-5x-3)$ 11a. _____

11b. _____

Name:
Instructor:

Date:
Section:

Do the Math Exercises 5.3
Multiplying Polynomials

In Problems 1–4, use the Distributive Property to find each product.

1. $3m(2m-7)$

2. $\dfrac{3}{5}b(15b-5)$

3. $4w(2w^2+3w-5)$

4. $(7r+3s)(2r^2s)$

1. _____

2. _____

3. _____

4. _____

In Problems 5–6, use the Distributive Property to find each product.

5. $(n-7)(n+4)$

6. $(5n-2y)(2n-3y)$

5. _____

6. _____

In Problems 7–10, find the product using the FOIL method.

7. $(x+3)(x+7)$

8. $(3z-2)(4z+1)$

9. $(x^2-5)(x^2-2)$

10. $(3r+5s)(6r+7s)$

7. _____

8. _____

9. _____

10. _____

Sullivan/Struve/Mazzarella, *Elementary & Intermediate Algebra*, 3e
Copyright © 2014 Pearson Education, Inc.

Do the Math Exercises 5.3

In Problems 11–12, find the product of the sum and difference of two terms.

11. $(6r-1)(6r+1)$

12. $(8a-5b)(8a+5b)$

11. _____

12. _____

In Problems 13–16, find the product.

13. $(x+4)^2$

14. $(6b-5)^2$

13. _____

14. _____

15. $(3x+2y)^2$

16. $\left(y-\dfrac{1}{3}\right)^2$

15. _____

16. _____

In Problems 17–20, find the product.

17. $(3a-1)(2a^2-5a-3)$

18. $-\dfrac{4}{3}k(k+7)(3k-9)$

17. _____

18. _____

19. $(2m^2-m+4)(-m^3-2m-1)$

20. $(2a-1)(a+4)(a+1)$

19. _____

20. _____

21. Subtract $2x+3$ from $(2x-5)^2$.

21. _____

Name:
Instructor:

Date:
Section:

Five-Minute Warm-Up 5.4
Dividing Monomials: The Quotient Rule and Integer Exponents

1. Find the product: $(7x^3y)^2$

    1. _____

2. Evaluate: $\left(\dfrac{6}{5}\right)^2$

    2. _____

3. Find the reciprocal:

    (a) -3 (b) $\dfrac{7}{8}$

    3a. _____

    3b. _____

4. Simplify: $\dfrac{72}{40}$

    4. _____

5. Simplify: $\dfrac{3}{8} - \dfrac{5}{6}$

    5. _____

Name:
Instructor:

Date:
Section:

Guided Practice 5.4
Dividing Monomials: The Quotient Rule and Integer Exponents

Objective 1: Simplify Exponential Expressions Using the Quotient Rule

1. To divide exponential expressions which have the same base, the quotient will have the common base, and to determine the exponent, _____ the exponent in the denominator from the exponent in the numerator.

1. _____

2. Simplify each expression. *(See textbook Examples 1 and 2)*

(a) $\dfrac{x^{11}}{x^4}$

(b) $\dfrac{-27a^5b^2}{12a^3b}$

2a. _____

2b. _____

Objective 2: Simplify Exponential Expressions Using the Quotient to a Power Rule

3. In your own words, describe the rule for raising a quotient to a power. State (a) if you should simplify the quotient first, if possible, and then raise this quotient to a power; (b) apply the power first, and then simplify, if possible; or (c) it does not matter which occurs first. Be sure to state any restrictions on your rule.

4. Simplify each expression. *(See textbook Examples 3 and 4)*

(a) $\left(-\dfrac{v}{2}\right)^3$

(b) $\left(\dfrac{28a^6}{16ab^4}\right)^2$

4a. _____

4b. _____

Objective 3: Simplify Exponential Expressions Using Zero as an Exponent

5. If a is a nonzero real number, $a^0 =$ _____.

5. _____

6. Simplify each expression. *(See textbook Example 5)*

(a) n^0

(b) $(9p^2)^0$

(c) $12x^0$

6a. _____

6b. _____

6c. _____

(d) -8^0

(e) $(-6x)^0$

(f) $-4x^0$ if $x = 2$

6d. _____

6e. _____

6f. _____

Sullivan/Struve/Mazzarella, *Elementary & Intermediate Algebra*, 3e
Copyright © 2014 Pearson Education, Inc.

Guided Practice 5.4

Objective 4: Simplify Exponential Expressions Involving Negative Exponents

7. If n is a positive integer and if a is a nonzero real number, then

   (a) $a^{-n} = $ _____   (b) $\dfrac{1}{a^{-n}} = $ _____   (c) $\left(\dfrac{a}{b}\right)^{-n} = $ _____

   7a. _____

   7b. _____

   7c. _____

From this point forward, assume all variables are nonzero. Always leave your final answer with only positive exponents.

8. Simplify each expression. *(See textbook Examples 6 – 9)*

   (a) $(-2)^{-5}$ (b) $6^{-2} - 3^{-2}$ (c) $5a^{-2}$

   8a. _____

   8b. _____

   8c. _____

Objective 5: Simplify Exponential Expressions Using the Laws of Exponents

9. Simplify: $\left(\dfrac{4}{3}x^{-2}y^{4}\right)\left(-18x^{5}y^{-7}\right)$ *(See textbook Example 13)*

Step 1: Rearrange factors.	Use the commutative and associative properties of multiplication to group together the coefficients, the variables with base x and the variables with base y:	$\left(\dfrac{4}{3}x^{-2}y^{4}\right)\left(-18x^{5}y^{-7}\right) = $ (a) _____
Step 2: Find each product.	Evaluate the product of the coefficients:	(b) _____
	Use $a^n \cdot a^m = a^{n+m}$ for x:	(c) _____
	Use $a^n \cdot a^m = a^{n+m}$ for y:	(d) _____
Step 3: Simplify. Write the product so that the exponents are positive.	Use $a^{-n} = \dfrac{1}{a^n}$ for y:	(e) _____

Name:
Instructor:

Date:
Section:

Do the Math Exercises 5.4
Dividing Monomials: The Quotient Rule and Integer Exponents

In Problems 1–4, use the Quotient Rule to simplify. All variables are nonzero.

1. $\dfrac{10^5}{10^2}$

2. $\dfrac{x^{20}}{x^{14}}$

1. _____

2. _____

3. $\dfrac{-36x^2y^5}{24xy^4}$

4. $\dfrac{-15r^9s^2}{-5r^8s}$

3. _____

4. _____

In Problems 5–8, use the Quotient to a Power Rule to simplify. All variables are nonzero.

5. $\left(\dfrac{4}{9}\right)^2$

6. $\left(\dfrac{7}{x^2}\right)^2$

5. _____

6. _____

7. $\left(-\dfrac{a^3}{b^{10}}\right)^5$

8. $\left(\dfrac{2mn^2}{q^3}\right)^4$

7. _____

8. _____

In Problems 9–12, use the Zero Exponent Rule to simplify. All variables are nonzero.

9. -100^0

10. $\left(\dfrac{2}{5}\right)^0$

9. _____

10. _____

Do the Math Exercises 5.4

11. $(-11xy)^0$ **12.** $-11xy^0$

11. _____

12. _____

In Problems 13–16, use the Negative Exponent Rules to simplify. Write answers with only positive exponents. All variables are nonzero.

13. $-7z^{-5}$ **14.** $\left(\dfrac{4}{p^2}\right)^{-2}$

13. _____

14. _____

15. $\dfrac{4}{b^{-3}}$ **16.** $\dfrac{9}{(4t)^{-1}}$

15. _____

16. _____

In Problems 17–20, use the Laws of Exponents to simplify. Write answers with positive exponents. All variables are nonzero.

17. $\left(\dfrac{3}{5}ab^{-3}\right)\left(\dfrac{5}{3}a^{-1}b^3\right)$ **18.** $\dfrac{30ab^{-4}}{15a^{-1}b^{-2}}$

17. _____

18. _____

19. $(5x^{-4}y^3)^{-2}$ **20.** $(5a^3b^{-4})\left(\dfrac{2ab^{-1}}{b^{-3}}\right)^2$

19. _____

20. _____

21. Explain why $\dfrac{6^5}{2^2} \neq 3^3$.

Name:
Instructor:

Date:
Section:

Five-Minute Warm-Up 5.5
Dividing Polynomials

1. Find the quotient: $\dfrac{-135x^5}{5x}$

    1. _____

2. Find the quotient: $\dfrac{12n}{108n^4}$

    2. _____

3. Find the product: $-9z(3z^2 + 2z - 1)$

    3. _____

4. Given the polynomial $5x + x^4 - 3x^2 - 7x^3 - x^4 + 13$,

    (a) state the degree.

    4a. _____

    (b) Write the polynomial in standard form.

    4b. _____

5. Simplify: $\dfrac{3}{4} + \dfrac{2}{4} - \dfrac{9}{4} = \dfrac{3 + 2 - 9}{4} =$

    5. _____

6. Subtract: $x^2 + 5x - 2 - (x^2 + 6x)$

    6. _____

Name:
Instructor:

Date:
Section:

Guided Practice 5.5
Dividing Polynomials

Objective 1: Divide a Polynomial by a Monomial

To divide a polynomial by a monomial, divide each of the terms of the polynomial numerator (dividend) by the monomial denominator (divisor). We will be using the Quotient Rule for Exponents and the fact that $\frac{a+b}{c} = \frac{a}{c} + \frac{b}{c}$. **You should continue to reduce fractions to lowest terms, use only positive exponents, and assume all variables are nonzero.**

1. Divide and simplify each expression. *(See textbook Examples 1 – 3)*

   (a) $\dfrac{8x^4 + 36x^3}{4x^2}$

   (b) $\dfrac{5x^2y + 20xy^3 - 5xy}{5xy}$

   1a. _____

   1b. _____

Objective 2: Divide a Polynomial by a Binomial

2. How do you check that long division has been done correctly?

To divide the polynomials, we use a process called long division. This is exactly the same process that you use to divide integers. Due to the tremendous popularity of hand-held calculators, this particular skill has diminished over the years. It is important to practice this process so that the steps are easy to recall while performing long division on polynomials.

3. Divide 8047 by 71. *(See textbook Example 4)*

   (a) Name the divisor.

   3a. _____

   (b) Name the dividend.

   3b. _____

   (c) Perform the division. What is the quotient?

   3c. _____

   (d) What is the remainder?

   3d. _____

   (e) Express the quotient as a mixed number.

   3e. _____

Guided Practice 5.5

4. Find the quotient and remainder when $(x^3 - 2x^2 + x + 6)$ is divided by $(x + 1)$. *(See textbook Example 5)*

Step 1: Divide the leading term of the dividend, x^3, by the leading term of the divisor, x. Enter the result over the term x^3.

(a) $\dfrac{x^3}{x} = $ _____

Step 2: Multiply ☐ by $x + 1$. Be sure to vertically align like terms.

(b) $x + 1 \overline{\smash{\big)}\, x^3 - 2x^2 + x + 6}$ with ☐ above

Step 3: Subtract your product from the dividend.

(c) $x + 1 \overline{\smash{\big)}\, x^3 - 2x^2 + x + 6}$ with x^2 above

Subtract and continue with Steps 4 and 5:

Step 4: Repeat Steps 1 – 3 treating $-3x^2 + x + 6$ as the dividend and dividing x into $-3x^2$ to obtain the next term in the quotient.

Step 5: Repeat Steps 1 – 3 treating $4x + 6$ as the dividend and dividing x into $4x$ to obtain the next term in the quotient.

When the degree of the remainder is less than the degree of the divisor, you are finished dividing.

Express the answer as the

$$\text{Quotient} + \dfrac{\text{Remainder}}{\text{Divisor}}.$$

(d) _____

Step 6: Check Verify that (Quotient)(Divisor) + Remainder = Dividend.

We leave it to you to verify the solution.

5. Before beginning long division, be sure that both the divisor and dividend are written in _____ form.

5. _____

6. *True or False* The following alternate approach correctly finds the quotient when a polynomial is divided by a binomial.

$$\dfrac{2x^4 - 8x}{2x + 4} = \dfrac{\cancel{2}x^4 - \cancel{8}x}{\cancel{2}x + \cancel{4}} = x^3 - 2x$$

6. _____

Do the Math Exercises 5.5
Dividing Polynomials

In Problems 1–8, divide and simplify.

1. $\dfrac{3x^3 - 6x^2}{3x^2}$

2. $\dfrac{16m^3 + 8m^2 - 4}{8m^2}$

1. _____

2. _____

3. $\dfrac{16a^5 - 12a^4 + 8a^3}{20a^4}$

4. $\dfrac{7p^4 + 21p^3 - 3p^2}{-6p^4}$

3. _____

4. _____

5. $\dfrac{-5y^2 + 15y^4 - 16y^5}{5y^2}$

6. $\dfrac{35xy + 20y}{-5y}$

5. _____

6. _____

7. $\dfrac{21y^2 + 35x^2}{-7x^2}$

8. $\dfrac{16m^2n^3 - 24m^4n^3}{-8m^3n^4}$

7. _____

8. _____

Do the Math Exercises 5.5

In Problems 9–14, find the quotient using long division.

9. $\dfrac{x^2+4x-32}{x-4}$

10. $\dfrac{x^3-x^2-40x+12}{x+6}$

9. _____

10. _____

11. $\dfrac{x^4-2x^3+x^2+x-1}{x-1}$

12. $\dfrac{x^3-7x^2+15x-11}{x-3}$

11. _____

12. _____

13. $\dfrac{9x^2-14}{2+3x}$

14. $\dfrac{64x^6-27}{4x^2-3}$

13. _____

14. _____

Name:
Instructor:

Date:
Section:

Five-Minute Warm-Up 5.6
Applying Exponent Rules: Scientific Notation

1. Find the product: $(2x^3)(9.5x^2)$

   1. _____

2. Find the product: $(1.6n^{-2})(1.6n^{-4})$

   2. _____

3. Find the quotient: $\dfrac{0.04p^{-3}}{0.2p^{-7}}$

   3. _____

4. Multiply each of the following:

   (a) 4.2×1000 (b) 3.05×0.001 (c) $7 \times 3 \times 0.1$

   4a. _____

   4b. _____

   4c. _____

5. Simplify each of the following:

   (a) $\dfrac{10^6}{10^{-2}}$ (b) $\dfrac{10^{-5}}{10^4}$ (c) $\dfrac{10^{-1}}{10^{-6}}$

   5a. _____

   5b. _____

   5c. _____

   (d) $10^{-5} \times 10^{-3}$ (e) $10^{-4} \times 10^9$

   5d. _____

   5e. _____

Sullivan/Struve/Mazzarella, *Elementary & Intermediate Algebra*, 3e
Copyright © 2014 Pearson Education, Inc.

Name: Date:
Instructor: Section:

Guided Practice 5.6
Applying Exponent Rules: Scientific Notation

Objective 1: Convert Decimal Notation to Scientific Notation

1. A number is written in scientific notation when it is of the form $x \times 10^N$.

    (a) x must be ____ 1a. _____

    (b) N is ____ 1b. _____

    (c) When 3,700 is written in scientific notation, the power of 10 is ____ (positive or negative). 1c. _____

    (d) When 0.002 is written in scientific notation, the power of 10 is ____ (positive or negative). 1d. _____

2. Write 45,000,000 in scientific notation. *(See textbook Example 1)*

    Step 1: The "understood" decimal point in 45,000,000 follows the last 0. Therefore, we will move the decimal point to the left until it is between the 4 and the 5. Do you see why? This requires that we move the decimal $N =$ ____ places. (a) _____

    Step 2: The original number is greater than 1, so the power of 10 is ____ (positive or negative). (b) _____

    Now write 45,000,000 in scientific notation. (c) _____

3. Write 0.00003 in scientific notation. *(See textbook Example 2)*

    Step 1: Because 0.00003 is less than 1, we shall move the decimal point until it is to the right of the 3. This requires that we move the decimal $N =$ ____ places. (a) _____

    Step 2: The original number is between 0 and 1, so the power of 10 is ____ (positive or negative). (b) _____

    Now write 0.00003 in scientific notation. (c) _____

Objective 2: Convert Scientific Notation to Decimal Notation

4. To convert a number from scientific notation to decimal notation, determine the exponent, N, on the number 10.

    (a) If the exponent is positive, move the decimal N decimal places to the ____. 4a. _____

    (b) If the exponent is negative, then move the decimal $|N|$ decimal places to the ____. 4b. _____

Guided Practice 5.6

5. Write 6.02×10^4 in decimal notation. *(See textbook Example 3)*

Step 1: Determine the exponent on the number 10. (a) _____

Step 2: Since the exponent is positive, we move the decimal point N places to the ____. (b) _____

Now write 6.02×10^4 in decimal notation. (c) _____

6. Write 9.1×10^{-3} in decimal notation. *(See textbook Example 4)*

Step 1: Determine the exponent on the number 10. (a) _____

Step 2: Since the exponent is negative, we move the decimal point $|N|$ places to the ____. (b) _____

Now write 9.1×10^{-3} in decimal notation. (c) _____

Objective 3: Use Scientific Notation to Multiply and Divide

7. Perform the indicated operation. Express the answer in scientific notation. *(See textbook Examples 5 and 6)*

(a) $(2 \times 10^5) \cdot (4 \times 10^8)$ (b) $(9 \times 10^{-15}) \cdot (6 \times 10^{10})$ 7a. _____

7b. _____

8. Perform the indicated operation. Express the answer in scientific notation. *(See textbook Example 7)*

(a) $\dfrac{(8 \times 10^{12})}{(2 \times 10^7)}$ (b) $\dfrac{(7.2 \times 10^{24})}{(8 \times 10^{-13})}$ 8a. _____

8b. _____

Do the Math Exercises 5.6
Applying Exponent Rules: Scientific Notation

In Problems 1–6, write each number in scientific notation.

1. 8,000,000,000

2. 0.0000001

3. 0.0000283

4. 401,000,000

5. 8

6. 120

1. _____

2. _____

3. _____

4. _____

5. _____

6. _____

In Problems 7–12, write each number in decimal notation.

7. 3.75×10^2

8. 6×10^6

9. 5×10^{-4}

10. 4.9×10^{-1}

7. _____

8. _____

9. _____

10. _____

Do the Math Exercises 5.6

11. 5.4×10^5

12. 5.123×10^{-3}

11. _____

12. _____

In Problems 13–18, perform the indicated operations. Express your answer in scientific notation.

13. $(3 \times 10^{-4})(8 \times 10^{-5})$

14. $(4 \times 10^7)(2.5 \times 10^{-4})$

13. _____

14. _____

15. $\dfrac{6 \times 10^3}{1.2 \times 10^5}$

16. $\dfrac{4.8 \times 10^7}{1.2 \times 10^2}$

15. _____

16. _____

17. $\dfrac{0.000275}{2500}$

18. $\dfrac{24,000,000,000}{0.00006 \times 2000}$

17. _____

18. _____

Name:
Instructor:

Date:
Section:

Five-Minute Warm-Up 6.1
Greatest Common Factor and Factoring by Grouping

1. Write as the product of prime numbers.
   (a) 36
   (b) 225

   1a. _____

   1b. _____

2. Use the Distributive Property to simplify: $-3(4x-9)$

   2. _____

3. Find the product: $(x+4)(2x-3)$

   3. _____

4. Find the product: $4x^2(x-3)(x+3)$

   4. _____

5. Given $\frac{1}{2} \cdot 10 = 5$,
   (a) list the factors.

   5a. _____

   (b) identify the product.

   5b. _____

6. Identify the missing factor: $10x^4 \cdot ? = 40x^6 y$

   6. _____

Name:
Instructor:
Date:
Section:

Guided Practice 6.1
Greatest Common Factor and Factoring by Grouping

Objective 1: Find the Greatest Common Factor of Two or More Expressions

1. The *greatest common factor (GCF)* of a list of polynomials is the largest expression that divides evenly into all of the polynomials.

Find the GCF of 27, 36 and 45. *(See textbook Examples 1 and 2)*

Step 1: Write each number as a product of prime factors.	(a) 27 = _____
	(b) 36 = _____
	(c) 45 = _____
Step 2: Determine the common prime factors.	(d) common factors are: _____
Step 3: Find the product of the common factors found in Step 2. This number is the GCF.	(e) GCF = _____

2. Find the greatest common factor (GCF) of each of the following. *(See textbook Examples 3 and 4)*

(a) $12x^2, 30x^6$ (b) $6a^2b^3, 15ab^4c, 27a^3bc^2$

2a. _____

2b. _____

Objective 2: Factor Out the Greatest Common Factor in Polynomials

3. Factor out the great common factor: $6m^4n^2 + 18m^3n^4 - 22m^2n^5$ *(See textbook Example 6)*

Step 1: Find the GCF.

(a) GCF = _____

Step 2: Rewrite each term as the product of the GCF and remaining factor.

(b) $6m^4n^2 + 18m^3n^4 - 22m^2n^5 = $ _____

Step 3: Factor out the GCF.

(c) $6m^4n^2 + 18m^3n^4 - 22m^2n^5 = $ _____

Step 4: Check   Distribute to verify that the factorization is correct.

4. Factor out the greatest common factor. *(See textbook Examples 9 and 10)*

(a) $-24x^5 - 9x^3$ (b) $4x(x+3) + 7(x+3)$

4a. _____

4b. _____

Sullivan/Struve/Mazzarella, *Elementary & Intermediate Algebra*, 3e
Copyright © 2014 Pearson Education, Inc.

Guided Practice 6.1

Objective 3: Factor Polynomials by Grouping

5. Factor by grouping is commonly used when a polynomial has _____ terms.   5. _____

6. *True or False* You may need to rearrange the terms in order to be able to identity a common binomial factor.   6. _____

7. Factor by grouping: $2x^2 - 4x + 3xy - 6y$ *(See textbook Example 11)*

Step 1: Group terms with common factors. In this problem the first two terms have a common factor and the last two terms have a common factor.

  (a) common factor of $2x^2 - 4x$: _____

  (b) common factor of $3xy - 6y$: _____

Step 2: In each grouping, factor out the common factor.

  (c) $(2x^2 - 4x) + (3xy - 6y) =$ _____

Step 3: Factor out the common factor that remains.

  (d) $2x^2 - 4x + 3xy - 6y =$ _____

Step 4: Check   Multiply to verify that the factorization is correct.

8. Factor by grouping: $9x - 18y - 4ax + 8ay$ *(See textbook Example 12)*   8. _____

9. In any factoring problem, the first step is to _____, provided that one exists.

10. Factor by grouping: $4x^3 + 6x^2 - 12x - 18$ *(See textbook Example 13)*   10. _____

Name:
Instructor:

Date:
Section:

Do the Math Exercises 6.1
Greatest Common Factor and Factoring by Grouping

In Problems 1–5, find the greatest common factor, GCF, of each group of expressions.

1. 35, 42, 63

2. $26xy^2, 39x^2y$

1. _____

2. _____

3. $2x^2yz, xyz^2, 5x^3yz^2$

4. $8(x+y)$ and $9(x+y)$

3. _____

4. _____

5. $6(a-b)$ and $15(a-b)^3$

5. _____

In Problems 6–11, factor the GCF from the polynomial.

6. $b^2 - 6b$

7. $8a^3b^2 + 12a^5b^2$

6. _____

7. _____

Do the Math Exercises 6.1

8. $5x^4 + 10x^3 - 25x^2$

9. $-2y^2 + 10y - 14$

8. _____

9. _____

10. $-22n^4 + 18n^2 + 14n$

11. $a(a-5) + 6(a-5)$

10. _____

11. _____

In Problems 12–15, factor by grouping.

12. $x^2 + ax + 2a + 2x$

13. $mn - 3n + 2m - 6$

12. _____

13. _____

14. $z^3 + 4z^2 + 3z + 12$

15. $x^3 - x^2 - 5x + 5$

14. _____

15. _____

16. Find the missing factor: $1 - 3x^{-1} + 2x^{-2} = x^{-2} \cdot ?$

16. _____

17. The area of the triangle is $6x^2 + 4x$. If the base is $4x$, write a polynomial that represents the height.

17. _____

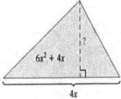

Name:
Instructor:

Date:
Section:

Five-Minute Warm-Up 6.2
Factoring Trinomials of the Form $x^2 + bx + c$

1. List all possible combinations of factors for the given product.
   (a) −12   (b) 36

   1a. _____

   1b. _____

In Problems 2 and 3, find (a) the sum of the integers and (b) the product of the integers.

2. −3 and −12   3. 8 and −2

   2a. _____

   2b. _____

   3a. _____

   3b. _____

4. Find two integers with the following properties.
   (a) sum of −5 and product of −36

   4a. _____

   (b) sum of −5 and product of 6

   4b. _____

   (c) sum of −4 and product of −5

   4c. _____

   (d) sum of 9 and product of 18

   4d. _____

5. Determine the coefficients of $-x^2 + 7x - 2$.

   5. _____

6. Find the product: $(x-6)(x-4)$

   6. _____

Name:
Instructor:

Date:
Section:

Guided Practice 6.2
Factoring Trinomials of the Form $x^2 + bx + c$

Objective 1: Factor Trinomials of the Form $x^2 + bx + c$

Step 1: Find the pair of integers whose product is c and whose sum is b. That is, determine m and n such that $mn = c$ and $m + n = b$.

Step 2: Write $x^2 + bx + c = (x + m)(x + n)$.

Step 3: Check your work by multiplying out the factored form.

1. Factor: $y^2 + 11y + 18$ *(See textbook Example 1)*

Step 1: We are looking for factors of $c = 18$ whose sum is $b = 11$. We begin by listing all factors of 18 and computing the sum of these factors.

(a)

Factors whose product is 18	1, 18	2, 9	3, 6	−1, −18	−2, −9	−3, −6
Sum of factors						

Which two factors sum to 11 and multiply to 18?   (b) _____

Step 2: We write the trinomial in the form $(y + m)(y + n)$.

(c) $y^2 + 11y + 18 =$ _____

Step 3: Check We multiply to verify our solution.

We leave it to you to verify the factorization.

2. Factor: $z^2 - 15z + 36$ *(See textbook Example 2)*

(a) Make a list of factors.

Factors whose product is 36	1, 36	2, 18	3, 12	4, 9	−1, −36	−2, −18	−3, −12	−4, −9
Sum of factors								

(b) Use the two factors whose sum is −15 to write the factorization of $z^2 - 15z + 36$.

2b. _____

Guided Practice 6.2

3. Factor: $x^2 + x - 56$ *(See textbook Example 3)*
(a) Make a list of factors.

Factors whose product is −56								
Sum of factors								

(b) Use the two factors whose sum is 1 to write the factorization of $x^2 + x - 56$.

3b. _____

4. Factor: $n^2 - 2n - 24$ *(See textbook Example 4)*
(a) Make a list of factors.

Factors whose product is −24							
Sum of factors							

(b) Use the two factors whose sum is −2 to write the factorization of $n^2 - 2n - 24$.

4b. _____

5. Show that $x^2 + 3x - 6$ is prime. *(See textbook Example 5)*

Objective 2: Factor Out the GCF, Then Factor $x^2 + bx + c$

6. Factor: $3p^3 - 9p^2 - 54p$ *(See textbook Example 8)*   6. _____

Do the Math Exercises 6.2
Factoring Trinomials of the Form $x^2 + bx + c$

In Problems 1–5, factor each trinomial completely. If the trinomial cannot be factored, say it is prime.

1. $n^2 + 12n + 20$

2. $y^2 - 8y - 9$

3. $y^2 + 6y - 40$

4. $t^2 + 2t - 38$

5. $x^2 - 14xy + 24y^2$

1. _____

2. _____

3. _____

4. _____

5. _____

In Problems 6–9, factor each trinomial completely by factoring out the GCF first and then factoring the resulting trinomial.

6. $4p^4 - 4p^3 - 8p^2$

7. $30x - 2x^2 - 100$

6. _____

7. _____

Sullivan/Struve/Mazzarella, *Elementary & Intermediate Algebra*, 3e

Do the Math Exercises 6.2

8. $-3x^2 - 18x - 15$

9. $-75x + x^3 + 10x^2$

8. _____

9. _____

In Problems 10–15, factor each polynomial completely. If the polynomial cannot be factored, say that it is prime.

10. $-2z^3 - 2z^2 + 24z$

11. $-r^2 - 12r - 36$

10. _____

11. _____

12. $x^2 - x + 6$

13. $-16x + x^3 - 6x^2$

12. _____

13. _____

14. $x^2 + 7xy + 12y^2$

15. $25 + 10x + x^2$

14. _____

15. _____

16. The answer key to your algebra exam said the factored form of $2 - 3x + x^2$ is $(x-1)(x-2)$. You have $(1-x)(2-x)$ on your paper. Is your answer correct or incorrect? Explain your reasoning.

16. _____

Name:
Instructor:

Date:
Section:

Five-Minute Warm-Up 6.3
Factoring Trinomials of the Form $ax^2 + bx + c$, $a \neq 1$

1. Determine the coefficients of $4n^2 + n - 1$?

   1. _____

2. Write as the product of prime numbers.
   (a) 18
   (b) 150

   2a. _____

   2b. _____

3. Find the product: $(2x+1)(3x-2)$

   3. _____

4. Factor by grouping: $4x - 4y + ax - ay$

   4. _____

5. Factor out the GCF: $12r^3s^2 + 3rs - 6rs^4$

   5. _____

6. Factor out the greatest common binomial factor: $7z(2z+1) - 4(2z+1)$

   6. _____

Sullivan/Struve/Mazzarella, *Elementary & Intermediate Algebra*, 3e
Copyright © 2014 Pearson Education, Inc.

Guided Practice 6.3
Factoring Trinomials of the Form $ax^2 + bx + c$, $a \neq 1$

Objective 1: Factor $ax^2 + bx + c$, $a \neq 1$ Using Grouping

1. Factor by grouping: $3x^2 - 13x + 12$ *(See textbook Example 1)*

    Step 1: Find the value of ac. (a) Identify the coefficients: $a =$ _____, $c =$ _____

    (b) The value of $a \cdot c =$ _____

    Step 2: Find the pair of integers, m and n, whose product is ac and whose sum is b. Since the sum is negative, only the pairs that contain 2 negative integers are listed here.

    (c)

Factors whose product is 36	−1, −36	−2, −18	−3, −12	−4, −9	−6, −6
Sum of factors					

    (d) Which two factors multiply to 36 and add to −13? _____

    Step 3: Write
    $ax^2 + bx + c = ax^2 + mx + nx + c$

    (e) $3x^2 - 13x + 12 =$ _____

    Step 4: Factor by expression in Step 3 by grouping. Factor out the common factor from each pair of terms in (e).

    (f) _____

    Factor out the common binomial factor.

    (g) _____

    Step 5: Check   Multiply to verify that the factorization is correct.

2. Factor by grouping: $13x^2 + 9x - 4$ *(See textbook Example 2)*   2. _____

Objective 2: Factor $ax^2 + bx + c$, $a \neq 1$ Using Trial and Error

3. Factor using trial and error: $3x^2 + 7x + 2$ *(See textbook Example 5)*

    Step 1: List the possibilities for the first terms of each binomial whose product is ax^2.

    (a) We list the possible ways of representing the first term, $3x^2$. Since 3 is a prime number, we have only one possiblity:

    $$(\underline{\quad} + ?)(\underline{\quad} + ?)$$

    Step 2: List the possibilities for the last terms of each binomial whose product is c.

    (b) The last term, 2, is also prime. List the choices of pairs of factors that multiply to 2.

    _____   _____

Guided Practice 6.3

	(c) However, did you notice that the coefficient of x, 7, is positive? To produce a positive sum, 7x, we must have two positive factors. What is the only option for factors of 2? _____
Step 3: Write out all the combinations of factors found in Steps 1 and 2. Multiply the binomials until a product is found that equals the trinomial.	(d) The order of the terms will change the product when the leading coefficient is not one. After testing the possibilities, which combination produces the required product? Write the factorization: $3x^2 + 7x + 2 =$ _____

4. Factor using trial and error: $18a^2 - 35ab + 12b^2$ (See textbook Example 6)   4. _____

5. Factor: $-6x^2 + 17x + 14$ (See textbook Example 10)   5. _____

Alternative Approach to Factor $ax^2 + bx + c$, $a \neq 1$, Using Synthetic Factoring

Another method for factoring trinomials of the form $ax^2 + bx + c$, $a \neq 1$, is synthetic factoring. When using this method, it is very important to first factor out any common factors.

EXAMPLE : How to Factor $ax^2 + bx + c$, $a \neq 1$, by Synthetic Factoring

Factor: $6x^2 + 11x + 3$

First, notice that $6x^2 + 11x + 3$ has no common factors. In this trinomial, $a = 6$, $b = 11$, and $c = 3$.

Step 1: Find the value of ac	The value of $a \cdot c$ is $6 \cdot 3 = 18$.
Step 2: Find the pair of integers, m and n, whose product is ac and whose sum is b.	We want to find integers whose product is 18 and whose sum is 11. Because both 18 and 11 are positive, we only list the positive factors of 18. <table><tr><td>**Factors of 18**</td><td>1, 18</td><td>2, 9</td><td>3, 6</td></tr><tr><td>**Sum**</td><td>19</td><td>11</td><td>9</td></tr></table>
Step 3: Write two fractions in the form $\dfrac{a}{m}$ and $\dfrac{a}{n}$.	We have $m = 2$, $n = 9$, and $a = 6$, so $\dfrac{a}{m} = \dfrac{6}{2}$ and $\dfrac{a}{n} = \dfrac{6}{9}$.

Guided Practice 6.3

Step 4: Write each fraction in lowest terms so that $\dfrac{a}{m} = \dfrac{p}{q}$ and $\dfrac{a}{n} = \dfrac{r}{s}$.	$\dfrac{a}{m} = \dfrac{6}{2} = \dfrac{3}{1} = \dfrac{p}{q}$ $\qquad$ $\dfrac{a}{n} = \dfrac{6}{9} = \dfrac{2}{3} = \dfrac{r}{s}$
Step 5: Write the factors $(px + q)(rx + s)$.	From Step 4, we have $p = 3$, $q = 1$, $r = 2$, and $s = 3$, so $$6x^2 + 11x + 3 = (3x + 1)(2x + 3)$$
Step 6: Check by multiplying the factors.	Multiply $(3x + 1)(2x + 3)$ to verify that the factorization is correct.

We summarize below the steps used in the example above.

Factoring $ax^2 + bx + c$, $a \ne 1$, By Synthetic Factoring, Where a, b, and c Have No Common Factors

Step 1: Find the value of ac
Step 2: Find the pair of integers, m and n, whose product is ac and whose sum is b.
Step 3: Write two fractions in the form $\dfrac{a}{m}$ and $\dfrac{a}{n}$.
Step 4: Write each fraction in lowest terms so that $\dfrac{a}{m} = \dfrac{p}{q}$ and $\dfrac{a}{n} = \dfrac{r}{s}$.
Step 5: Write the factors $(px + q)(rx + s)$.
Step 6: Check by multiplying the factors.

---

6. Factor using synthetic factoring: $-16x^2 - 4x + 6$

  (a) Factor out the common factor: _____

  (b) Determine each of the following using part (a): $a =$ _____, $b =$ _____, $c =$ _____, $a \cdot c =$ _____

Factors of –24							
Sum							

  (c) Which two factors multiply to $a \cdot c$ and add to b? _____

  (d) $\dfrac{a}{m} = \dfrac{?}{?}$ $\qquad$ $\dfrac{p}{q} = \dfrac{?}{?}$ $\qquad$ $p =$ _____ and $q =$ _____

  (e) $\dfrac{a}{n} = \dfrac{?}{?}$ $\qquad$ $\dfrac{r}{s} = \dfrac{?}{?}$ $\qquad$ $r =$ _____ and $s =$ _____

  (f) Write the factors $(px + q)(rx + s)$ _____

  (g) Write the factored form of $-16x^2 - 4x + 6$: _____

Guided Practice 6.3

Work Smart

Let's compare factoring by grouping and synthetic factoring by factoring $3x^2 + 10x + 8$. First, notice there is no GCF in the trinomial.

Grouping

Step 1: For $3x^2 + 10x + 8$, $a = 3$, $b = 10$, and $c = 8$, so $a \cdot c = 3 \cdot 8 = 24$.

Step 2: The factors of 24 whose sum is $b = 10$ are 6 and 4.

Step 3: Write $3x^2 + 10x + 8$ as
$$3x^2 + 10x + 8 = 3x^2 + 6x + 4x + 10$$

Step 4: Factor $3x^2 + 6x + 4x + 8$ by grouping.

$$3x^2 + 6x + 4x + 8 = (3x^2 + 6x) + (4x + 8)$$
$$= 3x(x + 2) + 4(x + 2)$$
$$= (x + 2)(3x + 4)$$

Synthetic Factoring

Step 1: For $3x^2 + 10x + 8$, $a = 3$, $b = 10$, and $c = 8$, so $a \cdot c = 3 \cdot 8 = 24$.

Step 2: The factors of 24 whose sum is $b = 10$ are 6 and 4.

Step 3: $\dfrac{a}{m} = \dfrac{3}{6}$ and $\dfrac{a}{n} = \dfrac{3}{4}$

Step 4: $\dfrac{a}{m} = \dfrac{3}{6} = \dfrac{1}{2} = \dfrac{p}{q}$ and $\dfrac{a}{n} = \dfrac{3}{4} = \dfrac{r}{s}$

Step 5: $(px + q)(rx + s) = (x + 2)(3x + 4)$, so

$$3x^2 + 10x + 8 = (x + 2)(3x + 4)$$

Both methods give the same result. Many people prefer trial and error while others prefer a more directed process such as grouping or synthetic factoring. You might know a process which has not been discussed here. Which method do you prefer and why?

7. Factor: $3x^2 + 22x + 7$ 7. _____

8. Factor: $-4 - 3x + 10x^2$ 8. _____

9. Factor: $21n^2 - 18n^3 + 9n$ 9. _____

Name:
Instructor:

Date:
Section:

Do the Math Exercises 6.3
Factoring Trinomials of the Form $ax^2 + bx + c$, $a \neq 1$

In Problems 1–6, factor each polynomial completely. Hint: None of the polynomials is prime.

1. $3x^2 + 16x - 12$

2. $5x^2 + 16x + 3$

1. _____

2. _____

3. $11p^2 - 46p + 8$

4. $3x^2 + 7xy + 2y^2$

3. _____

4. _____

5. $6x^2 - 14xy - 12y^2$

6. $-6x^2 - 3x + 45$

5. _____

6. _____

In Problems 7–12, factor each polynomial completely. Hint: None of the polynomials is prime.

7. $7n^2 - 27n - 4$

8. $25t^2 + 5t - 2$

7. _____

8. _____

Sullivan/Struve/Mazzarella, *Elementary & Intermediate Algebra*, 3e
Copyright © 2014 Pearson Education, Inc.

Do the Math Exercises 6.3

9. $20t^2 + 21t + 4$

10. $12p^2 - 23p + 5$

9. _____

10. _____

11. $18x^2 + 6xy - 4y^2$

12. $-10y^2 + 47y + 15$

11. _____

12. _____

In Problems 13–15, factor completely. If a polynomial cannot be factored, say it is prime.

13. $18x^2 + 88x - 10$

14. $9x^3y + 6x^2y + 3xy$

13. _____

14. _____

15. $8x^2 + 14x - 7$

16. $10x^2(x-1) - x(x-1) - 2(x-1)$

15. _____

16. _____

Name:
Instructor:

Date:
Section:

Five-Minute Warm-Up 6.4
Factoring Special Products

1. Evaluate: $(-10)^2$

   1. _____

2. Evaluate: $(-3)^3$

   2. _____

3. Find the product: $(4x^2)^3$

   3. _____

4. Find the product: $(2p+5)^2$

   4. _____

5. Find the product: $(4a+9b)(4a-9b)$

   5. _____

6. List the perfect squares that are less than 100. That is, find 1^2, 2^2, 3^2 etc.

   _____

7. List the perfect cubes that are less than 100. That is, find 1^3, 2^3, 3^3 etc.

   _____

Name:
Instructor:

Date:
Section:

Guided Practice 6.4
Factoring Special Products

Objective 1: Factor Perfect Square Trinomials

1. Find the product:

(a) $(A + B)^2 =$ _____

(b) $(A - B)^2 =$ _____

2. Factor completely: $z^2 - 8z + 16$ *(See textbook Example 1)*

Step 1: Determine whether the first term and the third term are perfect squares.

(a) What is the first term? _____ To be a perfect square, the first term must be of the form $(A)^2$. In this case case, $A =$ _____.

(b) What is the third term? _____ To be a perfect square, the third term must be of the form $(B)^2$. In this case case, $B =$ _____.

Step 2: Determine whether the middle term is 2 times or –2 times the product of the expression being squared in the first and last term.

(c) What is the middle term? _____ Is this equal to the product of 2 or -2 times A times B? _____

Step 3: Use $A^2 - 2AB + B^2 = (A - B)^2$

(d) $z^2 - 8z + 16 =$ _____

3. Factor each perfect square trinomial. *(See textbook Examples 2 and 4)*

(a) $p^2 + 22p + 121$ (b) $12m^2n - 36mn^2 + 27n^3$

3a. _____

3b. _____

Objective 2: Factor the Difference of Two Squares

4. Find the product: $(A - B)(A + B) =$ _____

5. Factor each difference of two squares completely. *(See textbook Examples 5 – 7)*

(a) $x^2 - 49$ (b) $25m^6 - 36n^4$

5a. _____

5b. _____

(c) $a^8 - 256$ (d) $12x^2y - 75y^3$

5c. _____

5d. _____

Guided Practice 6.4

Objective 3: Factor the Sum or Difference of Two Cubes

6. Find the product:

(a) $(A + B)(A^2 - AB + B^2) =$ _____

(b) $(A - B)(A^2 + AB + B^2) =$ _____

(c) $(A + B)^3 =$ _____

7. Factor the sum or difference of two cubes. *(See textbook Examples 8 and 9)*
   (a) $p^3 - 64$ (b) $8x^6 + 27y^3$

   7a. _____

   7b. _____

8. Factor completely. Remember to factor out the GCF first.
   (a) $24x^4 + 375x$ (b) $250x^4y - 16xy^4$

   8a. _____

   8b. _____

Name:
Instructor:

Date:
Section:

Do the Math Exercises 6.4
Factoring Special Products

In Problems 1–4, factor each perfect square trinomial completely.

1. $m^2 + 12m + 36$

2. $9a^2 - 12a + 4$

1. _____

2. _____

3. $16y^2 - 72y + 81$

4. $4a^2 + 20ab + 25b^2$

3. _____

4. _____

In Problems 5–6, factor each difference of two squares completely.

5. $36m^2 - 25n^2$

6. $a^4 - 16$

5. _____

6. _____

In Problems 7–10, factor each sum or difference of two cubes completely.

7. $64r^3 - 125s^3$

8. $m^9 - 27n^6$

7. _____

8. _____

Do the Math Exercises 6.4

9. $125y^3 + 27z^6$ **10.** $40x^3 + 135y^6$ 9. _____

10. _____

In Problems 11–15, factor completely. If the polynomial is prime, state so.

11. $12n^3 - 36n^2 + 27n$ **12.** $32a^3 + 4b^6$ 11. _____

12. _____

13. $x^4 - 225x^2$ **14.** $12m^2 - 14m + 21$ 13. _____

14. _____

15. $3x^2 + 18x + 27$ **16.** $(x+3)^2 - 25$ 15. _____

16. _____

Five-Minute Warm-Up 6.5
Summary of Factoring Techniques

In Problems 1 – 2, find each product.

1. $(4x + 3y)(5x - 2y)$

    1. _____

2. $(a + 4b)^2$

    2. _____

In Problems 3 – 5, factor completely.

3. $-27a^3 + 9a^2 - 18a$

    3. _____

4. $6m(2n - 1) - 9(2n - 1)$

    4. _____

5. $4p^2 - 8pq + 3p - 6q$

    5. _____

Name: Date:
Instructor: Section:

Guided Practice 6.5
Summary of Factoring Techniques

Objective 1: Factor Polynomials Completely

1. Review the **Steps for Factoring** listed at the beginning of textbook Section 6.5. Step 1 should always be

   _____, if possible.

2. Factor: $4x^2 + 16x - 84$ *(See textbook Example 1)*

Step 1: Factor out the greatest common factor (GCF), if any exists.	Determine the GCF:	**(a)** GCF = _____
	Factor out the GCF:	**(b)** $4x^2 + 16x - 84 =$ _____
Step 2: Identify the number of terms in the polynomial in parentheses.	How many terms are in the polynomial in parentheses?	**(c)** _____
Step 3: We concentrate on the trinomial in parentheses. It is not a perfect square trinomial. The trinomial has a leading coefficient of 1 so we try $(x + m)(x + n)$, where $mn = c$ and $m + n = b$.	Factor the trinomial:	**(d)** _____
Step 4: Check		We leave it to you to multiply the binomials and then distribute to verify that the factorization is correct.

3. Factor: $64a^2 - 81b^2$

Step 1: Factor out the greatest common factor (GCF), if any exists.	There is no GCF.	
Step 2: Identify the number of terms in the polynomial.	How many terms are in the polynomial?	**(a)** _____
Step 3: Because the first term $64a^2 = (8a)^2$, and the second term, $81b^2 = (9b)^2$, are both perfect squares, we have the difference of two squares.	Factor the binomial:	**(b)** _____
Step 4: Check		We leave it to you to verify that the factorization is correct.

Sullivan/Struve/Mazzarella, *Elementary & Intermediate Algebra*, 3e
Copyright © 2014 Pearson Education, Inc.

Guided Practice 6.5

4. Factor: $8a^2 + 24ab + 18b^2$ *(See textbook Example 3)*

Step 1: Factor out the greatest common factor (GCF), if any exists.

What is the GCF of 8, 24, and 18? **(a)** _____

Factor out the GCF: **(b)** _____

Step 2: Identify the number of terms in the polynomial in parentheses.

How many terms are in the polynomial? **(c)** _____

Step 3: We concentrate on the polynomial in parentheses. Is it a perfect square trinomial?

(d) What is the first term? ____ To be a perfect square, the first term must be of the form $(A)^2$. In this case case, $A =$ ____.

(e) What is the third term? ____ To be a perfect square, the third term must be of the form $(B)^2$. In this case case, $B =$ ____.

(f) What is the middle term? ____ Is this equal to the product of 2 or -2 times A times B? _____

Use $A^2 + 2AB + B^2 = (A + B)^2$ Write the factorization: **(g)** _____

Step 4: Check   We leave it to you to verify that the factorization is correct.

5. Factor completely: $27p^6 + 1$ *(See textbook Example 4)*   5. _____

6. Factor completely: $2n^3 - 10n^2 - 6n + 30$ *(See textbook Example 6)*   6. _____

7. Factor completely: $-3xy^2 - 12xy - 9x$ *(See textbook Example 7)*   7. _____

Do the Math Exercises 6.5
Summary of Factoring Techniques

In Problems 1–15, factor completely. If a polynomial cannot be factored, say it is prime.

1. $x^2 - 256$

2. $1 - y^9$

1. _____

2. _____

3. $x^2 + 5x - 6$

4. $2x^3 - x^2 - 18x + 9$

3. _____

4. _____

5. $100x^2 - 25y^2$

6. $6s^2t^2 + st - 1$

5. _____

6. _____

7. $8x^6 + 125y^3$

8. $48m - 3m^9$

7. _____

8. _____

Do the Math Exercises 6.5

9. $4t^4 + 16t^2$

10. $8n^3 - 18n$

9. _____

10. _____

11. $n^2 + 2n + 8$

12. $24p^3q + 81q^4$

11. _____

12. _____

13. $x^2(x+y) - 16(x+y)$

14. $-9x - 3x^3 - 12x^2$

13. _____

14. _____

15. $6a^3 - 5a^2b + ab^2$

16. $x^4 + 3x^2 + 2$

15. _____

16. _____

Five-Minute Warm-Up 6.6
Solving Polynomial Equations by Factoring

1. Solve: $2x - 1 = 0$

    1. _____

2. Solve: $-3(x - 2) + 15 = 0$

    2. _____

3. Evaluate $3n^2 + 4n - 5$ for
   (a) $n = 2$ (b) $n = -3$

    3a. _____

    3b. _____

In Problems 4 – 7, factor completely.

4. $x^2 + 2x - 63$

    4. _____

5. $4p^2 + 4p - 3$

    5. _____

6. $x^2 - 81$

    6. _____

7. $n^2 - 16n + 64$

    7. _____

Name:
Instructor:
Date:
Section:

Guided Practice 6.6
Solving Polynomial Equations by Factoring

Objective 1: Solve Quadratic Equations Using the Zero-Product Property

1. State the **Zero-Product Property**. _____
   _____
   _____

2. Solve $(x+1)(2x-3) = 0$ *(See textbook Example 1)*   2. _____

3. A second-degree equation is also called a _____

4. What does it mean to write a quadratic equation in *standard form*? _____
   _____

5. Solve: $x^2 + 5x - 24 = 0$ *(See textbook Example 2)*

Step 1: Is the quadratic equation in standard form? Yes, it is written in the form $ax^2 + bx + c = 0$.	$x^2 + 5x - 24 = 0$
Step 2: Factor the expression on the left side of the equation.	(a) _____
Step 3: Set each factor to 0. (b) _____	(c) _____
Step 4: Solve each first-degree equation. (d) _____	(e) _____
Step 5: Check Substitute your values into the original equation.	We leave it to you to verify the solutions.
Write the solution set:	(f) _____

6. Solve: $3t^2 + 11t = 4$ *(See textbook Example 3)*

   (a) Write the quadratic equation in standard form: _____

   (b) Factor the polynomial on the left side of the equation: _____

   (c) Solve and state the solution set. _____

Sullivan/Struve/Mazzarella, *Elementary & Intermediate Algebra*, 3e
Copyright © 2014 Pearson Education, Inc.

Guided Practice 6.6

7. Solve: $(x-2)(x-3) = 56$ *(See textbook Example 5)*

   (a) To solve this equation, the first step is _____

   (b) Next, write the quadratic equation in _____

   (c) Factor the polynomial on the left side of the equation: _____

   (d) Solve and state the solution set. _____

8. Solve: $4x^2 + 36 = 24x$ *(See textbook Example 6)*

   (a) Write the quadratic equation in standard form: _____

   (b) Factor the polynomial on the left side of the equation: _____

   (c) Solve and state the solution set. _____

Objective 2: Solve Polynomial Equations of Degree Three or Higher Using the Zero-Product Property

9. Solve: $p^3 + 2p^2 - 9p - 18 = 0$ *(See textbook Example 9)*

Step 1: Write the equation in standard form.	The polynomial is already in standard form.	$p^3 + 2p^2 - 9p - 18 = 0$
Step 2: Factor the expression on the left side of the equation. Begin with the GCF, if any.	Does the equation contain a GCF?	(a) _____
	How many terms are in the polynomial?	(b) _____
	What technique should you use?	(c) _____
	Factor the polynomial:	(d) _____
Step 3: Set each factor to 0.	(e) _____ (f) _____	(g) _____
Step 4: Solve each first-degree equation.	(h) _____ (i) _____	(j) _____
Step 5: Check Substitute your values into the original equation.		We leave it to you to verify the solutions.
	Write the solution set:	(k) _____

Do the Math Exercises 6.6
Solving Polynomial Equations by Factoring

In Problems 1–2, solve each equation using the Zero-Product Property.

1. $3x(x+9) = 0$

2. $(4z-3)(z+4) = 0$

1. _____

2. _____

In Problems 3–4, identify each equation as a linear equation or a quadratic equation.

3. $2x + 1 - (x+7) = 3x + 1$

4. $(x+2)(x-2) = 14$

3. _____

4. _____

In Problems 5–12, solve each quadratic equation by factoring.

5. $p^2 - 5p - 24 = 0$

6. $14x - 49x^2 = 0$

5. _____

6. _____

7. $3x^2 + x - 14 = 0$

8. $k^2 + 12k + 36 = 0$

7. _____

8. _____

Do the Math Exercises 6.6

9. $a^2 - 6a = 16$

10. $m^2 - 30 = 7m$

9. _____

10. _____

11. $4p - 3 = -4p^2$

12. $(x+5)(x-3) = 9$

11. _____

12. _____

In Problems 13–14, solve each polynomial equation by factoring.

13. $3x^3 + x^2 - 14x = 0$

14. $m^3 + 2m^2 - 9m - 18 = 0$

13. _____

14. _____

15. **Tossing a Ball** A ball is thrown vertically upward from the ground with an initial velocity of 64 feet per second. Solve the equation $-16t^2 + 64t = 48$ to find the time t (in seconds) at which the ball is 48 feet from the ground.

15. _____

Five-Minute Warm-Up 6.7
Modeling and Solving Problems with Quadratic Equations

1. Solve: $x(x+4) = 45$ 1. _____

2. Solve: $p^2 + 12p + 36 = 0$ 2. _____

3. Solve: $6x^2 + 17x + 5 = 0$ 3. _____

4. The lengths of the two legs of a right triangle are 9 and 12 inches. Find the length of the hypotenuse.   4. _____

5. Find the product: $(x-5)^2$ 5. _____

6. Factor completely: $-18h^3 + 27h^2 - 9h$ 6. _____

Name:
Instructor:

Date:
Section:

Guided Practice 6.7
Modeling and Solving Problems with Quadratic Equations

Objective 1: Model and Solve Problems Involving Quadratic Equations

1. A ball is thrown off a cliff from a height of 64 feet above sea level. The height h of the ball above the water (in feet) at any time t (in seconds) can be modeled by the equation
$$h = -16t^2 + 48t + 64$$
(See textbook Example 1)

    (a) When will the height of the ball be 64 feet above sea level? 1a. _____

    (b) When will the ball strike the water? 1b. _____

2. The length and width of two sides of a rectangle are consecutive odd integers. The area of the rectangle is 255 square centimeters. Find the dimensions of the rectangle. *(See textbook Example 2)*

Step 1: Identify This is a geometry problem involving the area of a rectangle. It also involves consecutive odd integers.

    (a) If n represents one of the odd integers, express the next consecutive odd integer: 2a. _____

    (b) In general, what formula do we use to calculate the area of a rectangle? 2b. _____

    (c) Step 2: Name Use the variables from (a) to identify an expression that represents the length and width of the rectangle.
2c.
width = _____
length = _____

    (d) Step 3: Translate Write an equation that will model the area of this rectangle. 2d. _____

    (e) Step 4: Solve the equation from step 3. 2e. _____

Step 5: Check Is your answer reasonable? Does it meet the necessary conditions?

    (f) Step 6: Answer the question. _____

Sullivan/Struve/Mazzarella, *Elementary & Intermediate Algebra*, 3e
Copyright © 2014 Pearson Education, Inc.

Guided Practice 6.7

Objective 2: Model and Solve Problems Using the Pythagorean Theorem

3. In your own words, state the Pythagorean Theorem.

4. If x and y are the lengths of the legs and z is the length of the hypotenuse, write an equation which uses these variables to state the Pythagorean Theorem.

4. _____

5. Use the Pythagorean Theorem to find the missing length in the right triangle. *(See textbook Example 4)*

   (a) one leg = 18 cm; second leg = 24 cm, find the hypotenuse.

   (b) one leg is 24 ft; hypotenuse is 25 ft, find the second leg.

   5a. _____

   5b. _____

6. In a right triangle, the hypotenuse measures $(2x-3)$ meters. If one leg measures x meters and the other leg measures $(x+3)$ meters,

   (a) solve for x.

   6a. _____

   (b) What are the lengths of the sides of the right triangle?

   6b. _____

Name:
Instructor:

Date:
Section:

Do the Math Exercises 6.7
Modeling and Solving Problems with Quadratic Equations

In Problems 1–2, use the given area to find the missing sides of the rectangle.

1.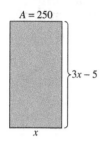

2.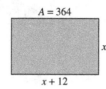

1. _____

2. _____

In Problems 3–4, use the given area to find the height and base of the triangle.

3.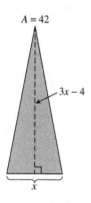

4.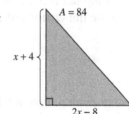

3. _____

4. _____

In Problems 5–6, use the given area to find the dimensions of the quadrilateral.

5.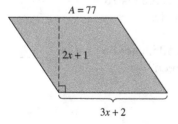

6.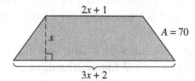

5. _____

6. _____

Sullivan/Struve/Mazzarella, *Elementary & Intermediate Algebra*, 3e
Copyright © 2014 Pearson Education, Inc.

Do the Math Exercises 6.7

In Problems 7–8, use the Pythagorean Theorem to find the lengths of the sides of the triangle.

7.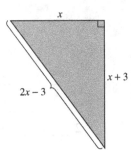

8.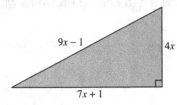

7. _____

8. _____

9. **Projectile Motion** If $h = -16t^2 + 96t$ represents the height of a rocket, in feet, t seconds after it was fired, when will the rocket be 144 feet high?

9. _____

10. **Rectangular Room** The width of a rectangular room is 4 feet less than the length. If the area of the room is 21 square feet, find the dimensions of the room.

10. _____

11. **Triangle** The base of a triangle is 2 meters more than the height. If the area of the triangle is 24 square meters, find the base and height of the triangle.

11. _____

12. **Playing Field Dimensions** The length of a rectangular playing field is 3 yards less than twice the width. If the area of the field is 104 square yards, what are its dimensions?

12. _____

13. **Jasper's Big-Screen TV** Jasper purchased a new big-screen TV. The TV screen is 10 inches wider than it is tall and is surrounded by a casing that is 2 inches wide.

(a) Jasper lost his tape measure but sees on the box that the TV measures 50 inches on the diagonal, including the casing. What are the dimensions of the TV screen?

13a. _____

(b) Jasper knows that the size of the opening where he wants the TV installed is 29 by 42 inches. Will this TV fit into his space?

13b. _____

Name:
Instructor:

Date:
Section:

Five-Minute Warm-Up 7.1
Simplifying Rational Expressions

1. Evaluate $\dfrac{-4a + 2b}{3}$ for $a = -2$ and $b = -7$.

   1. _____

2. Factor: $4x^2 + x - 3$

   2. _____

3. Solve: $2x^2 - x - 15 = 0$

   3. _____

4. Write $\dfrac{30}{63}$ in lowest terms.

   4. _____

5. Divide: $\dfrac{24x^2 y}{36xy^3}$

   5. _____

Name:
Instructor:

Date:
Section:

Guided Practice 7.1
Simplifying Rational Expressions

Objective 1: Evaluate a Rational Expression

1. Evaluate the following rational expressions. *(See textbook Example 1)*

   (a) $\dfrac{3}{2x+4}$ for $x = 3$

   (b) $\dfrac{5x-3}{x+1}$ for $x = 7$

   1a. _____

   1b. _____

2. Evaluate the following rational expressions. *(See textbook Examples 2 and 3)*

   (a) $\dfrac{x+9}{x^2+8x-12}$ for $x = -3$

   (b) $\dfrac{-3a+2b}{-a-b}$ for $a = 6$ and $b = 2$

   2a. _____

   2b. _____

Objective 2: Determine Values for Which a Rational Expression is Undefined

3. To determine the undefined values of a rational expression, find the values where _____.

4. Find the value(s) of x for which the expression is undefined. *(See textbook Examples 4 and 5)*

   (a) $\dfrac{x-3}{x+8}$

   (b) $\dfrac{x^2-2x-3}{x^2-9x+18}$

   4a. _____

   4b. _____

Objective 3: Simplify Rational Expressions

5. To simplify a rational expression always use the following steps:

   (a) Completely _____ the numerator and denominator of the rational expression.

   (b) _____ _____ the common factors.

Guided Practice 7.1

6. Simplify: $\dfrac{6x-18}{x^2-5x+6}$ (See textbook Examples 6 and 7)

Step 1: Completely factor the numerator and the denominator. Factor each expression: (a) $\dfrac{6x-18}{x^2-5x+6} =$ _____

Step 2: Divide out common factors. The rational expression is completely simplified if the numerator and the denominator share no common factor other than 1.

Divide out common factors: (b) _____

Write the simplified expression: (c) _____

7. Simplify: $\dfrac{2x^2+12x+10}{4x^2+28x+40}$ (See textbook Example 8)

7. _____

8. Simplify: $\dfrac{50-2x^2}{4x^2+8x-60}$ (See textbook Example 9)

8. _____

Name:
Instructor:

Date:
Section:

Do the Math Exercises 7.1
Simplifying Rational Expressions

In Problems 1–4, evaluate each expression for the given values.

1. $\dfrac{x}{x+4}$
   (a) $x = 8$
   (b) $x = -4$
   (c) $x = 0$

2. $\dfrac{2m-1}{m}$
   (a) $m = 4$
   (b) $m = 1$
   (c) $m = -1$

1a. _____

1b. _____

1c. _____

2a. _____

2b. _____

2c. _____

3. $\dfrac{a^2 - 2a}{a - 4}$
   (a) $a = 5$
   (b) $a = -1$
   (c) $a = -4$

4. $\dfrac{b^2 - a^2}{(a-b)^2}$
   (a) $a = 3, b = 2$
   (b) $a = 5, b = 4$
   (c) $a = -2, b = 2$

3a. _____

3b. _____

3c. _____

4a. _____

4b. _____

4c. _____

In Problems 5–8, find the value(s) of the variable for which the rational expression is undefined.

5. $\dfrac{5m^3}{m+8}$

6. $\dfrac{12}{4a-3}$

5. _____

6. _____

7. $\dfrac{2x^2}{x^2 + x - 2}$

8. $\dfrac{3h+2}{h^3 + 5h^2 + 4h}$

7. _____

8. _____

Do the Math Exercises 7.1

In Problems 9–16, simplify each rational expression. Assume that no variable has a value which results in a denominator with a value of zero.

9. $\dfrac{6p^2 + 3p}{3p}$

10. $\dfrac{4-z}{z-4}$

9. _____

10. _____

11. $\dfrac{x^2 - 9}{x^2 + 5x + 6}$

12. $\dfrac{x^3 + 3x^2 + 2x}{x^3 - 2x^2 - 3x}$

11. _____

12. _____

13. $\dfrac{a-b}{b^2 - a^2}$

14. $\dfrac{z^3 + 3z^2 + 2z + 6}{z^2 + 5z + 6}$

13. _____

14. _____

15. $\dfrac{2x^2 + 5x - 3}{4x^2 - 8x + 3}$

16. $\dfrac{5 + 4x - x^2}{4x^2 - 17x - 15}$

15. _____

16. _____

17. Explain why we can divide out only common *factors*, not common *terms*.

Name:
Instructor:

Date:
Section:

Five-Minute Warm-Up 7.2
Multiplying and Dividing Rational Expressions

1. Find the product: $\dfrac{21}{4} \cdot \dfrac{18}{49}$

    1. _____

2. Find the reciprocal of $-\dfrac{5}{2}$.

    2. _____

3. Find the quotient: $\dfrac{15}{36} \div \dfrac{70}{24}$

    3. _____

4. Factor: $-5x^2 + 25x$

    4. _____

5. Simplify the rational expression: $\dfrac{x^2 - 9}{x^2 - 3x - 18}$

    5. _____

Sullivan/Struve/Mazzarella, *Elementary & Intermediate Algebra*, 3e
Copyright © 2014 Pearson Education, Inc.

Name:
Instructor:

Date:
Section:

Guided Practice 7.2
Multiplying and Dividing Rational Expressions

Objective 1: Multiply Rational Expressions

1. Multiply $\dfrac{x+3}{8} \cdot \dfrac{4x-12}{x^2-9}$. Simplify the result, if possible. *(See textbook Example 1)*

   Step 1: Completely factor the polynomials in each numerator and denominator.

   $\dfrac{x+3}{8} \cdot \dfrac{4x-12}{x^2-9} =$

   Factor the numerator and denominator: **(a)** _____

   Step 2: Multiply using $\dfrac{a}{b} \cdot \dfrac{c}{d} = \dfrac{ac}{bd}$

   Write the product: **(b)** _____

   Step 3: Divide out common factors in the numerator and denominator.

   Divide out common factors and express the answer as a simplified rational expression in factored form. **(c)** _____

   Write the simplified result: **(d)** _____

2. Multiply $\dfrac{x^2-4x}{x^2-4} \cdot \dfrac{x^2-x-6}{x-4}$. Simplify the result, is possible. *(See textbook Examples 2 and 3)*

   (a) Completely factor the polynomials in each numerator and denominator. **(a)** _____

   (b) Multiply. **(b)** _____

   (c) Divide out common factors in the numerator and denominator. **(c)** _____

   (d) Express the answer as a simplified rational expression in factored form. **(d)** _____

3. Find the product and simplify: $\dfrac{x^2-y^2}{13x^2-13xy} \cdot \dfrac{26y-26x}{x^2+2xy+y^2}$ *(See textbook Example 4)* 3. _____

Sullivan/Struve/Mazzarella, *Elementary & Intermediate Algebra*, 3e
Copyright © 2014 Pearson Education, Inc.

Guided Practice 7.2

Objective 2: Divide Rational Expressions

4. Divide $\dfrac{3x+15}{36} \div \dfrac{5x+25}{24}$. Simplify the result, is possible. *(See textbook Example 5)*

Step 1: Multiply the dividend by the reciprocal of the divisor.

$\dfrac{3x+15}{36} \div \dfrac{5x+25}{24} =$

Rewrite the quotient as a product: **(a)** _____

Step 2: Completely factor the polynomials in each numerator and denominator.

Factor each expression: **(b)** _____

Step 3: Multiply.

Write the product: **(c)** _____

Step 4: Divide out common factors in the numerator and denominator.

Divide out the common factors: **(d)** _____

Express the answer as a simplified rational expression in factored form: **(e)** _____

5. Find the quotient and simplify: $\dfrac{x^2-4x+3}{2x^2-5x+3} \div (x-3)$ *(See textbook Example 7)*   5. _____

6. Find the quotient and simplify: $\dfrac{\dfrac{3y^2-3x^2}{x^2+2xy+y^2}}{\dfrac{6x^2-6xy}{3y+3x}}$ *(See textbook Example 8)*   6. _____

Name:
Instructor:

Date:
Section:

Do the Math Exercises 7.2
Multiplying and Dividing Rational Expressions

In Problems 1–4, multiply.

1. $\dfrac{x-4}{4} \cdot \dfrac{12x}{x^2-16}$

2. $\dfrac{8n-8}{n^2-3n+2} \cdot \dfrac{n+2}{12}$

3. $\dfrac{z^2-4}{3z-2} \cdot \dfrac{3z^2+7z-6}{z^2+z-6}$

4. $(x-3) \cdot \dfrac{x-2}{x^2-5x+6}$

1. _____

2. _____

3. _____

4. _____

In Problems 5–8, divide.

5. $\dfrac{p+7}{3p+21} \div \dfrac{p}{6}$

6. $\dfrac{z^2-25}{10z} \div \dfrac{z^2-10z+25}{5z}$

7. $\dfrac{3y^2+6y}{8} \div \dfrac{y+2}{12y-12}$

8. $\dfrac{4z^2+12z+9}{8z+16} \div \dfrac{4z^2+12z+9}{12z+24}$

5. _____

6. _____

7. _____

8. _____

Do the Math Exercises 7.2

In Problems 9–16, perform the indicated operation.

9. $\dfrac{x-4}{12x-18} \cdot \dfrac{6}{x^2-16}$

10. $\dfrac{x+5}{3} \div \dfrac{30x}{4x+20}$

9. _____

10. _____

11. $\dfrac{2r^2+rs-3s^2}{r^2-s^2} \cdot \dfrac{r^2-2rs-3s^2}{2r+3s}$

12. $\dfrac{2x^2-5x+3}{x^2-1} \cdot \dfrac{x^2+1}{2x^2-x-3}$

11. _____

12. _____

13. $\dfrac{(x-3)^2}{8xy^2} \div \dfrac{x^2-9}{(4x^2y)^2}$

14. $\dfrac{p^2-49}{p^2-5p-14} \cdot \dfrac{p-7}{14-5p-p^2}$

13. _____

14. _____

15. $\dfrac{1}{b^2+b-12} \div \dfrac{1}{b^2-5b-36}$

16. $\dfrac{\dfrac{t^2}{t^2-16}}{\dfrac{t^2-3t}{t^2-t-12}}$

15. _____

16. _____

Five-Minute Warm-Up 7.3
Adding and Subtracting Rational Expressions with a Common Denominator

1. Write $\dfrac{27}{45}$ in lowest terms.

    1. _____

2. Find each sum and simplify:

    (a) $\dfrac{8}{3} + \dfrac{5}{3}$

    (b) $\dfrac{11}{8} + \dfrac{9}{8}$

    2a. _____

    2b. _____

3. Find each difference and simplify:

    (a) $\dfrac{9}{2} - \dfrac{7}{2}$

    (b) $\dfrac{13}{24} - \left(-\dfrac{2}{24}\right)$

    3a. _____

    3b. _____

4. Determine the additive inverse of $\dfrac{1}{2}$.

    4. _____

5. Simplify: $-(2x - 3)$

    5. _____

Name:
Instructor:
Date:
Section:

Guided Practice 7.3
Adding and Subtracting Rational Expressions with a Common Denominator

Objective 1: Add Rational Expressions with a Common Denominator

1. Find the sum and simplify, if possible: $\dfrac{2x}{x^2 - 3x} + \dfrac{5x}{x^2 - 3x}$ (See textbook Example 1)

Step 1: Add the numerators and write the result over the common denominator.

Use $\dfrac{a}{c} + \dfrac{b}{c} = \dfrac{a+b}{c}$: (a) _____

Step 2: Simplify the rational expression.

Factor the numerator and denominator: (b) _____

Divide out the common factors: (c) _____

Express the answer in simplified form: (d) _____

2. Find the sum and simplify, if possible: $\dfrac{3x - 10}{x^2 - 25} + \dfrac{2x - 15}{x^2 - 25}$ (See textbook Example 2)   2. _____

Objective 2: Subtract Rational Expressions with a Common Denominator

3. Find the difference and simplify, if possible: $\dfrac{x^2}{x^2 - 5x - 6} - \dfrac{1}{x^2 - 5x - 6}$ (See textbook Example 3)

Step 1: Subtract the numerators and write the result over the common denominator.

Use $\dfrac{a}{c} - \dfrac{b}{c} = \dfrac{a-b}{c}$: (a) _____

Step 2: Simplify the rational expression.

Factor the numerator and denominator: (b) _____

Divide out the common factors: (c) _____

Express the answer in simplified form: (d) _____

Sullivan/Struve/Mazzarella, *Elementary & Intermediate Algebra*, 3e
Copyright © 2014 Pearson Education, Inc.

Guided Practice 7.3

4. Find the difference and simplify if possible: $\dfrac{2x^2 + 3x}{x - 2} - \dfrac{x^2 + 3x + 4}{x - 2}$

4. _____

(See textbook Example 4)

Objective 3: Add or Subtract Rational Expressions with Opposite Denominators

5. Find the sum and simplify, if possible: $\dfrac{6n - 5}{n - 5} + \dfrac{2n - n^2}{5 - n}$ (See textbook Example 5)

5. _____

6. Find the difference and simplify, if possible: $\dfrac{x^2 + 1}{x^2 - 9} - \dfrac{4x + 2}{9 - x^2}$

6. _____

(See textbook Example 6)

Do the Math Exercises 7.3
Adding and Subtracting Rational Expressions with a Common Denominator

In Problems 1–4, add the rational expressions.

1. $\dfrac{4n+1}{8} + \dfrac{12n-1}{8}$

2. $\dfrac{4p-1}{p-1} + \dfrac{p+1}{p-1}$

3. $\dfrac{3x^2+4}{3x-6} + \dfrac{3x^2-28}{3x-6}$

4. $\dfrac{2x}{2x^2-x-15} + \dfrac{5}{2x^2-x-15}$

1. _____

2. _____

3. _____

4. _____

In Problems 5–8, subtract the rational expressions.

5. $\dfrac{4a^2}{3a-1} - \dfrac{a^2+a}{3a-1}$

6. $\dfrac{7x+13}{3x} - \dfrac{x+1}{3x}$

7. $\dfrac{2z^2+7z}{z^2-1} - \dfrac{z^2-6}{z^2-1}$

8. $\dfrac{2n^2+n}{n^2-n-2} - \dfrac{n^2+6}{n^2-n-2}$

5. _____

6. _____

7. _____

8. _____

Do the Math Exercises 7.3

In Problems 9–12, add or subtract the rational expressions.

9. $\dfrac{4}{x-1} + \dfrac{8x}{1-x}$

10. $\dfrac{x^2+3}{x-1} - \dfrac{x+3}{1-x}$

9. _____

10. _____

11. $\dfrac{x^2+6x}{x^2-1} + \dfrac{4x+3}{1-x^2}$

12. $\dfrac{m^2+2mn}{n^2-m^2} - \dfrac{n^2}{m^2-n^2}$

11. _____

12. _____

In Problems 13–16, perform the indicated operation.

13. $\dfrac{3n^2}{2n-1} - \dfrac{3n^2}{2n-1}$

14. $\dfrac{x^2}{x^2-9} + \dfrac{6x+9}{x^2-9}$

13. _____

14. _____

15. $\dfrac{2s-3t}{s-t} + \dfrac{6s-4t}{t-s}$

16. $\dfrac{n+3}{2n^2-n-3} - \dfrac{-n^2-3n}{2n^2-n-3}$

15. _____

16. _____

Name:
Instructor:

Date:
Section:

Five-Minute Warm-Up 7.4
Finding the Least Common Denominator and Forming Equivalent Rational Expressions

1. Write $\dfrac{7}{15}$ as a fraction with 45 as its denominator.

    1.

2. Find the least common denominator (LCD) of $\dfrac{9}{20}$ and $\dfrac{2}{45}$.

    2.

3. Write each of the rational expressions in Problem 2 as equivalent expressions using the LCD.

    3.

4. Identify the missing factor: $12x^2y \cdot ? = 36x^2y^3$

    4.

5. Factor: $x^2 - 4x + 4$

    5.

6. Factor: $-3x^2 - 9x$

    6. _____

Sullivan/Struve/Mazzarella, *Elementary & Intermediate Algebra*, 3e
Copyright © 2014 Pearson Education, Inc.

Guided Practice 7.4
Finding the Least Common Denominator and Forming Equivalent Rational Expressions

Objective 1: Find the Least Common Denominator of Two or More Rational Expressions

1. Find the least common denominator (LCD) of $\frac{7}{24}$ and $\frac{3}{20}$. *(See textbook Example 1)*

Step 1: Write each denominator as the product of prime factors.

Write 24 as the product of prime factors using exponential form for repeating factors: **(a)** _____

Write 20 as the product of prime factors using exponential form for repeating factors: **(b)** _____

Step 2: Find the product of each prime factor the greatest number of times it appears in any factorization.

List every different factor in (a) and (b), using the highest exponent that appears: **(c)** _____

Multiply the factors in (c) and state the LCD of $\frac{7}{24}$ and $\frac{3}{20}$: **(d)** _____

2. Find the least common denominator (LCD) of the rational expressions $\frac{7}{6x^2}$ and $\frac{5}{9x}$.

(See textbook Example 2)

Step 1: Write each denominator as the product of prime factors. Write the factored form using exponents, if necessary.

Write $6x^2$ as the product of prime factors: **(a)** _____

Write $9x$ as the product of prime factors: **(b)** _____

Step 2: Find the product of each prime factor the greatest number of times it appears in any factorization.

List every different factor in (a) and (b), using the highest exponent that appears: **(c)** _____

Multiply the factors in (c) and state the LCD of $\frac{7}{6x^2}$ and $\frac{5}{9x}$: **(d)** _____

Sullivan/Struve/Mazzarella, *Elementary & Intermediate Algebra*, 3e
Copyright © 2014 Pearson Education, Inc.

Guided Practice 7.4

3. Find the LCD of the rational expressions: $\dfrac{7c}{8a^3b}$ and $\dfrac{11d}{10a^2b^2}$ (See textbook Example 3)   3. _____

4. Find the LCD of the rational expressions: $\dfrac{8}{7a^2 - 14a}$ and $\dfrac{5}{2a^2}$ (See textbook Example 4)   4. _____

5. Find the LCD of the rational expressions: $\dfrac{4x}{x^2 - 9}$ and $\dfrac{x^2}{x^2 - 6x + 9}$ 5. _____
(See textbook Example 5)

Objective 2: Write a Rational Expression Equivalent to a Given Rational Expression

6. Write the rational expression $\dfrac{x-2}{3x^2 - x}$ as an equivalent rational expression with denominator $9x^3 - 3x^2$.
(See textbook Example 8)

Step 1: Write each denominator in factored form.	Write $3x^2 - x$ as the product of prime factors:	(a) _____
	Write $9x^3 - 3x^2$ as the product of prime factors:	(b) _____
Step 2: Determine the "missing factor(s)".	Compare (a) and (b). What is (are) the missing factor(s)?	(c) _____
Step 3: Multiply the original rational expression by 1.	Write 1 as $\dfrac{\text{missing factor}}{\text{missing factor}}$.	(d) _____
Step 4: Find the product. Leave the denominator in factored form.	Multiply the expression in (d) and the original rational expression. This product is the equivalent rational expression.	(e) _____

Objective 3: Use the LCD to Write Equivalent Rational Expressions

7. Find the LCD of the rational expressions $\dfrac{4}{9xy^2}$ and $\dfrac{5}{12x^2}$. Then rewrite each rational expression with the LCD. (See textbook Example 9)   7. _____

_____

Name: Date:
Instructor: Section:

Do the Math Exercises 7.4
Finding the Least Common Denominator and Forming Equivalent Rational Expressions

In Problems 1–6, find the LCD of the given rational expressions.

1. $\dfrac{3}{7y^2}; \dfrac{4}{49y}$

2. $\dfrac{2}{6x-18}; \dfrac{3}{8x-24}$

1. _____

2. _____

3. $\dfrac{5}{2x^2-12x+18}; \dfrac{4}{4x^2-36}$

4. $\dfrac{8}{x^2-1}; \dfrac{3}{x^2-2x+1}$

3. _____

4. _____

5. $\dfrac{-1}{c^2-49}; \dfrac{5}{21-3c}$

6. $\dfrac{3z}{(z+5)(z-4)}; \dfrac{-z}{(4-z)(5+z)}$

5. _____

6. _____

In Problems 7–10, write an equivalent rational expression with the given denominator.

7. $\dfrac{7a+1}{abc}$ with denominator a^2b^2c

8. $\dfrac{3}{x+2}$ with denominator x^2+5x+6

7. _____

8. _____

9. $\dfrac{7a}{3a^2+9a+6}$ denominator $6a^2+18a+12$

10. 7 with denominator t^2+1

9. _____

10. _____

Do the Math Exercises 7.4

In Problems 11–18, find the LCD of the rational expressions. Then rewrite each as an equivalent rational expression with the LCD.

11. $\dfrac{4}{5n^2}; \dfrac{3}{7n}$

12. $\dfrac{2p^2-1}{6p}; \dfrac{3p^3+2}{8p^2}$

11. _____

12. _____

13. $\dfrac{x+2}{x}; \dfrac{x}{x+2}$

14. $\dfrac{3b}{7a}; \dfrac{2a}{7a+14b}$

13. _____

14. _____

15. $\dfrac{5}{m-6}; \dfrac{2m}{6-m}$

16. $\dfrac{4a}{a^2-1}; \dfrac{7}{a+1}$

15. _____

16. _____

17. $\dfrac{4n}{n^2-n-6}; \dfrac{2}{n^2+4n+4}$

18. $\dfrac{3}{2n^2-7n+3}; \dfrac{2n}{n-3}$

17. _____

18. _____

Name:
Instructor:

Date:
Section:

Five-Minute Warm-Up 7.5
Adding and Subtracting Rational Expressions with Unlike Denominators

1. Determine the least common denominator (LCD) and then write the rational expressions as equivalent expressions using the LCD.

   (a) $\dfrac{3}{2}$ and $\dfrac{5}{3}$

   (b) $\dfrac{7}{12}$ and $\dfrac{5}{28}$

   1a. _____

   1b. _____

2. Factor: $6z^2 - 7z + 2$

   2. _____

3. Simplify: $\dfrac{8x^2 - 4x}{4x}$

   3. _____

4. Simplify: $\dfrac{16 - x^2}{2x^2 - 8x}$

   4. _____

5. Find the sum: $\dfrac{3}{8x} + \dfrac{1}{8x}$

   5. _____

Name: Date:
Instructor: Section:

Guided Practice 7.5
Adding and Subtracting Rational Expressions with Unlike Denominators

Objective 1: Add and Subtract Rational Expressions with Unlike Denominators

1. Find the sum and simplify, if possible: $\dfrac{7}{15} + \dfrac{10}{25}$ *(See textbook Example 1)*

Step 1: Find the least common denominator of 15 and 25.

Write 15 as the product of prime factors: (a) _____

Write 25 as the product of prime factors: (b) _____

Determine the LCD: (c) _____

Step 2: Write each fraction as an equivalent fraction with the LCD.

Find the equivalent fraction: (d) $\dfrac{7}{15} \cdot \dfrac{}{} = \dfrac{}{}$

Find the equivalent fraction: (e) $\dfrac{10}{25} \cdot \dfrac{}{} = \dfrac{}{}$

Step 3: Find the sum of the numerators, written over the common denominator.

Use $\dfrac{a}{c} + \dfrac{b}{c} = \dfrac{a+b}{c}$: (f) _____

Step 4: Simplify, if possible.

What is the sum, reduced to lowest terms? (g) _____

2. Find the sum and simplify, if possible: $\dfrac{5}{12x} + \dfrac{7x}{18x^2}$ *(See textbook Example 2)*

Step 1: Find the least common denominator.

Write $12x$ as the product of prime factors: (a) _____

Write $18x^2$ as the product of prime factors: (b) _____

Determine the LCD: (c) _____

Step 2: Write each rational expression as an equivalent expression with the LCD.

Find the equivalent expression: (d) $\dfrac{5}{12x} \cdot \dfrac{}{} = \dfrac{}{}$

Find the equivalent expression: (e) $\dfrac{7x}{18x^2} \cdot \dfrac{}{} = \dfrac{}{}$

Step 3: Add the rational expressions found in Step 2.

Use $\dfrac{a}{c} + \dfrac{b}{c} = \dfrac{a+b}{c}$: (f) _____

Step 4: Simplify, if possible.

What is the sum, simplified to lowest terms? (g) _____

Sullivan/Struve/Mazzarella, *Elementary & Intermediate Algebra*, 3e
Copyright © 2014 Pearson Education, Inc.

Guided Practice 7.5

3. Find the sum and simplify, if possible: $\dfrac{4}{x+2} + \dfrac{2}{x+1}$ (See textbook Example 3)

3. _____

4. Find the difference and simplify, if possible: $\dfrac{2}{x} - \dfrac{4}{x-2}$ (See textbook Example 5)

Step 1: Find the least common denominator.	Determine the LCD:	(a) _____
Step 2: Write each rational expression as an equivalent expression with the LCD.	Find the equivalent expression:	(b) $\dfrac{2}{x} \cdot \dfrac{}{} = \dfrac{}{}$
	Find the equivalent expression:	(c) $\dfrac{4}{x-2} \cdot \dfrac{}{} = \dfrac{}{}$
Step 3: Subtract the rational expressions found in Step 2.	Use $\dfrac{a}{c} - \dfrac{b}{c} = \dfrac{a-b}{c}$:	(d) _____
Step 4: Simplify, if possible.	What is the difference, simplified to lowest terms?	(e) _____

5. Find the sum and simplify, if possible: $\dfrac{y-5}{y-3} + \dfrac{y^2-5y-6}{9-y^2}$

5. _____

(See textbook Example 8)

6. Find the difference and simplify, if possible: $3 - \dfrac{x-2}{x-4}$ (See textbook Example 9)

6. _____

Do the Math Exercises 7.5
Adding and Subtracting Rational Expressions with Unlike Denominators

In Problems 1–6, find each sum and simplify.

1. $\dfrac{7}{12} + \dfrac{3}{4}$

2. $\dfrac{5}{2x} + \dfrac{6}{5}$

3. $\dfrac{2}{x-1} + \dfrac{x-1}{x+1}$

4. $\dfrac{7}{n-4} + \dfrac{8}{4-n}$

5. $\dfrac{2x-6}{x^2-x-6} + \dfrac{x+4}{x+2}$

6. $\dfrac{3}{x-4} + \dfrac{x+4}{x^2-16}$

1. _____
2. _____
3. _____
4. _____
5. _____
6. _____

In Problems 7–10, find each difference and simplify, if necessary.

7. $\dfrac{8}{21} - \dfrac{6}{35}$

8. $\dfrac{x-2}{x+2} - \dfrac{x+2}{x-2}$

7. _____
8. _____

Do the Math Exercises 7.5

9. $\dfrac{x}{x+4} - \dfrac{-4}{x^2+8x+16}$

10. $\dfrac{p+1}{p^2-2p} - \dfrac{2p+3}{2-p}$

9. _____

10. _____

In Problems 11–16, find the LCD of the rational expressions. Then rewrite each as an equivalent rational expression with the LCD.

11. $\dfrac{5}{4n} - \dfrac{3}{n^2}$

12. $\dfrac{x-2}{x+3} + 2$

11. _____

12. _____

13. $\dfrac{-1}{3n-n^2} + \dfrac{1}{3n^2-9n}$

14. $\dfrac{a-5}{2a^2-6a} - \dfrac{6}{12a^2-4a^3}$

13. _____

14. _____

15. $\dfrac{x+1}{x-3} + \dfrac{x+2}{x-2} - \dfrac{x^2+3}{x^2-5x+6}$

16. $\dfrac{7}{w-3} - \dfrac{5}{w} - \dfrac{2w+6}{w^2-9}$

15. _____

16. _____

Five-Minute Warm-Up 7.6
Complex Rational Expressions

1. Factor: $2x^2 - 5x - 12$ 1. _____

2. Factor: $3x^2 - 75$ 2. _____

3. Find the quotient: $\dfrac{x-2}{6} \div \dfrac{x^2-4}{8}$ 3. _____

4. Find the product: $\dfrac{x-1}{x-2} \cdot (x^2 - 3x + 2)$ 4. _____

5. Find the product: $\left(\dfrac{3}{x} + \dfrac{2}{x^2}\right) \cdot x^3$ 5. _____

Name: Date:
Instructor: Section:

Guided Practice 7.6
Complex Rational Expressions

Objective 1: Simplify a Complex Rational Expression by Simplifying the Numerator and Denominator Separately (Method I)

1. In your own words, define a *complex rational expression*.

2. Simplify $\dfrac{\dfrac{5}{x} - \dfrac{x}{5}}{\dfrac{1}{5} - \dfrac{5}{x^2}}$ using Method I. *(See textbook Examples 2 and 3)*

Step 1: Write the numerator of the complex fraction as a single rational expression.

Determine the LCD of x and 5: **(a)** _____

Write the equivalent rational expressions on the LCD and then use $\dfrac{a}{c} - \dfrac{b}{c} = \dfrac{a-b}{c}$: **(b)** _____

Step 2: Write the denominator of the complex fraction as a single rational expression.

Determine the LCD of x^2 and 5: **(c)** _____

Write the equivalent rational expressions on the LCD and then add use $\dfrac{a}{c} - \dfrac{b}{c} = \dfrac{a-b}{c}$: **(d)** _____

Step 3: Rewrite the complex rational expression using the rational expressions determined in Steps 1 and 2.

(e) _____

Step 4: Simplify the rational expression using the techniques for dividing rational expressions from textbook Section 7.2.

Rewrite the division problem as a multiplication problem: **(f)** _____

Divide out common factors and express the answer as a simplified rational expression in factored form. **(g)** _____

Guided Practice 7.6

Objective 2: Simplify a Complex Rational Expression Using the Least Common Denominator (Method II)

3. Method II uses the LCD to simplify complex rational expressions. We use several of the properties of real numbers to simplify the complex rational expression; that is, to find an equivalent rational expression which has a single fraction bar. State the property of real numbers that is illustrated below.

(a) $\dfrac{x-9}{1} \cdot \dfrac{1}{x-9} = \dfrac{x-9}{x-9} = 1$ 3a. _____

(b) $\dfrac{2}{x} \cdot \dfrac{x-9}{x-9} = \dfrac{2}{x}$ 3b. _____

(c) $\dfrac{2}{x} \cdot \dfrac{x-9}{x-9} = \dfrac{2x-18}{x^2-9x}$ 3c. _____

4. Simplify $\dfrac{1+\dfrac{1}{x}}{1-\dfrac{1}{x^2}}$ using Method II. *(See textbook Examples 5 and 6)*

Step 1: Find the least common denominator among all the denominators in the complex rational expression.

Determine the LCD of x and x^2: (a) _____

Step 2: Multiply both the numerator and denominator of the complex rational expression by the LCD found in Step 1.

(b) _____

Distribute the LCD to each term: (c) _____

Step 3: Simplify the rational expression.

Divide out the common factors: (d) _____

Simplify the expression: (e) _____

Do the Math Exercises 7.6
Complex Rational Expressions

In Problems 1–4, simplify the complex rational expression using Method I.

1. $\dfrac{\dfrac{4}{t^2}-1}{\dfrac{t+2}{t^3}}$

2. $\dfrac{\dfrac{2}{x}+\dfrac{3}{x^2}}{\dfrac{2x+3}{x}}$

1. _____

2. _____

3. $\dfrac{\dfrac{5}{n-1}+3}{n-\dfrac{2}{n-1}}$

4. $\dfrac{\dfrac{2}{a+b}}{\dfrac{1}{a}+\dfrac{1}{b}}$

3. _____

4. _____

In Problems 5–8, simplify the complex rational expression using Method II.

5. $\dfrac{\dfrac{3c}{4}+\dfrac{3d}{10}}{\dfrac{3c}{2}-\dfrac{6d}{5}}$

6. $\dfrac{\dfrac{a}{2}+4}{2-\dfrac{a}{2}}$

5. _____

6. _____

7. $\dfrac{\dfrac{x}{x+1}}{1+\dfrac{1}{x-1}}$

8. $\dfrac{6x+\dfrac{3}{y}}{\dfrac{9x+3}{y}}$

7. _____

8. _____

Do the Math Exercises 7.6

In Problems 9–14, simplify the complex rational expression using either Method I or Method II.

9. $\dfrac{1 - \dfrac{4}{x^2}}{1 - \dfrac{1}{x} - \dfrac{6}{x^2}}$

10. $\dfrac{\dfrac{-6}{y^2 + 5y + 6}}{\dfrac{2}{y+3} - \dfrac{3}{y+2}}$

9. _____

10. _____

11. $\dfrac{2 - \dfrac{3}{x} - \dfrac{2}{x^2}}{1 - \dfrac{5}{x} + \dfrac{6}{x^2}}$

12. $\dfrac{1 - \dfrac{n^2}{25m^2}}{1 - \dfrac{n}{5m}}$

11. _____

12. _____

13. $\dfrac{\dfrac{x}{x+y} - 1}{\dfrac{y}{x+y} - 1}$

14. $\dfrac{\dfrac{5}{2b+3} + \dfrac{1}{2b-3}}{\dfrac{6b}{8b^2 - 18}}$

13. _____

14. _____

15. **Finding the Mean** The arithmetic mean of a set of numbers is found by adding the numbers and then dividing by the number of entries on the list. Write a complex rational expression to find the arithmetic mean of the expressions $\dfrac{z}{2}, \dfrac{z-3}{4}$, and $\dfrac{2z-1}{6}$ and then simplify the complex rational expression.

15. _____

Five-Minute Warm-Up 7.7
Rational Equations

1. Solve: $-8 + 4(x + 6) = 10x$

    1. _____

2. Factor: $x^2 - 3x - 28$

    2. _____

3. Solve: $4p^2 - 1 = 0$

    3. _____

4. Find the values for which the expression $\dfrac{x-2}{x^2 + 6x - 16}$ is undefined.

    4. _____

5. Solve for P: $A = P + Prt$

    5. _____

Name:
Instructor:

Date:
Section:

Guided Practice 7.7
Rational Equations

Objective 1: Solve Equations Containing Rational Expressions

1. When solving rational equations it is important to identify the restrictions on the variable. We exclude all values of the variable that result in _____

2. Solve: $\dfrac{x+5}{x-7} = \dfrac{x-3}{x+7}$ *(See textbook Example 2)*

Step 1: Determine the value(s) of the variable that cause the rational expressions(s) in the equation to be undefined.

For what values of x will the denominator be equal to zero? **(a)** _____

Step 2: Determine the least common denominator (LCD) of all the denominators.

LCD: **(b)** _____

Step 3: Multiply both sides of the equation by the LCD and simplify the expressions on each side.

Multiply both sides by the LCD: **(c)** _____

Divide out common factors: **(d)** _____

Multiply: **(e)** _____

Step 4: Solve the resulting equation.

Solve for x: **(f)** _____

Step 5: Check Verify your solution using the original equation.

Write the solution set: **(g)** _____

3. Solve: $\dfrac{9}{x} + \dfrac{4}{5x} = \dfrac{7}{10}$ *(See textbook Example 3)*

    (a) Determine the excluded values of the variable.   3a. _____

    (b) Determine the LCD of all of the denominators.   3b. _____

    (c) Multiply both sides of the equation by the LCD. What is the resulting equation?   3c. _____

    (d) Solve the resulting equation.   3d. _____

Sullivan/Struve/Mazzarella, *Elementary & Intermediate Algebra*, 3e
Copyright © 2014 Pearson Education, Inc.

Guided Practice 7.7

4. Solve: $\dfrac{6}{x^2-1} = \dfrac{5}{x-1} - \dfrac{3}{x+1}$ (See textbook Example 6)

   (a) Determine the excluded values of the variable. 4a. _____

   (b) Determine the LCD of all of the denominators. 4b. _____

   (c) Multiply both sides of the equation by the LCD and solve the resulting equation. 4c. _____
   What is the solution to this equation?

   (d) What is the solution set? 4d. _____

Objective 2: Solve for a Variable in a Rational Equation

5. Solve: $\dfrac{6}{z} = \dfrac{2}{x} + \dfrac{1}{y}$ for y (See textbook Example 8)

Step 1: Determine the value(s) of the variable(s) that result in an undefined rational expression.	For what values of the variables will the denominator be equal to zero?	(a) _____
Step 2: Determine the least common denominator (LCD) of all the denominators.	LCD:	(b) _____
Step 3: Multiply both sides of the equation by the LCD and simplify the expression on each side of the equation.	Multiply both sides by the LCD:	(c) _____
	Divide out common factors:	(d) _____
Step 4: Solve the resulting equation for y.	Subtract $2yz$ from both sides:	(e) _____
	Factor out y:	(f) _____
	Divide by the coefficient of y:	(g) _____
	Simplify, if possible. Write the solution:	(h) _____

Do the Math Exercises 7.7
Rational Equations

In Problems 1–10, solve each equation and state the solution set. Remember to identify the values of the variable for which the expressions in each rational equation are undefined.

1. $\dfrac{4}{p} - \dfrac{5}{4} = \dfrac{5}{2p} + \dfrac{3}{8}$

2. $\dfrac{4}{x-4} = \dfrac{5}{x+4}$

1. _____

2. _____

3. $\dfrac{x-2}{4x} - \dfrac{x+2}{3x} = \dfrac{1}{2x} - \dfrac{1}{2}$

4. $\dfrac{6}{t} - \dfrac{2}{t-1} = \dfrac{2-4t}{t^2-t}$

3. _____

4. _____

5. $\dfrac{4}{x-3} - \dfrac{3}{x-2} = \dfrac{2x+1}{x^2-5x+6}$

6. $\dfrac{2x+3}{x-1} - 2 = \dfrac{3x-1}{4x-4}$

5. _____

6. _____

7. $x = \dfrac{6-5x}{6x}$

8. $\dfrac{2x}{x+4} = \dfrac{x+1}{x+2} - \dfrac{7x+12}{x^2+6x+8}$

7. _____

8. _____

Sullivan/Struve/Mazzarella, *Elementary & Intermediate Algebra*, 3e
Copyright © 2014 Pearson Education, Inc.

Do the Math Exercises 7.7

9. $\dfrac{2}{b+2} - \dfrac{5b+6}{b^2-b-6} = \dfrac{-b}{b-3}$

10. $\dfrac{5}{x-2} - \dfrac{2}{2-x} = \dfrac{4}{x+1}$

9. _____

10. _____

In Problems 11–14, solve the equation for the indicated variable.

11. $\dfrac{a}{b+2} = c$ for b

12. $\dfrac{3}{i} - \dfrac{4}{j} = \dfrac{8}{k}$ for j

11. _____

12. _____

13. $X = \dfrac{ab}{a-b}$ for a

14. For what value of k will the solution set of $\dfrac{1}{2} + \dfrac{3x}{k} = 1 + \dfrac{x}{3}$ be $\{3\}$?

13. _____

14. _____

15. **Drug Concentration** The concentration C of a drug in a patient's bloodstream in milligrams per liter t hours after ingestion is modeled by $C = \dfrac{40t}{t^2+3}$. When will the concentration of the drug be 10 milligrams per liter?

15. _____

16. **Average Cost** Suppose that the average daily cost $\overline{C}$ in dollars of manufacturing x bicycles is given by the equation $\overline{C} = \dfrac{x^2 + 75x + 5000}{x}$. Determine the level of production for which the average daily cost will be $240.

16. _____

Five-Minute Warm-Up 7.8
Models Involving Rational Equations

1. Solve: $\dfrac{20}{r} = \dfrac{15}{r-3}$

1. _____

2. Solve: $\dfrac{1}{x} + \dfrac{1}{x-2} = \dfrac{3}{4}$

2. _____

3. Solve: $\dfrac{2}{r} + \dfrac{8}{r+3} = 2$

3. _____

Name:
Instructor:
Date:
Section:

Guided Practice 7.8
Models Involving Rational Equations

Objective 1: Model and Solve Ratio and Proportion Problems

1. Write an example of a ratio.   1. _____

2. Write an example of a proportion.   2. _____

3. Solve: $\dfrac{5}{k+4} = \dfrac{2}{k-1}$ *(See textbook Example 1)*   3. _____

4. **Flight Accidents** According to the Statistical Abstract of the United States, in 2001, there were 1.22 fatal airplane accidents per 100,000 flight hours. If there were a total of 321 fatal accidents in 2001, to the nearest million, how many flight hours were flown that year? *(See textbook Example 2)*   4. _____

Objective 2: Model and Solve Problems with Similar Figures

5. In your own words, what does it mean if two geometric figures are *similar*?

6. Suppose that a 6-foot-tall man casts a shadow of 3.2 feet. At the same time of day, a tree casts a shadow of 8 feet. How tall is the tree? *(See textbook Example 5)*

    (a) Write a proportion that can be used to solve this problem.
         6a. _____

    (b) Solve the proportion. How tall is the tree?
         6b. _____

Objective 3: Model and Solve Work Problems

7. Problems of this type involve completing a job or a task when working at a constant rate. We convert the time t that it takes to complete the job into a unit rate. That is, if it takes t hours to complete a job, then $\dfrac{1}{t}$ of the job is completed per hour.

    (a) If it takes 15 minutes to complete a job, what part of the job is completed per minute?   7a. _____

    (b) If it takes t hours to complete a job, what part of the job is completed per hour?   7b. _____

    (c) If it takes $t + 3$ hours to complete a job, what part of the job is competed per hour?   7c. _____

Sullivan/Struve/Mazzarella, *Elementary & Intermediate Algebra*, 3e
Copyright © 2014 Pearson Education, Inc.

Guided Practice 7.8

8. Josh can clean the math building on his campus in 3 hours. Ken takes 5 hours to clean the same building. If they work together, how long will it take for Josh and Ken to clean the math building?
(See textbook Example 6)

Step 1: Identify We want to know how long it will take Josh and Ken working together to clean the building.

Step 2: Name We let t represent the time (in hours) that it takes to clean the building when working together.

Step 3: Translate What fraction of the job is completed in one hour when working individually and when working together? Write the following ratios:

(a) Part of the job completed by Josh in one hour: 8a. _____

(b) Part of the job completed by Ken in one hour: 8b. _____

(c) Part of the job completed when working together in one hour: 8c. _____

(d) Write the model for this problem:

8d. _____

Step 4: Solve the equation from Step 3.

Step 5: Check Is your answer reasonable?

(e) **Step 6: Answer** the question. 8e. _____

Objective 4: Model and Solve Uniform Motion Problems

9. A small plane can travel 1000 miles with the wind in the same time it can go 600 miles against the wind. If the speed of the plane in still air is 180 mph, what is the speed of the wind? *(See textbook Example 8)*

(a) If w is the speed of the wind, what is the rate (in mph) when travelling with the wind: 9a. _____

(b) If w is the speed of the wind, what is the rate when travelling against the wind: 9b. _____

(c) Use $t = \dfrac{d}{r}$ to write a rational expression for the time travelled with the wind: 9c. _____

(d) Use $t = \dfrac{d}{r}$ to write a rational expression for the time travelled against the wind: 9d. _____

(e) Write an equation that can be used to solve this problem: 9e. _____

(f) Solve your equation and answer the question. 9f. _____

Name:
Instructor:

Date:
Section:

Do the Math Exercises 7.8
Models Involving Rational Equations

In Problems 1–6, solve the proportion.

1. $\dfrac{6}{5} = \dfrac{8}{3x}$

2. $\dfrac{k}{k+3} = \dfrac{6}{15}$

1. _____

2. _____

3. $\dfrac{n+6}{3} = \dfrac{n+4}{5}$

4. $\dfrac{n-2}{4n+7} = \dfrac{-2}{n+1}$

3. _____

4. _____

5. $\dfrac{4}{x^2} = \dfrac{1}{x+3}$

6. $\dfrac{1}{z+3} = \dfrac{2z+1}{z^2+10z+21}$

5. _____

6. _____

In Problems 7 and 8, write an algebraic expression that represents each phrase.

7. If Brian worked for n days and Kellen worked for half as many days, write an algebraic expression that represents the number of days Kellen worked.

7. _____

8. If the rate of the current in a stream is 3 mph and Betsy can paddle her canoe at a rate of r mph in still water, write an algebraic expression that represents Betsy's rate when she paddles downstream.

8. _____

Do the Math Exercises 7.8

In Problems 9–14, write an equation that could be used to model each of the following. SOLVE THE EQUATION.

9. **Buying Candy** If 2 lb of candy costs $3.50, how much candy can be purchased for $8.75?

   9. _____

   _____

10. **Comparing Pesos and Pounds** If 50 Mexican pesos are worth approximately 2.5 British pounds and if Sean takes 80 pounds as spending money in Mexico City, how many pesos will he have?

    10. _____

    _____

11. **Telephone Pole Shadow** A bush that is 4 m tall casts a shadow that is 1.75 m long. How tall is a telephone pole if its shadow is 8.75 m?

    11. _____

    _____

12. **Pruning Trees** Martin can prune his fruit trees in 4 hours. His neighbor can prune the same trees for him in 7 hours. If they work together on this job, to the nearest tenth of an hour, how long will it take to prune the trees?

    12. _____

    _____

13. **River Trip** A stream has a current of 3 mph. Find the speed of Marty's boat in still water if she can go 10 miles downstream in the same time it takes to go 4 miles upstream.

    13. _____

    _____

14. **A Bike Trip** A bicyclist rides his bicycle 12 miles up a hill and then 16 miles on level terrain. His speed on level ground is 3 miles per hour faster than his speed going uphill. The cyclist rides for the same amount of time going uphill and on level ground. Find his speed going uphill.

    14. _____

    _____

Five-Minute Warm-Up 8.1
Graphs of Equations

1. Plot the following points on the real number line: $\frac{3}{2}, -2, 0, 3$.

   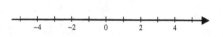

2. Determine which of the following are solutions to the equation $4x - 5(x+1) = 6$.

   (a) $x = -1$ (b) $x = -11$ (c) $x = 1$ (d) $x = 11$

   2. _____

3. Evaluate the expression $-x^2 - 3x + 4$ for the given values of the variable.

   (a) $x = 3$ (b) $x = -2$ (c) $x = 0$

   3a. _____

   3b. _____

   3c. _____

4. Solve the equation $4x - 3y = -12$ for y.

   4. _____

5. Evaluate each of the following absolute values.

   (a) $|-12|$ (b) $|0|$ (c) $|125|$

   5a. _____

   5b. _____

   5c. _____

Guided Practice 8.1
Graphs of Equations

1. Label each quadrant and axis in the rectangular or Cartesian coordinate system.

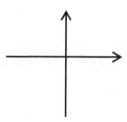

2. What name do we give the ordered pair $(0, 0)$? _____

In Problems 3 and 4, circle one answer for each underlined choice.

3. To plot the ordered pair $(-3, 2)$, you would move 3 units <u>up, down, left or right</u> from the origin.

4. To plot the ordered pair $(-6, -2)$, you would move <u>6 or 2</u> units down from the origin.

Objective 1: Graph an Equation Using the Point-Plotting Method *(See textbook Example 1)*

5. Graph the equation $y = 3x - 1$ by plotting points.

Step 1: We want to find all the points (x, y) that satisfy the equation. To determine these points we choose values of x and use the equation to determine the corresponding values of y.

	x	$y = 3x - 1$	(x, y)
(a)	-2	$3(\_\_\_) - 1 = \_\_\_$	$(-2, \_\_\_)$
(b)	-1	$3(\_\_\_) - 1 = \_\_\_$	$(-1, \_\_\_)$
(c)	0	$3(\_\_\_) - 1 = \_\_\_$	$(0, \_\_\_)$
(d)	1	$3(\_\_\_) - 1 = \_\_\_$	$(1, \_\_\_)$
(e)	2	$3(\_\_\_) - 1 = \_\_\_$	$(2, \_\_\_)$

Step 2: Draw the axes in the Cartesian plane and plot the points listed in the third column. Now connect the points to obtain the graph of the equation (a line).

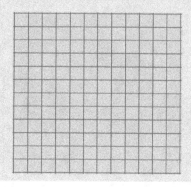

Guided Practice 8.1

Objective 2: Identify the Intercepts from the Graph of an Equation *(See textbook Example 4)*

6. The point(s), if any, where a graph crosses or touches the *x*-axis is called the *x*-intercept.   6._____
Suppose a graph crosses the *x*-axis at a point *a* and touches the *y*-axis at a point *b*. Write
these points as ordered pairs.   _____

7. The point(s), if any, where a graph crosses or touches the *y*-axis is called the *y*-intercept.   7._____
Suppose a graph crosses the *y*-axis at a point *c*. Write this point as an ordered pair.

Objective 3: Interpret Graphs

8. (1, 5) is a point on the graph of $3x - y = -2$. If the *x*-axis represents the number of picnic tables manufactured and sold and the *y*-axis represents the profit (in tens of dollars) from the sale of those tables, describe the meaning of the ordered pair (1, 5).

_____

9. (2, 8) is a point on the graph of $y = -x^2 + 3x + 6$. If the *x*-axis represents the time (in seconds) after a ball leaves the hand of a thrower and the *y*-axis represents the height (in feet) above the ground, describe the meaning of the ordered pair (2, 8).

_____

Name:
Instructor:

Date:
Section:

Do the Math Exercises 8.1
Graphs of Equations

1. Determine the coordinates of each of the points plotted. Tell in which quadrant or on what coordinate axis each point lies.

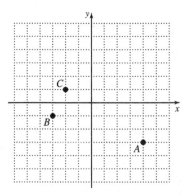

1.
A _____

B _____

C _____

2. Use the graph above to plot each of the following points. Tell in which quadrant or on which coordinate axis each point lies.

   $D(0, 5)$ $E(2, 3)$ $F(1, 0)$

2.
D _____

E _____

F _____

3. Determine which of the following are points on the graph of the equation $-4x + 3y = 18$.

(a) $(1, 7)$ **(b)** $(0, 6)$ **(c)** $(-3, 10)$ **(d)** $\left(\dfrac{3}{2}, 4\right)$

3. _____

4. Determine which of the following are points on the graph of the equation $x^2 + y^2 = 1$.

(a) $(0, 1)$ **(b)** $(1, 1)$ **(c)** $\left(\dfrac{1}{2}, \dfrac{1}{2}\right)$ **(d)** $\left(\dfrac{\sqrt{3}}{2}, \dfrac{1}{2}\right)$

4. _____

Do the Math Exercises 8.1

5. The graph of the equation is given. List the intercepts of the graph.

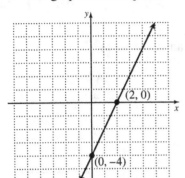

5. *x*-intercept:

_____

y-intercept:

_____

In Problems 6 – 9, graph each equation by plotting points.

6. $y = x - 2$

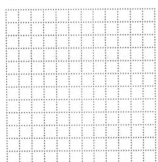

7. $y = x^2 - 2$

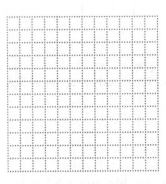

8. $y = |x| + 1$

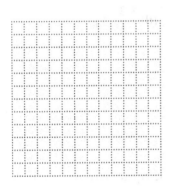

9. $3x + y = -2$

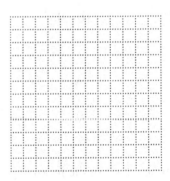

10. If $(a, -2)$ is a point on the graph of $y - 3x = 5$, what is a?

10. _____

11. If $(-2, b)$ is a point on the graph of $y = -2x^2 + 3x + 1$, what is b?

11. _____

Name:
Instructor:

Date:
Section:

Five-Minute Warm-Up 8.2
Relations

1. Plot the points in the rectangular coordinate system.
   $A(-3, 2); B(0, -2); C(4, -1); D(2, 0); E(-1, -3)$

   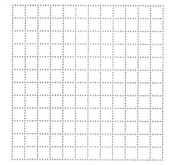

2. Graph: $y = -x$

   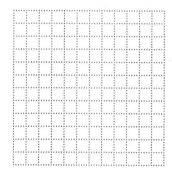

3. Graph: $4x - 2y = -8$

   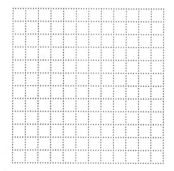

4. Graph: $y = -2x^2 + 3$

   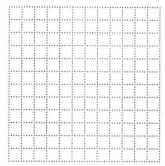

Sullivan/Struve/Mazzarella, *Elementary & Intermediate Algebra*, 3e
Copyright © 2014 Pearson Education, Inc.

Name:
Instructor:
Date:
Section:

Guided Practice 8.2
Relations

Objective 1: Understand Relations

1. In your own words, write a definition for a *relation*.

2. Represent the relation shown in the figure below as a set of ordered pairs. Then state the domain and the range of the relation. *(See textbook Examples 1 and 2)*

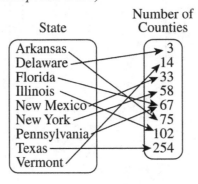

_____

_____

Objective 2: Find the Domain and the Range of a Relation

3. The *domain* is the set of all _____ and is the set of _____ -coordinates for the relation which is defined by the set of ordered pairs (x, y).

4. The *range* is the set of all _____ and is the set of _____ -coordinates for the relation which is defined by the set of ordered pairs (x, y).

In Problems 5 and 6, the figure shows the graph of a relation. Determine (a) the domain and (b) the range of the relation. *(See textbook Example 4)*

5.
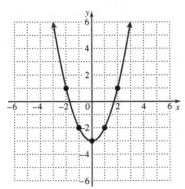

6.
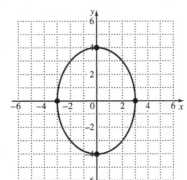

5a. _____

5b. _____

6a. _____

6b. _____

Sullivan/Struve/Mazzarella, *Elementary & Intermediate Algebra*, 3e
Copyright © 2014 Pearson Education, Inc.

Guided Practice 8.2

Objective 3: Graph a Relation Defined by an Equation

7. Graph the relation $y = x^2 - 3$. Use the graph to determine **(f)** the domain and **(g)** the range of the relation. *(See textbook Example 5)*

	x	y	(x, y)
(a)	-2		
(b)	-1		
(c)	0		
(d)	1		
(e)	2		

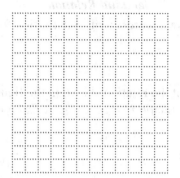

7f. _____

7g. _____

Do the Math Exercises 8.2
Relations

In Problems 1–2, represent each relation as a set of ordered pairs. Then state the domain and range of each relation.

1.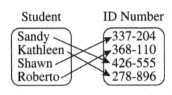

2.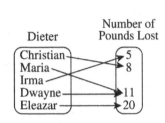

1. _____

2. _____

In Problems 3–4, use the set of ordered pairs to represent the relation as a map. Then state the domain and range of each relation.

3. {(−2, 3); (−4, 3); (−2, 2); (1, 1)}

4. {(3, 0); (−1, −6); (−1, 6); (1, 0)}

3. _____

4. _____

In Problems 5–8, identify the domain and range of the relation shown in the figure.

5.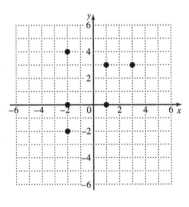

6.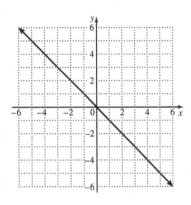

5. _____

6. _____

7.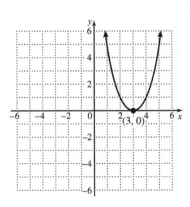

8.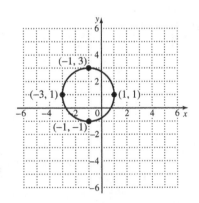

7. _____

8. _____

Sullivan/Struve/Mazzarella, *Elementary & Intermediate Algebra*, 3e
Copyright © 2014 Pearson Education, Inc.

307

Do the Math Exercises 8.2

In Problems 9–14, graph each relation. Use the graph to identify the domain and range of the relation.

9. $y = x + 4$

10. $y = 3x - 2$

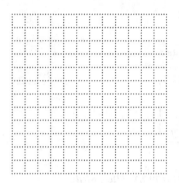
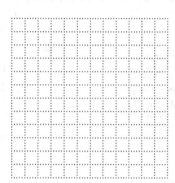

9. _____

_____

10. _____

_____

11. $y = -x^2 + 4x - 4$

12. $y = 3x^2 + 6x$

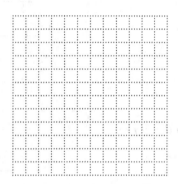
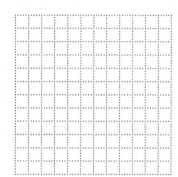

11. _____

_____

12. _____

_____

13. $y = -x^2 + 5x + 2$

14. $y = x^2 - 5x + 1$

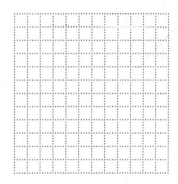
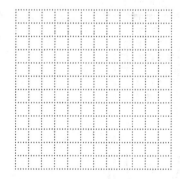

13. _____

_____

14. _____

_____

Five-Minute Warm-Up 8.3
An Introduction to Functions

1. Evaluate the expression $2x^2 - 4x + 5$ for
   (a) $x = -4$ (b) $x = 2$

   1a. _____

   1b. _____

2. Evaluate the expression $-\frac{2}{3}x + 2$ for
   (a) $x = 9$ (b) $x = -\frac{5}{2}$

   2a. _____

   2b. _____

3. Evaluate the expression $|3x - 10|$ for $x = -\frac{4}{9}$.

   3. _____

4. Evaluate the expression $\sqrt{9 - x^2}$ for $x = -1$.

   4. _____

5. State (a) the domain and (b) the range: $\{(-2, 4), (-1, 2), (2, 4), (3, 6)\}$

   5a. _____

   5b. _____

Name:
Instructor:
Date:
Section:

Guided Practice 8.3
An Introduction to Functions

Objective 1: Determine Whether a Relation Expressed as a Map or Ordered Pairs Represents a Function

1. In your own words, write a definition for a *function*.

2. Determine whether the relation represents a function. If the relation is a function, then state its domain and range. *(See textbook Example 2)*

   (a) $\{(-1, -1), (4, 4), (5, 5)\}$ _____

   (b) $\{(2, -1), (3, -2), (2, -3)\}$ _____

Objective 2: Determine Whether a Relation Expressed as an Equation Represents a Function

3. The symbol $\pm$ is a shorthand device and is read "plus or minus." Write the two equations that are represented by $y = \pm 2x$ and then determine whether the equation shows y is a function of x.

   (a) $y = \pm 2x$ means $y = $ _____ and also $y = $ _____ . (b) Is y a function of x? _____

4. Determine whether each equation shows y as a function of x. *(See textbook Examples 3 and 4)*

   (a) $y = 9$ (b) $x = -1$ 4a. _____

          4b. _____

   (c) $y = \pm \sqrt{x - 2}$ (d) $4x - 3y = -24$ 4c. _____

          4d. _____

Objective 3: Determine Whether a Relation Expressed as a Graph Represents a Function

5. We use the Vertical Line Test to determine whether the graph of a relation is a function. In your own words, state the *Vertical Line Test (VLT)*.

Guided Practice 8.3

6. Which of the graphs are graphs of functions? *(See textbook Example 5)* 6. _____

(a) 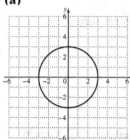   (b) 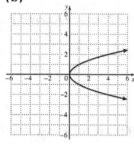   (c) 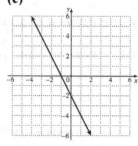   (d)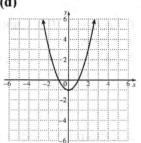

Objective 4: Find the Value of a Function

In Problems 7 – 10, find the value of each function. *(See textbook Examples 6 and 7)*

7. $f(x) = -2x^2 + 3x$

(a) $f(-4)$ (b) $f\left(\dfrac{3}{2}\right)$

8. $g(x) = \dfrac{3}{2}x + 2$

(a) $g(8)$ (b) $g(-2)$

7a. _____

7b. _____

8a. _____

8b. _____

9. $h(t) = 4$

(a) $h(0)$ (b) $h(3)$

10. $F(z) = z^2 + 4$

(a) $F(-2)$ (b) $F(-5)$

9a. _____

9b. _____

10a. _____

10b. _____

Do the Math Exercises 8.3
An Introduction to Functions

In Problems 1–2, determine whether each relation represents a function. If the relation is a function, state its domain and range.

1.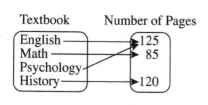

2.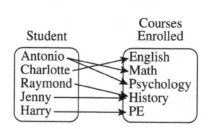

1. _____

2. _____

In Problems 3–4, determine whether each relation represents a function. If the relation is a function, state its domain and range.

3. $\{(3, 1); (2, 1); (0, 1); (-1, 1)\}$

4. $\{(5, 0); (3, 1); (-1, 2); (3, 2)\}$

3. _____

4. _____

In Problems 5–8, determine whether each equation represents y as a function of x.

5. $x = -4$

6. $y = 0$

5. _____

6. _____

7. $y = x^2 - 3x + 2$

8. $y = \sqrt{x}$

7. _____

8. _____

In Problems 9–10, determine whether the graph is that of a function.

9.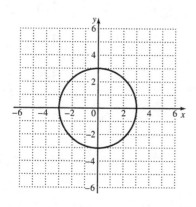

10.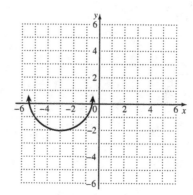

9. _____

10. _____

Do the Math Exercises 8.3

In Problems 11–15, find the following values for each function:

(a) $f(0)$ (b) $f(3)$ (c) $f(-2)$

11. $f(x) = 2x + 4$

12. $f(x) = 1 - x$

11a. _____

11b. _____

11c. _____

12a. _____

12b. _____

12c. _____

13. $f(x) = -1$

14. $f(x) = x^2 + 4$

13a. _____

13b. _____

13c. _____

14a. _____

14b. _____

14c. _____

15. $f(x) = -2x^2 + x - 3$

15a. _____

15b. _____

15c. _____

Name:
Instructor:

Date:
Section:

Five-Minute Warm-Up 8.4
Functions and Their Graphs

In Problems 1 and 2, solve each equation.

1. $-4x + 2 = 0$

2. $-18 + 3y = 0$

1. _____

2. _____

3. Graph the equation $y = -\dfrac{4}{3}x + 2$.

4. Graph the equation $y = x^2 - 4$ by plotting points.

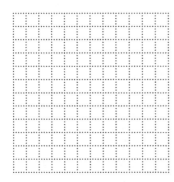

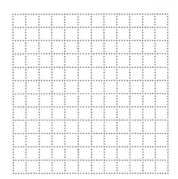

In Problems 5 and 6, determine the (a) domain and (b) the range from the graph.

5.

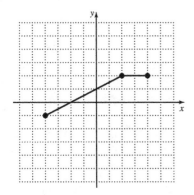

6.

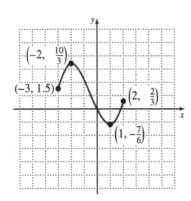

5a. _____

5b. _____

6a. _____

6b. _____

Name:
Instructor:

Date:
Section:

Guided Practice 8.4
Functions and Their Graphs

Objective 1: Graph a Function

1. Complete the table and graph the function $f(x) = |2x - 4|$. *(See textbook Example 1)*

x	$f(x)$	$(x, f(x))$
−2		
−1		
0		
1		
2		
3		
4		

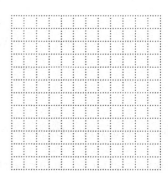

Objective 2: Obtain Information from the Graph of a Function

2. Use the graph of the function to answer parts (a) – (d). *(See textbook Example 2)*

   (a) Determine the domain of the function.   2a. _____

   (b) Determine the range of the function.   2b. _____

   (c) Identify the *x*-intercept(s).   2c. _____

   (d) Identify the *y*-intercept(s).   2d. _____

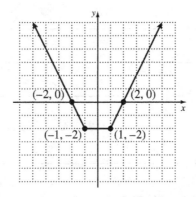

Sullivan/Struve/Mazzarella, *Elementary & Intermediate Algebra*, 3e
Copyright © 2014 Pearson Education, Inc.

Guided Practice 8.4

3. Consider the function $f(x) = -\dfrac{5}{2}x + 8$. *(See textbook Example 4)*

    (a) Is the point $(4, -2)$ on the graph of the function? 3a. _____

    (b) If $x = 6$, what is $f(x)$? What point is on the graph of the function? 3b. _____

    _____

    (c) If $f(x) = 3$, what is x? What point is on the graph of f? 3c. _____

    _____

4. The zeros of a function are also the ___-intercepts of the graph of the function. To find the zeros of a function we set the function equal to _____ and solve for x.

5. Find the zeros of the function f whose graph is shown below. *(See textbook Example 5)*   5. _____

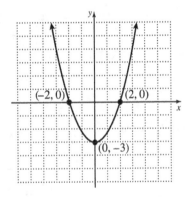

Objective 3: Know Properties and Graphs of Basic Functions

6. List the 6 basic functions described in Table 3 and briefly describe each graph.

    (a) _____

    (b) _____

    (c) _____

    (d) _____

    (e) _____

    (f) _____

Name:
Instructor:

Date:
Section:

Do the Math Exercises 8.4
Functions and Their Graphs

In Problems 1–5, find the domain of each function.

1. $G(x) = -8x + 3$

2. $H(x) = \dfrac{x+5}{2x+1}$

3. $s(t) = 2t^2 - 5t + 1$

4. $H(q) = \dfrac{1}{6q+5}$

5. $f(x) = \dfrac{4x-9}{7}$

1. _____
2. _____
3. _____
4. _____
5. _____

In Problems 6 and 7, graph each function.

6. $F(x) = x^2 + 1$

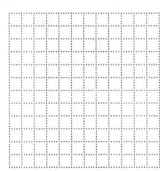

7. $H(x) = |x+1|$

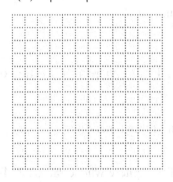

*In Problems 8 and 9, find **(a)** the domain, **(b)** the range, **(c)** the intercepts, if any, and **(d)** the zeros, if any.*

8.

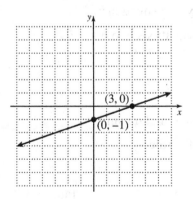

9.

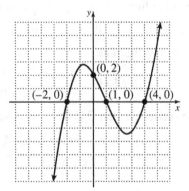

8a. _____
8b. _____
8c. _____
8d. _____
9a. _____
9b. _____
9c. _____
9d. _____

Sullivan/Struve/Mazzarella, *Elementary & Intermediate Algebra*, 3e
Copyright © 2014 Pearson Education, Inc.

Do the Math Exercises 8.4

10. Use the table of values for the function G to answer the following:

x	G(x)
-5	-3
-4	0
0	5
3	8
7	5

(a) What is $G(3)$? 10a. _____

(b) For what number(s) is $G(x) = 5$? 10b. _____

(c) What is the x-intercept of the graph of G? 10c. _____

(d) What is the y-intercept of the graph of G? 10d. _____

(e) Are there any zeros of the function G? If so, name the zero(s). 10e. _____

11. Use the function $f(x) = 3x + 5$ to answer the following:

(a) Is the point (-2, 1) on the graph of the function?

(b) If $x = 4$, what is $f(x)$? What point is on the graph of the function?

11a. _____

11b. _____

(c) If $f(x) = -4$, what is x? What point is on the graph of the function?

(d) Is $x = 0$ a zero of the function?

11c. _____

11d. _____

12. **Geometry** The volume V of a sphere as a function of its radius r is given by $V(r) = \frac{4}{3}\pi r^3$.

(a) What is the domain of this function? 12a. _____

(b) Find the volume of the sphere whose radius is $4\frac{1}{2}$ cm. Round your answer to one decimal place. 12b. _____

13. **A piecewise-defined function** is a function defined by more than one equation. For example, the absolute value function $f(x) = |x|$ is actually defined by two equations: $f(x) = x$ if $x \geq 0$ and $f(x) = -x$ if $x < 0$. We can combine these equations into one piecewise-defined function written as

$$f(x) = \begin{cases} x & \text{for } x \geq 0 \\ -x & \text{for } x < 0 \end{cases}$$

To evaluate $f(3)$, we recognize that $3 \geq 0$, so we use the rule $f(x) = x$ and obtain $f(3) = 3$. To evaluate $f(-4)$, we recognize that $-4 < 0$, so we use the rule $f(x) = -x$ and obtain $f(-4) = -(-4) = 4$.

Given $f(x) = \begin{cases} x + 3 & \text{for } x < 0 \\ -2x + 1 & \text{for } x \geq 0 \end{cases}$, find each of the following:

(a) $f(3)$ (b) $f(-2)$ (c) $f(0)$

13a. _____

13b. _____

13c. _____

Five-Minute Warm-Up 8.5
Linear Functions and Models

In Problems 1 – 2, graph each linear equation.

1. $\dfrac{2}{3}x - \dfrac{1}{2}y = -2$

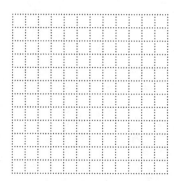

2. $x = -1$

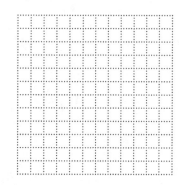

3. Find the equation of the line through $(-1, 2)$ and $(2, -7)$.

3. _____

In Problems 4 and 5, solve for the given variable.

4. $6.5 - 1.5(x - 4) + 4x = 10.25$

5. $-\dfrac{7}{8}(2y + 3) \leq \dfrac{5}{4}(y - 2)$

4. _____

5. _____

Sullivan/Struve/Mazzarella, *Elementary & Intermediate Algebra*, 3e

Name:
Instructor:

Date:
Section:

Guided Practice 8.5
Linear Functions and Models

Objective 1: Graph Linear Functions

1. A _____ function is a function of the form $f(x) = mx + b$ where m and b are real numbers. The graph of a linear function is called a _____.

2. To graph a linear function, we use the same technique used to graph a linear equation written in slope-intercept form, $y = mx + b$, where m is the _____ and $(0,b)$ is the _____.

3. Graph the linear function $f(x) = \dfrac{3}{2}x - 2$. *(See textbook Example 1)*

   (a) Identify the y-intercept. 3a. _____

   (b) Identify the slope. 3b. _____

   (c) Graph the function.

Objective 2: Find the Zero of a Linear Function

4. To find the zero of a linear function $f(x) = mx + b$, we solve the equation _____.

5. Perimeter of a Rectangle In a given rectangle, the length is 3 ft less than twice the width. If x represents the width of the rectangle, the perimeter can be calculated by the function $P(x) = 2x + 2(2x - 3)$. *(See textbook Example 3)*

   (a) What is the implied domain of the function? 5a. _____

   (b) What is the perimeter of a rectangle whose width is 12 ft? 5b. _____

   (c) What is the width of a rectangle whose perimeter is 84 ft? 5c. _____

   (d) For what width of the rectangle will the perimeter exceed 12 feet? 5d. _____

Sullivan/Struve/Mazzarella, *Elementary & Intermediate Algebra*, 3e
Copyright © 2014 Pearson Education, Inc.

Guided Practice 8.5

Objective 3: Build Linear Models from Verbal Descriptions

6. The linear cost function is $C(x) = ax + b$, where b represents the _____ costs of operating a business and _____ represents the costs associated with manufacturing one additional item. *(See textbook Example 4)*

7. Some companies use *straight-line depreciation* to depreciate their assets so that the value of the asset declines by a constant amount each year. To calculate the amount the asset depreciates each year, divide the total cost by the number of years of useful life.

A cab company bought a new car for $22,500 and plans to drive it until there is no scrap value. The life of a car in the cab fleet is 5 years. *(See textbook Example 5)*

(a) By how much does the car depreciate each year? 7a. _____

(b) This rate can be expressed as the slope of a linear function. Is this slope positive or negative? 7b. _____

(c) The *book value* is the value of the asset at a particular time. To find the book value, we take the original value and deduct the amount of depreciation after a given time.

Write a linear function that expresses the book value V of the car as a function of its age, x. 7c. _____

(d) What is the implied domain of the linear function? 7d. _____

(e) What is the book value after 4 years? 7e. _____

(f) When will the book value of the car be $10,125? 7f. _____

(g) What is the independent variable? 7g. _____

(h) What is the dependent variable? 7h. _____

Objective 4: Build Linear Models from Data

8. How many points are needed to determine the equation of a line? 8. _____

9. If the data appears to be linearly related, we select two points on the line of best fit. The linear model can be determined by finding the equation of the line through the two points as described in Section 1.6, Example 12.

(a) To find the equation of the line, you must first determine the _____.

(b) Second, use the point-slope form of a line, _____, to find the equation.

Name:
Instructor:

Date:
Section:

Do the Math Exercises 8.5
Linear Functions and Models

In Problems 1 – 4, graph each linear function.

1. $f(x) = 2x - 4$

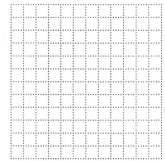

2. $g(x) = -\dfrac{3}{2}x + 1$

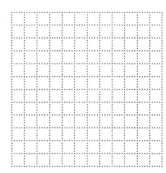

3. $h(x) = \dfrac{1}{4}x + 2$

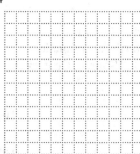

4. $F(x) = 3$

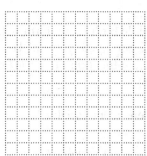

In Problems 5 and 6, find the zeros of the linear function.

5. $f(x) = 3x - 24$

6. $g(x) = -\dfrac{3}{2}x + 6$

5. _____

6. _____

7. Suppose that $f(x) = \dfrac{4}{3}x + 5$ and $g(x) = \dfrac{1}{3}x + 1$.

   (a) Solve $f(x) = g(x)$.

   7a. _____

   (b) What is the value of f at the solution?

   7b. _____

   (c) What is the value of g at the solution?

   7c. _____

   (d) Solve $f(x) > g(x)$.

   7d. _____

Do the Math Exercises 8.5

8. Graph $f(x) = \frac{4}{3}x + 5$ and $g(x) = \frac{1}{3}x + 1$ on the same Cartesian plane. Label the intersection point.

9. (a) Find a linear function g such that $g(1) = 5$ and $g(5) = 17$. (b) What is $g(-3)$?

9a. _____

9b. _____

10. **Birth Rate** A multiple birth is any birth with 2 or more children born. The birth rate is the number of births per 1,000 women. The birth rate B of multiple births as a function of age a is given by the function $B(a) = 1.73a - 14.56$ for $15 \leq a \leq 44$. [Source: Centers for Disease Control]

   (a) What is the independent variable?

   10a. _____

   (b) What is the dependent variable?

   10b. _____

   (c) What is the domain of this linear function?

   10c. _____

   (d) What is the multiple birth rate of women who are 22 years of age according to the model?

   10d. _____

   (e) What is the age of women whose multiple birth rate is 49.45?

   10e. _____

11. A strain of *E. coli* Beu 397-recA441 is placed into a Petri dish at 30° Celsius and allowed to grow. The population is estimated by means of an optical device in which the amount of light that passes through the Petri dish is measured. The data below was collected. Do you think that a linear function could be used to describe the relation between the two variables? Why or Why not?

   11. _____

Time, x	Population, y
0	0.09
2.5	0.18
3.5	0.26
4.5	0.35
6	0.50

Name:
Instructor:

Date:
Section:

Five-Minute Warm-Up 8.6
Compound Inequalities

1. Use **(a)** set-builder notation and **(b)** interval notation to list the set of all real numbers x such that $-4 \leq x < -1$. **(c)** Graph the inequality.

   1a.

   1b. _____

   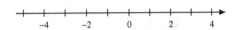

2. Graph the inequality $x < -3$.

   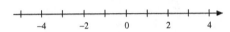

3. Use interval notation to express the inequality shown in the graph.

   3. _____

4. Solve $2(x+4) = x - (3x+4)$.

   4. _____

5. Solve $\dfrac{3}{8}x + 1 \leq -\dfrac{5}{4}$.

   5. _____

6. Solve $-2(1-x) + 2x < 7x + 4$.

   6. _____

Name: Date:
Instructor: Section:

Guided Practice 8.6
Compound Inequalities

Objective 1: Determine the Intersection or Union of Two Sets

1. The *intersection* of two sets A and B, denoted _____, is the set of all elements that belong to both set A and set B. The word _____ implies intersection.

2. The *union* of two sets A and B, denoted _____, is the set of all elements that are in set A or in set B or in both A and B. The word _____ implies union.

3. Suppose $A = \{x | x > -2\}$, $B = \{x | x \leq 1\}$, and $C = \{x | x \geq 3\}$. *(See textbook Example 2)*
   (a) Determine $A \cap B$. Graph the solution set on a real number line.

   Step 1: Graph $A = \{x | x > -2\}$
   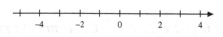

   Step 2: Graph $B = \{x | x \leq 1\}$
   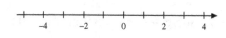

   Step 3: Identify where the graph of the inequalities overlap.

   (b) Determine $B \cup C$. Graph the solution set on a real number line.

   Step 1: Graph $B = \{x | x \leq 1\}$

   Step 2: Graph $C = \{x | x \geq 3\}$
   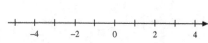

   Step 3: Identify the solution as the set that is in either B or C.

Objective 2: Solve Compound Inequalities Involving "and" (See textbook Example 3)

4. Solve $5x - 1 < 9$ and $5x \geq -20$. Graph the solution set.

Step 1: Solve each inequality separately.

$5x - 1 < 9$

Add 1 to both sides: (a) _____

Divide both sides by 5: (b) _____

$5x \geq -20$

Divide both sides by 5: (c) _____

Step 2: Find the intersection of the solution sets, which will represent the solution set to the compound inequality.

Graph (b): (d)

Graph (c): (e)

Graph the intersection (overlap): (f)

Sullivan/Struve/Mazzarella, *Elementary & Intermediate Algebra*, 3e
Copyright © 2014 Pearson Education, Inc.

Guided Practice 8.6

5. If $a < b$, then we can write $a < x$ and $x < b$ more compactly as _____.

6. Solve $-8 \leq 5x - 3 \leq 7$ and graph the solution set. *(See textbook Example 6)*

Our goal is to get the variable by itself in the "middle" with a coefficient of 1. Remember that the inequality is always written in order, the smaller number on left and the larger number on the right side of the inequality.

$$-8 \leq 5x - 3 \leq 7$$

(a) Add 3 to all three parts: $\quad -8 \;\underline{\quad} \leq 5x - 3 \;\underline{\quad} \leq 7 \;\underline{\quad}$

(b) Simplify: _____

(c) Divide all three parts by 5 and simplify: _____

(d) Graph the solution set:

Objective 3: Solve Compound Inequalities Involving "or" *(See textbook Example 7)*

7. Solve $-\dfrac{3}{2}x + 6 > 9$ or $7x - 10 > 4$. Write the solution set in interval notation.

Step 1: Solve each inequality separately.

$$-\dfrac{3}{2}x + 6 > 9$$

Subtract 6 from both sides: (a) _____

Multiply both sides by $-\dfrac{2}{3}$:
Don't forget to reverse the direction of the inequality. (b) _____

$$7x - 10 > 4$$

Add 10 to both sides: (c) _____

Divide both sides by 7: (d) _____

Step 2: Find the union of the solution sets, which will represent the solution set to the compound inequality.

Graph (b): (e)

Graph (d): (f)

Graph the union: (g)

Write the solution in interval notation: (h) _____

Do the Math Exercises 8.6
Compound Inequalities

In Problems 1 – 3, use $A = \{4, 5, 6, 7, 8, 9\}$, $B = \{1, 5, 7, 9\}$, and $C = \{2, 3, 4, 6\}$ to find each set.

1. $A \cup C$

2. $A \cap C$

3. $B \cap C$

1. _____

2. _____

3. _____

In Problems 4 and 5, use $E = \{x \mid x \leq 2\}$ and $F = \{x \mid x \geq -2\}$ to find each of the following.

4. $E \cap F$

5. $E \cup F$

4. _____

5. _____

In Problems 6 – 17, solve each compound inequality. Express your answer in interval notation.

6. $x \leq 5$ and $x > 0$

7. $x < 0$ or $x \geq 6$

6. _____

7. _____

8. $7x + 2 \geq 9$ and $4x + 3 \leq 7$

9. $x + 3 \leq 5$ or $x - 2 \geq 3$

8. _____

9. _____

10. $-12 < 7x + 2 \leq 6$

11. $3x \geq 7x + 8$ or $x < 4x - 9$

10. _____

11. _____

12. $-\dfrac{4}{5}x - 5 > 3$ or $7x - 3 > 4$

13. $0 < \dfrac{3}{2}x - 3 \leq 3$

12. _____

13. _____

Do the Math Exercises 8.6

14. $x - \dfrac{3}{2} \leq \dfrac{5}{4}$ and $-\dfrac{2}{3}x - \dfrac{2}{9} < \dfrac{8}{9}$

15. $-3 < -4x + 1 < 17$

14. _____

15. _____

16. $-4 \leq \dfrac{4x-3}{3} < 3$

17. $-15 < -3(x+2) \leq 1$

16. _____

17. _____

In Problem 18, use the Addition Property and/or Multiplication Properties to find a and b.

18. If $-4 < x < 3$, then $a < 2x - 7 < b$.

18. _____

19. **Diastolic Blood Pressure** Blood pressure is measured using two numbers. One of the numbers measures diastolic blood pressure. The diastolic blood pressure represents the pressure while the heart is resting between beats. In a healthy person, the diastolic blood pressure should be greater than 60 and less than 90. If we let the variable x represent a person's diastolic blood pressure, express the diastolic blood pressure of a healthy person using compound inequality.

19. _____

20. **Electric Bill** In North Carolina, Duke Energy charges $42.41 plus $0.084192 for each additional kilowatt hour (kwh) used during the months from November through June for usage in excess of 350 kwh. Suppose one homeowner's electric bill ranged from a low of $55.04 to a high of $89.56 during this time period. Over what range (in kwh) did the usage vary?

20. _____

Name:
Instructor:

Date:
Section:

Five-Minute Warm-Up 8.7
Absolute Value Equations and Inequalities

1. Evaluate each expression.

   (a) $|-12|$ (b) $|0|$ (c) $\left|\dfrac{3}{4}\right|$ (d) $|-5.2|$

   1a. _____

   1b. _____

   1c. _____

   1d. _____

2. Express the distance between the origin, 0, and 45 as an absolute value.

   2. _____

3. Express the distance between the origin, 0, and -12 as an absolute value.

   3. _____

4. Solve each equation.

   (a) $-3x + 7 = -5$ (b) $4(x+1) = x + 5x - 10$

   4a. _____

   4b. _____

5. Solve each inequality.

   (a) $6x - 10 < 8x + 2$ (b) $\dfrac{1}{2}(3x - 1) \leq \dfrac{2}{3}(x + 3)$

   5a. _____

   5b. _____

Sullivan/Struve/Mazzarella, *Elementary & Intermediate Algebra*, 3e
Copyright © 2014 Pearson Education, Inc.

Name: Date:
Instructor: Section:

Guided Practice 8.7
Absolute Value Equations and Inequalities

Objective 1: Solve Absolute Value Equations

1. If a is a positive real number and if u is any algebraic expression, then $|u| = a$ is equivalent to _____ or _____.

2. When solving absolute value equations the first step is to _____.

3. Solve the equation $|3x - 1| - 5 = -3$. *(See textbook Example 2)*

Step 1: Isolate the expression containing the absolute value.

$|3x - 1| - 5 = -3$

Add 5 to both sides: **(a)** _____

Step 2: Rewrite the absolute value equation as two equations: $u = a$ and $u = -a$, where u is the algebraic expression in the absolute value symbol. Here $u = 3x - 1$ and $a = 2$.

(b) _____ or _____

Step 3: Solve each equation.

(c) $x =$ _____ or $x =$ _____

Step 4: Check. Verify each solution.

Substitute your values for x into the original equation. If the statement is true, then the value is a solution of the absolute value equation. If the statement is false, delete the value from the solution set.

(d) solution set: _____

4. If u and v are any algebraic expressions, then $|u| = |v|$ is equivalent to _____ or _____.

Objective 2: Solve Absolute Value Inequalities Involving $<$ or $\leq$

5. If a is a positive real number and if u is any algebraic expression, then

$|u| < a$ is equivalent to _____.

$|u| \leq a$ is equivalent to _____.

Sullivan/Struve/Mazzarella, *Elementary & Intermediate Algebra*, 3e
Copyright © 2014 Pearson Education, Inc.

Guided Practice 8.7

6. Solve the inequality $|4x - 3| \leq 9$. Write the solution set in interval notation. *(See textbook Example 6)*

Step 1: The inequality is in the form $|u| \leq a$ where $u = 4x - 3$ and $a = 9$. We rewrite the inequality as a compound inequality that does not involve absolute value.

$|4x - 3| \leq 9$

Use the fact that $|u| \leq a$ means $-a \leq u \leq a$:

(a) _____

Step 2: Solve the resulting compound inequality.

Add 3 to all three parts:

(b) _____

Divide all three parts of the inequality by 4:

(c) _____

Graph the solution:

(d) _____

Write the solution in interval notation:

(e) _____

Objective 3: Solve Absolute Value Inequalities Involving > or ≥

7. If a is a positive real number and if u is any algebraic expression, then

   $|u| > a$ is equivalent to _____ or _____.

   $|u| \geq a$ is equivalent to _____ or _____.

8. Solve the inequality $3|8x + 3| > 9$. Write the solution set in interval notation. *(See textbook Example 9)*

Step 1: The inequality is in the form $|u| > a$ where $u = 8x + 3$ and $a = 3$. We rewrite the inequality as a compound inequality that does not involve absolute value.

$3|8x + 3| > 9$

Isolate the absolute value:

(a) _____

Rewrite the inequality:

(b) _____

Step 2: Solve each inequality separately.

(c) _____ or _____

Step 3: Find the union of the solution sets of each inequality.

Graph the solution:

(d)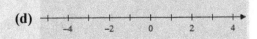

Write the solution in interval notation:

(e) _____

Name:
Instructor:

Date:
Section:

Do the Math Exercises 8.7
Absolute Value Equations and Inequalities

In Problems 1 and 2, solve each absolute value equation.

1. $\left|\dfrac{2x-3}{5}\right| = 2$

2. $3|y - 4| + 4 = 16$

1. _____

2. _____

In Problems 3 – 6, solve each absolute value inequality. Express your answer in set-builder notation.

3. $|y + 4| < 6$

4. $|-3x + 2| - 7 \leq -2$

3. _____

4. _____

5. $|x + 4| \geq 7$

6. $3|z| + 8 > 2$

5. _____

6. _____

In Problems 7 – 14, solve each absolute value equation or inequality. For absolute value inequalities, express your answer in interval notation.

7. $|2x + 1| = x - 3$

8. $|3 - 5x| < |-7|$

7. _____

8. _____

9. $|3x - 4| = -9$

10. $|-9x + 2| \geq -1$

9. _____

10. _____

Do the Math Exercises 8.7

11. $|4x+3|=1$

12. $|4x-3|>1$

11. _____

12. _____

13. $|7x+5|+4<3$

14. $|4y+3|\geq -3$

13. _____

14. _____

In Problems 15 and 16, write each statement as an absolute value inequality.

15. x differs from -4 by less than 2

16. twice x differs from 7 by more than 3

15. _____

16. _____

17. **Gestation Period** The length of human pregnancy is about 266 days. It can be shown that a mother whose gestation period x satisfies the inequality $\left|\dfrac{x-266}{16}\right|>1.96$ has an unusual length of pregnancy. Determine the length of pregnancy that would be considered unusual.

17. _____

18. Explain why the solution set of $|5x-3|>-5$ is the set of all real numbers.

Name:
Instructor:

Date:
Section:

Five-Minute Warm-Up 8.8
Variation

1. Solve each equation.

   (a) $-75 = 15x$

   (b) $-\dfrac{2}{3}k = \dfrac{8}{27}$

   1a. _____

   1b. _____

2. Solve: $-12 = \dfrac{k}{4}$

   2. _____

3. Graph the equation $y = \dfrac{1}{2}x$.

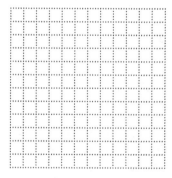

4. Graph the equation $y = 2x$.

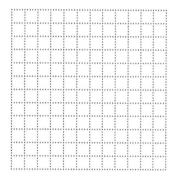

Sullivan/Struve/Mazzarella, *Elementary & Intermediate Algebra*, 3e
Copyright © 2014 Pearson Education, Inc.

Name: Date:
Instructor: Section:

Guided Practice 8.8
Variation

Objective 1: Model and Solve Direct Variation Problems

1. We say that y varies directly with x, or y is directly proportional to x, if there is a nonzero number k such that _____.

2. The number k is called the _____.

3. If y varies directly with x, then y is a _____ function of x and has a y-intercept of _____.

4. Suppose that y is directly proportional to x and when $x = -12$, $y = 5$. Find y when $x = 20$. Write the direct variation equation, calculate the constant k, and then use the given values to solve the unknown.

4. _____

Objective 2: Model and Solve Inverse Variation Problems

5. We say that y varies inversely with x, or y is inversely proportional to x, if there is a nonzero number k such that _____.

6. Suppose that y varies inversely with x. When $x = 4$, $y = 12$. Find y when $x = 18$. Write the inverse variation equation, calculate the constant, k, and then use the given values to solve for the unknown. *(See textbook Example 3)*

6. _____

Guided Practice 8.8

Objective 3: Model and Solve Joint Variation and Combined Variation Problems

7. Suppose that r varies jointly with s and t. When $r = 12$, $s = 8$ and $t = 3$. Find r when $s = 14$ and $t = 6$. Write the joint variation equation, calculate the constant, k, and then use the given values to solve for the unknown. *(See textbook Example 4)*

7. _____

8. **Gas Laws** The volume V of an ideal gas varies directly with the temperature T and inversely with the pressure P. If a cylinder contains oxygen at a temperature of 300 kelvin (K) and a pressure of 15 atmospheres (Atm) in a volume of 100 liters, what is the constant of proportionality k? If a piston is lowered into the cylinder, decreasing the volume occupied by the gas to 70 liters and raising the temperature to 315 K, what is the pressure?

(a) Write the equation that shows the relationship between the variables, V, T, and P.

8a. _____

(b) To calculate the constant of proportionality, substitute each of the following variables into your equation from 8a.

8b.
$T = $ _____

$P = $ _____

$V = $ _____

(c) Solve your equation to determine the constant of proportionality, k.

8c. _____

(d) Substitute your value of k into 8a to determine the function relating the variables.

8d. _____

(e) Substitute the values of V and T in the last sentence into your equation from 8d to answer the question.

8e. _____

Name:
Instructor:

Date:
Section:

Do the Math Exercises 8.8
Variation

In Problems 1 and 2, **(a)** find the constant of proportionality, k, **(b)** write the linear function relating the two variables, and **(c)** find the quantity indicated.

1. Suppose that y varies directly with x. When $x = 3$, then $y = 15$. Find y when $x = 5$.

2. Suppose that y is directly proportional to x. When $x = 20$, then $y = 4$. Find y when $x = 35$.

1a. _____

1b. _____

1c. _____

2a. _____

2b. _____

2c. _____

In Problems 3 and 4, find the quantity indicated.

3. A is directly proportional to B. If A is 360 when B is 72, find B when A is 400.

4. m varies directly with r. If r is 24 when m is 9, find r when m is 24.

3. _____

4. _____

5. **Mortgage Payments** The monthly payment p on a mortgage varies directly with the amount borrowed b. Suppose that you decide to borrow $120,000 using a 15-year mortgage at 5.5% interest. You are told that your payment is $980.50. Assume that you have decided to buy a more expensive home that requires you borrow $150,000. What will your monthly payment be?

5. _____

6. **Buying Gasoline** The cost to purchase a tank of gasoline varies directly with the number of gallons purchased. You notice that the person in front of you spent $34.50 on 15 gallons of gas. If your SUV needs 35 gallons of gas, how much will you spend?

6. _____

Do the Math Exercises 8.8

7. **Falling Objects** The velocity of a falling object (ignoring air resistance) v is directly proportional to the time t of the fall. If, after 2 seconds, the velocity of the object is 64 feet per second, what will its velocity be after 3 seconds?

7. _____

In Problems 8 – 10, (a) find the constant, k; (b) write the function relating the two variables; and (c) find the quantity indicated.

8. Suppose that y is inversely proportional to x. When $x = 20$, then $y = 4$. Find y when $x = 35$.

9. Suppose that y varies jointly with x and z. When $y = 20$, $x = 6$ and $z = 10$. Find y when $x = 8$ and $z = 15$.

8a. _____

8b. _____

8c. _____

9a. _____

9b. _____

9c. _____

10. Suppose that Q varies directly with x and inversely with y. When $Q = \frac{14}{5}$, $x = 4$ and $y = 3$. Find Q when $x = 8$ and $y = 3$.

10a. _____

10b. _____

10c. _____

11. **Resistance** The current, i, in a circuit is inversely proportional to its resistance, R, measured in ohms. Suppose that when the current in a circuit is 30 amperes, the resistance is 8 ohms. Find the current in the same circuit when the resistance is 10 ohms.

11. _____

12. **Intensity of Light** The intensity, I, of light (measured in foot-candles) varies inversely with the square of the distance from the bulb. Suppose the intensity of a 100-watt light bulb at a distance of 2 meters is 0.075 foot-candles. Determine the intensity of the bulb at a distance of 3 meters.

12. _____

Name:
Instructor:

Date:
Section:

Five-Minute Warm-Up 9.1
Square Roots

In Problems 1 – 4, identify which numbers in the set $\left\{\dfrac{-2}{3}, 5, -2.3, \dfrac{21}{7}, 1.\overline{6}, \dfrac{0}{11}, -10, \dfrac{11}{0}, 2.1010010001...\right\}$ belong to the sets listed below.

1. Integer _____

2. Rational _____

3. Irrational _____

4. Real _____

In Problems 5 – 10, evaluate the expressions.

5. 5^2

6. 11^2

7. $(0.8)^2$

8. $(1.2)^2$

9. $\left(\dfrac{4}{7}\right)^2$

10. $\left(\dfrac{13}{6}\right)^2$

5. _____

6. _____

7. _____

8. _____

9. _____

10. _____

Name:
Instructor:

Date:
Section:

Guided Practice 9.1
Square Roots

Objective 1: Evaluate Square Roots of Perfect Squares

1. Fill in the blank:
   (a) Every positive real number has two square roots, one _____ and one _____.

   1a. _____
   _____

   (b) We use the symbol $\sqrt{}$, called a _____, to denote the nonnegative square root.

   1b. _____

   (c) The nonnegative square root is called the _____ square root.

   1c. _____

   (d) The number under the radical is called the _____.

   1d. _____

2. Evaluate each square root. *(See textbook Example 1)*
   (a) $\sqrt{64}$ (b) $\sqrt{400}$

   2a. _____

   2b. _____

   (c) $\sqrt{\dfrac{4}{25}}$ (d) $\left(\sqrt{17}\right)^2$

   2c. _____

   2d. _____

3. Evaluate each expression. *(See textbook Example 2)*
   a) $-3\sqrt{49}$ (b) $\sqrt{1+16+64}$

   3a. _____

   3b. _____

   (c) $\sqrt{1}+\sqrt{16}+\sqrt{64}$ (d) $\sqrt{49+3\cdot 4\cdot(-2)}$

   3c. _____

   3d. _____

Objective 2: Determine Whether a Square Root is Rational, Irrational, or Not a Real Number

4. Fill in the blank:
   (a) The square root of a perfect square is a(n) _____ number.

   4a. _____

   (b) The square root of a positive rational number that is not a perfect square, such as $\sqrt{20}$, is a(n) _____ number.

   4b. _____

   (c) The square root of a negative real number, such as $\sqrt{-2}$ is not a _____ number.

   4c. _____

Sullivan/Struve/Mazzarella, *Elementary & Intermediate Algebra*, 3e
Copyright © 2014 Pearson Education, Inc.

Guided Practice 9.1

5. Determine if each square root is rational, irrational, or not a real number. Then evaluate each real square root. For each square root that is irrational, express the square root as a decimal rounded to two decimal places. *(See textbook Example 4)*

(a) $\sqrt{48}$

(b) $\sqrt{-289}$

5a. _____

5b. _____

(c) $\sqrt{256}$

(d) $-\sqrt{225}$

5c. _____

5d. _____

Objective 3: Find Square Roots of Variable Expressions

For a variable expression, where the variable can represent either positive or negative number, the square root of the expression squared is the absolute value of the expression. So, $\sqrt{4x^2} = |2x|$ or $2|x|$.

6. Evaluate each square root. *(See textbook Example 5)*

(a) $\sqrt{(-23)^2}$

(b) $\sqrt{\left(\dfrac{17}{68}\right)^2}$

6a. _____

6b. _____

6c. _____

(c) $\sqrt{(3y-2)^2}$

(d) $\sqrt{25x^2 - 30x + 9}$

6d. _____

Do the Math Exercises 9.1
Square Roots

In Problems 1 – 4, evaluate each square root.

1. $\sqrt{9}$

2. $\sqrt{\dfrac{4}{81}}$

3. $\sqrt{0.16}$

4. $\left(\sqrt{3.7}\right)^2$

1. _____

2. _____

3. _____

4. _____

In Problems 5 – 8, tell if the square root is rational, irrational, or not a real number. If the square root is rational, find the exact value; if the square root is irrational, write the approximate value rounded to two decimal places.

5. $\sqrt{-50}$

6. $\sqrt{\dfrac{49}{100}}$

7. $\sqrt{24}$

8. $\sqrt{12}$

5. _____

6. _____

7. _____

8. _____

Do the Math Exercises 9.1

In Problems 9 – 12, simplify each square root.

9. $\sqrt{(-13)^2}$

10. $\sqrt{w^2}$

9. _____

10. _____

11. $\sqrt{(x-8)^2}$

12. $\sqrt{9z^2 - 24z + 16}$

11. _____

12. _____

In Problems 13 – 15, simplify each expression.

13. $2\sqrt{\dfrac{9}{4}} - \sqrt{4}$

14. $\sqrt{9^2 - 4 \cdot 1 \cdot 20}$

13. _____

14. _____

15. $\sqrt{(-3)^2 - 4 \cdot 3 \cdot 2}$

15. _____

Name:
Instructor:

Date:
Section:

Five-Minute Warm-Up 9.2
nth Roots and Rational Exponents

In Problems 1 – 4, simplify each expression.

1. $\sqrt{0.25}$

2. $\left(\sqrt{\dfrac{4}{9}}\right)^2$

1. _____

2. _____

3. $\sqrt{(4x+3)^2}$

4. $\sqrt{x^2 - 2xy + y^2}$

3. _____

4. _____

In Problems 5 – 8, simplify each expression.

5. 7^{-2}

6. $\dfrac{1}{x^{-2}}$

5. _____

6. _____

7. $\left(\dfrac{3x}{y^2}\right)^2$

8. $\left(\dfrac{4a^2 b^{-1}}{8a^2 b^3}\right)^{-3}$

7. _____

8. _____

Name:
Instructor:

Date:
Section:

Guided Practice 9.2
*n*th Roots and Rational Exponents

Objective 1: Evaluate nth Roots

1. In the notation $\sqrt[3]{8} = 2$, 3 is called the _____, 8 is called the _____ and 2 is the _____.

2. If the index is even, then the radicand must be _____ in order for the radical to simplify to a real number.

   If the index is odd, then the radicand can be _____ and the expression will simplify to any real number.

3. Since $(-4)^2 = 16$ and $4^2 = 16$, it could be interpreted that $\sqrt{16} = -4$ or 4. In fact, $\sqrt{16} = 4$ only, because when the index even, we use the _____ _____, which must be ≥ 0.

4. Evaluate each root without using a calculator. *(See textbook Example 1)*

   (a) $\sqrt[3]{-27}$ (b) $\sqrt{-36}$ (c) $\sqrt[4]{\dfrac{16}{81}}$

   4a. _____

   4b. _____

   4c. _____

5. Write $\sqrt[4]{64}$ as a decimal rounded to two decimal places. *(See textbook Example 2)*   5. _____

Objective 2: Simplify Expressions of the form $\sqrt[n]{a^n}$

6. Simplify: $\sqrt[4]{(2x-1)^4}$ *(See textbook Example 3)*   6. _____

Objective 3: Evaluate Expressions of the Form $a^{\frac{1}{n}}$

7. Write each of the following expressions as a radical and simplify, if possible. *(See textbook Example 4)*

   (a) $144^{\frac{1}{2}}$ (b) $(-64)^{\frac{1}{2}}$ (c) $2x^{\frac{1}{3}}$

   7a. _____

   7b. _____

   7c. _____

Guided Practice 9.2

Objective 4: Evaluate Expressions of the Form $a^{\frac{m}{n}}$

8. If a is a real number and $\frac{m}{n}$ is a rational number in lowest terms with $n \geq 2$, then

$$a^{\frac{m}{n}} = \underline{} \text{ or } \underline{}, \text{ provided that } \sqrt[n]{a} \text{ exists.}$$

9. Evaluate each of the following expressions, if possible. *(See textbook Example 6)*
   (a) $36^{\frac{3}{2}}$ (b) $-16^{\frac{5}{2}}$ (c) $(-64)^{\frac{2}{3}}$

9a. _____

9b. _____

9c. _____

10. Write $25^{\frac{2}{3}}$ as a decimal rounded to two decimal places. *(See textbook Example 7)*

10. _____

11. Write each radical expression with a rational exponent. *(See textbook Example 8)*
    (a) $\sqrt[4]{(2x)^3}$ (b) $\left(\sqrt[3]{2x^2 y}\right)^2$

11a. _____

11b. _____

12. Rewrite each of the following with positive exponents, and completely simplify, if possible. *(See textbook Example 9)*
    (a) $64^{-\frac{1}{2}}$ (b) $\dfrac{2}{16^{-\frac{1}{2}}}$ (c) $(4x)^{-\frac{5}{2}}$

12a. _____

12b. _____

12c. _____

Name:
Instructor:

Date:
Section:

Do the Math Exercises 9.2
nth Roots and Rational Exponents

In Problems 1 – 7, simplify each radical.

1. $\sqrt[3]{216}$

2. $\sqrt[3]{-64}$

3. $-\sqrt[4]{256}$

4. $\sqrt[3]{\dfrac{8}{125}}$

5. $\sqrt[4]{6^4}$

6. $\sqrt[5]{n^5}$

7. $\sqrt[6]{(2x-3)^6}$

1. _____

2. _____

3. _____

4. _____

5. _____

6. _____

7. _____

In Problems 8 – 16, evaluate each expression, if possible.

8. $16^{\frac{1}{2}}$

9. $-25^{\frac{1}{2}}$

10. $-81^{\frac{1}{4}}$

11. $(-81)^{\frac{1}{2}}$

12. $-100^{\frac{5}{2}}$

13. $-(-32)^{\frac{3}{5}}$

14. $121^{-\frac{1}{2}}$

15. $\dfrac{1}{49^{-\frac{3}{2}}}$

16. $27^{-\frac{4}{3}}$

8. _____

9. _____

10. _____

11. _____

12. _____

13. _____

14. _____

15. _____

16. _____

Sullivan/Struve/Mazzarella, *Elementary & Intermediate Algebra*, 3e
Copyright © 2014 Pearson Education, Inc.

Do the Math Exercises 9.2

In Problems 17 – 19, rewrite each of the following radicals with a rational exponent.

17. $\sqrt[4]{x^3}$

18. $\left(\sqrt[5]{3x}\right)^2$

17. _____

18. _____

19. $\sqrt[4]{(3pq)^7}$

19. _____

In Problems 20 – 21, use a calculator to write each expression as a decimal rounded to two decimal places.

20. $\sqrt[3]{85}$

21. $100^{0.25}$

20. _____

21. _____

In Problems 22 – 26, evaluate each expression, if possible.

22. $100^{\frac{3}{2}}$

23. $\sqrt[4]{-1}$

22. _____

23. _____

24. $125^{-\frac{1}{3}}$

25. $100^{\frac{1}{2}} - 4^{\frac{3}{2}}$

24. _____

25. _____

26. $(-125)^{-\frac{1}{3}}$

26. _____

27. Explain why $(-9)^{\frac{1}{2}}$ is not a real number, but $-9^{\frac{1}{2}}$ is a real number.

Name:
Instructor:

Date:
Section:

Five-Minute Warm-Up 9.3
Simplifying Expressions Using the Laws of Exponents

In Problems 1 – 8, simplify each expression.

1. $9z^6 \cdot 6z$

2. $\dfrac{18u^5}{12u^3}$

3. 25^{-2}

4. $\left(\dfrac{4}{3}\right)^{-3}$

5. $\dfrac{9}{7}x^5 y^{-2} \cdot \dfrac{28}{12}xy$

6. $\left(-5p^3\right)^{-2}$

7. $\left(\dfrac{8a^{-1}}{b^{-5}}\right)^2$

8. $\left(\dfrac{2x^{-2}y}{6x^{-3}y^2}\right)^{-1} \cdot \left(\dfrac{xy^{-2}}{9xy^2}\right)^2$

9. Evaluate: $\sqrt{\dfrac{81}{49}}$

1. _____

2. _____

3. _____

4. _____

5. _____

6. _____

7. _____

8. _____

9. _____

Name:
Instructor:

Date:
Section:

Guided Practice 9.3
Simplifying Expressions Using the Laws of Exponents

Objective 1: Simplify Expressions Involving Rational Exponents

1. List four characteristics of an expression with rational exponents which is completely simplified.

2. Simplify each of the following expressions involving rational exponents. Express the answer with positive rational exponents, if necessary. *(See textbook Examples 1 and 2)*

(a) $6^{\frac{12}{5}} \cdot 6^{-\frac{2}{5}}$ (b) $\dfrac{\left(-8a^4\right)^{\frac{1}{3}}}{a^{\frac{5}{6}}}$

2a. _____

2b. _____

3. Simplify the following expression involving rational exponents. Express the answer with positive rational exponents, if necessary. *(See textbook Example 3)*

$\left(xy^{-\frac{3}{8}}\right) \cdot \left(x^{\frac{1}{2}}y^{-2}\right)^{\frac{3}{4}}$

3. _____

Sullivan/Struve/Mazzarella, *Elementary & Intermediate Algebra*, 3e

Guided Practice 9.3

Objective 2: Simplify Radical Expressions

4. Rewrite the radical as an expression involving a rational exponent and simplify. Express the answer with a simplified radical, if necessary. *(See textbook Example 4)*

(a) $\sqrt[6]{81^3}$ (b) $\sqrt[4]{128x^8y^4}$ (c) $\dfrac{\sqrt[4]{x^3}}{\sqrt{x}}$ (d) $\sqrt[3]{\sqrt{p}}$

4a. _____

4b. _____

4c. _____

4d. _____

Objective 3: Factor Expressions Containing Rational Exponents

5. Simplify $6x^{\frac{2}{3}} + 2x^{\frac{1}{3}}(5x - 2)$ by factoring out $2x^{\frac{1}{3}}$. Express the answer with positive rational exponents, if necessary. *(See textbook Example 5)*

5. _____

Name:
Instructor:

Date:
Section:

Do the Math Exercises 9.3
Simplifying Expressions Using the Laws of Exponents

In Problems 1 – 9, simplify each of the following expressions.

1. $3^{\frac{1}{3}} \cdot 3^{\frac{5}{3}}$

2. $\dfrac{10^{\frac{7}{5}}}{10^{\frac{2}{5}}}$

3. $\dfrac{y^{\frac{1}{5}}}{y^{\frac{9}{10}}}$

4. $\left(36^{-\frac{1}{4}} \cdot 9^{\frac{3}{4}}\right)^{-2}$

5. $\left(a^{\frac{5}{4}} \cdot b^{\frac{3}{2}}\right)^{\frac{2}{5}}$

6. $\left(a^{\frac{4}{3}} \cdot b^{-\frac{1}{2}}\right)\left(a^{-2} \cdot b^{\frac{5}{2}}\right)$

7. $\left(25 p^{\frac{2}{5}} q^{-1}\right)^{\frac{1}{2}}$

8. $\left(\dfrac{64 m^{\frac{1}{2}} n}{m^{-2} n^{\frac{4}{3}}}\right)^{\frac{1}{2}}$

9. $\left(\dfrac{27 x^{\frac{1}{2}} y^{-1}}{y^{-\frac{2}{3}} x^{-\frac{1}{2}}}\right)^{\frac{1}{3}} - \left(\dfrac{4 x^{\frac{1}{3}} y^{\frac{4}{9}}}{x^{-\frac{1}{3}} y^{\frac{2}{3}}}\right)^{\frac{1}{2}}$

In Problems 10 –12, distribute and simplify.

10. $x^{\frac{1}{3}}\left(x^{\frac{5}{3}} + 4\right)$

11. $3a^{-\frac{1}{2}}(2 - a)$

12. $8p^{\frac{2}{3}}\left(p^{\frac{4}{3}} - 4p^{-\frac{2}{3}}\right)$

1. _____

2. _____

3. _____

4. _____

5. _____

6. _____

7. _____

8. _____

9. _____

10. _____

11. _____

12. _____

Do the Math Exercises 9.3

In Problems 13 – 17, use rational exponents to simplify each radical. Assume all variables are positive.

13. $\sqrt[3]{x^6}$

14. $\sqrt[9]{125^6}$

15. $\sqrt{25x^4 y^6}$

16. $\sqrt[4]{p^3} \cdot \sqrt[3]{p}$

17. $\sqrt{5} \cdot \sqrt[3]{25}$

13. _____

14. _____

15. _____

16. _____

17. _____

In Problems 18 – 21, simplify by factoring out the given expression.

18. $3(x-5)^{\frac{1}{2}}(3x+1) + 6(x-5)^{\frac{3}{2}}; (x-5)^{\frac{1}{2}}$

19. $x^{-\frac{2}{3}}(3x+2) + 9x^{\frac{1}{3}}; x^{-\frac{2}{3}}$

20. $4(x+3)^{\frac{1}{2}} + (x+3)^{-\frac{1}{2}}(2x+1); (x+3)^{-\frac{1}{2}}$

21. $24x(x^2-1)^{\frac{1}{3}} + 9(x^2-1)^{\frac{4}{3}}; (x^2-1)^{\frac{1}{3}}$

18. _____

19. _____

20. _____

21. _____

In Problems 22 – 25, simplify each expression.

22. $\sqrt[6]{27^2}$

23. $25^{\frac{3}{4}} \cdot 25^{\frac{3}{4}}$

24. $\left(8^4\right)^{\frac{5}{12}}$

25. $\sqrt[9]{a^6} - \dfrac{\sqrt[6]{a^5}}{\sqrt[6]{a}}$

22. _____

23. _____

24. _____

25. _____

Five-Minute Warm-Up 9.4
Simplifying Radical Expressions Using Properties of Radicals

1. List the perfect squares that are less than 150.   _____

2. List the perfect cubes that are less than 150.   _____

3. List the perfect fourths that are less than 150.   _____

In Problems 4 – 9, simplify each radical.

4. $\sqrt{-4}$

5. $\sqrt{10^2}$

4. _____

5. _____

6. $\sqrt{x^2}$

7. $\sqrt{(4x+1)^2}$

6. _____

7. _____

8. $\sqrt{4p^2 - 4p + 1}$

9. $\sqrt{144 + 25}$

8. _____

9. _____

Name:
Instructor:

Date:
Section:

Guided Practice 9.4
Simplifying Radical Expressions Using Properties of Radicals

Objective 1: Use the Product Property to Multiply Radical Expressions

1. The Product Property of Radicals states that if $\sqrt[n]{a}$ and $\sqrt[n]{b}$ are real numbers and $n \geq 2$ is an integer, then $\sqrt[n]{a} \cdot \sqrt[n]{b} = \sqrt[n]{ab}$. Use this property to multiply each of the following. *(See textbook Example 1)*

(a) $\sqrt{5} \cdot \sqrt{7}$ (b) $\sqrt{x-5} \cdot \sqrt{x+5}$ (c) $\sqrt[3]{7x} \cdot \sqrt[3]{2x}$

1a. _____

1b. _____

1c. _____

Objective 2: Use the Product Property to Simplify Radical Expressions

2. A radical expression is *simplified* when _____.

3. Simplify: $\sqrt{75}$ *(See textbook Example 2)*

Step 1: Since the index is 2, we write each factor of the radicand as the product of two factors, one of which is a perfect square. What perfect square is a factor of 75? (a) _____

Step 2: Write the radical as the product of two radicals, one of which contains the perfect square. $\sqrt{75} =$ (b) _____

Step 3: Take the square root of each perfect power. (c) _____

4. Simplify: $\dfrac{-6 + \sqrt{48}}{2}$ *(See textbook Example 4)*

$\dfrac{-6 + \sqrt{48}}{2} =$

Step 1: Since the index is 2, write each factor of the radicand as the product of the factors, one of which is a perfect square. (a) _____

Step 2: Use $\sqrt{a \cdot b} = \sqrt{a} \cdot \sqrt{b}$. (b) _____

Step 3: Take the square root of each perfect power. (c) _____

Step 4: Factor out the 2 in the numerator. (d) _____

Step 5: Divide out the common factor. (e) _____

Sullivan/Struve/Mazzarella, *Elementary & Intermediate Algebra*, 3e
Copyright © 2014 Pearson Education, Inc.

Guided Practice 9.4

5. Explain how to simplify an expression with an index that does not divide evenly into the exponent on the variable in the radicand.

_____

Objective 3: Use the Quotient Property to Simplify Radical Expressions

6. Simplify $\sqrt{\dfrac{12x^2}{121}}$. Assume $x \geq 0$. *(See textbook Example 8)*   6. _____

7. Simplify $\dfrac{\sqrt{72a}}{\sqrt{2a^5}}$. Assume $a > 0$. *(See textbook Example 9)*   7. _____

Objective 4: Multiply Radicals with Unlike Indices

8. Multiply: $\sqrt{6} \cdot \sqrt[4]{48}$ *(See textbook Example 10)*

(a) What is the first step to multiply $\sqrt{6} \cdot \sqrt[4]{48}$?   8a. _____

(b) Determine the LCD of 2 and 4.   8b. _____

(c) Use $a^{\frac{n}{m}} = \left(a^{\frac{1}{m}}\right)^n$ to rewrite each factor: 8c. _____

(d) Use $a^r \cdot b^r = (a \cdot b)^r$:   8d. _____

(e) Multiply: 8e. _____

(f) Write the expression using radicals and simplify: 8f. _____

Name:
Instructor:

Date:
Section:

Do the Math Exercises 9.4
Simplifying Radical Expression Using Properties of Radicals

In Problems 1 – 3, use the Product Rule to multiply. Assume that all variables can be any real number.

1. $\sqrt[4]{6a^2} \cdot \sqrt[4]{7b^2}$

2. $\sqrt{p-5} \cdot \sqrt{p+5}$

3. $\sqrt[3]{\dfrac{-9x^2}{4}} \cdot \sqrt[3]{\dfrac{4}{3x}}$

1. _____

2. _____

3. _____

In Problems 4 – 11, simplify each radical using the Product Property. Assume that all variables can be any real number.

4. $\sqrt[4]{162}$

5. $\sqrt{20a^2}$

6. $\sqrt[3]{-64p^3}$

7. $4\sqrt{27b}$

8. $\sqrt{s^9}$

9. $\sqrt[5]{x^{12}}$

10. $\sqrt[3]{-54q^{12}}$

11. $\sqrt{75x^6 y}$

4. _____

5. _____

6. _____

7. _____

8. _____

9. _____

10. _____

11. _____

In Problems 12 and 13, simplify each expression.

12. $\dfrac{5 - \sqrt{100}}{5}$

13. $\dfrac{-6 + \sqrt{48}}{8}$

12. _____

13. _____

Sullivan/Struve/Mazzarella, *Elementary & Intermediate Algebra*, 3e
Copyright © 2014 Pearson Education, Inc.

Do the Math Exercises 9.4

In Problems 14 – 17, multiply and simplify. Assume that all variables are greater than or equal to zero.

14. $\sqrt{3} \cdot \sqrt{12}$

15. $\sqrt{6x} \cdot \sqrt{30x}$

14. _____

15. _____

16. $\sqrt[3]{9a} \cdot \sqrt[3]{6a^2}$

17. $\sqrt[3]{16m^2n} \cdot \sqrt[3]{27m^2n}$

16. _____

17. _____

In Problems 18 – 23, simplify. Assume that all variables are greater than zero.

18. $\sqrt{\dfrac{5}{36}}$

19. $\sqrt[4]{\dfrac{5x^4}{16}}$

18. _____

19. _____

20. $\sqrt[3]{\dfrac{-27x^9}{64y^{12}}}$

21. $\dfrac{\sqrt[4]{64}}{\sqrt[4]{4}}$

20. _____

21. _____

22. $\dfrac{\sqrt{54y^5}}{\sqrt{3y}}$

23. $\dfrac{\sqrt[3]{-128x^8}}{\sqrt[3]{2x^{-1}}}$

22. _____

23. _____

In Problems 24 and 25, multiply and simplify.

24. $\sqrt{2} \cdot \sqrt[3]{7}$

25. $\sqrt[4]{3} \cdot \sqrt[8]{5}$

24. _____

25. _____

Name:
Instructor:

Date:
Section:

Five-Minute Warm-Up 9.5
Adding, Subtracting, and Multiplying Radical Expressions

1. Add: $(2z^3 - 7z^2 + 1) + (z^3 + 2z^2 - 4z - 1)$

1. _____

2. Subtract: $(-2a^2b^2 + ab - 3b^2) - (a^2b^2 + a^2 - 5ab - 2b^2)$

2. _____

3. Multiply: $-4x^2(2x^2 + 3xy - 5y^2)$

3. _____

4. Multiply: $\dfrac{3}{2}x^2\left(\dfrac{4}{27}x^3 - \dfrac{8}{21}x^2 + \dfrac{2}{3}\right)$

4. _____

5. Multiply: $(9c + 2)(2c - 3)$

5. _____

6. Multiply: $(ab - 2)(ab + 2)$

6. _____

7. Multiply: $(7n - 3)^2$

7. _____

Sullivan/Struve/Mazzarella, *Elementary & Intermediate Algebra*, 3e
Copyright © 2014 Pearson Education, Inc.

Name:
Instructor:

Date:
Section:

Guided Practice 9.5
Adding, Subtracting, and Multiplying Radical Expressions

Objective 1: Add or Subtract Radical Expressions

1. Describe the characteristics of *like* radicals.

2. Add or subtract, as indicated. Assume all variables are greater than or equal to zero.
(See textbook Example 1)

(a) $8\sqrt{x} + \sqrt{x}$

(b) $13\sqrt[3]{3p} - 6\sqrt[3]{3p} + \sqrt[3]{3p}$

2a. _____

2b. _____

3. Simplify the radicals and then perform the indicated operations. Assume all variables are greater than or equal to zero. *(See textbook Examples 2 and 3)*

(a) $3\sqrt{8} - 4\sqrt{32}$

(b) $3n^2\sqrt{54n^4} - 2n\sqrt{150n^6} + 2\sqrt{24n^9}$

3a. _____

3b. _____

Sullivan/Struve/Mazzarella, *Elementary & Intermediate Algebra*, 3e
Copyright © 2014 Pearson Education, Inc.

Guided Practice 9.5

Objective 2: Multiply Radical Expressions

4. Multiply and simplify: $(9 + 2\sqrt{6})(2 - 3\sqrt{2})$ (See textbook Example 4)

4. _____

5. Multiply and simplify: $(7 - \sqrt{3})(7 + \sqrt{3})$ (See textbook Example 5)

Step 1: What special products formula can be used to multiply $(7 - \sqrt{3})(7 + \sqrt{3})$?

5a. _____

Step 2: Use the formula from part (a) to multiply and simplify $(7 - \sqrt{3})(7 + \sqrt{3})$.

5b. _____

6. Use **Heron's Formula** for finding the area of triangle whose sides are known. Heron's Formula states that the area A of a triangle with sides a, b, and c, is

$$A = \sqrt{s(s-a)(s-b)(s-c)}$$

where

$$s = \frac{1}{2}(a+b+c)$$

Find the area of the shaded region by computing the difference in the areas of the triangles. That is, compute "area of larger triangle minus area of smaller triangle." Write your answer as a radical in simplified form.

6. _____

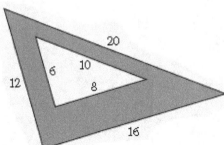

Name:
Instructor:

Date:
Section:

Do the Math Exercises 9.5
Adding, Subtracting, and Multiplying Radical Expressions

In Problems 1 – 9, add or subtract as indicated. Assume all variables are positive or zero.

1. $6\sqrt{3} + 8\sqrt{3}$

2. $12\sqrt[4]{z} - 5\sqrt[4]{z}$

1. _____

2. _____

3. $4\sqrt[3]{5} - 3\sqrt{5} + 7\sqrt[3]{5} - 8\sqrt{5}$

4. $6\sqrt{3} + \sqrt{12}$

3. _____

4. _____

5. $7\sqrt[4]{48} - 4\sqrt[4]{243}$

6. $2\sqrt{48z} - \sqrt{75z}$

5. _____

6. _____

7. $3\sqrt{63z^3} + 2z\sqrt{28z}$

8. $\sqrt{48y^2} - 4y\sqrt{12} + \sqrt{108y^2}$

7. _____

8. _____

9. $-2\sqrt[3]{5x^3} + 4x\sqrt[3]{40} - \sqrt[3]{135}$

9. _____

In Problems 10 – 18, multiply and simplify. Assume all variables are positive or zero.

10. $\sqrt{5}(5 + 3\sqrt{3})$

11. $\sqrt[3]{6}(\sqrt[3]{2} + \sqrt[3]{12})$

10. _____

11. _____

12. $(5 + \sqrt{5})(3 + \sqrt{6})$

13. $(9 + 5\sqrt{10})(1 - 3\sqrt{10})$

12. _____

13. _____

Do the Math Exercises 9.5

14. $(\sqrt{6}-2\sqrt{2})(2\sqrt{6}+3\sqrt{2})$ 　　　　**15.** $(2-\sqrt{3})^2$

14. _____

15. _____

16. $(\sqrt{3}-1)(\sqrt{3}+1)$ 　　　　**17.** $(6+3\sqrt{2})(6-3\sqrt{2})$

16. _____

17. _____

18. $(\sqrt[3]{y}-6)(\sqrt[3]{y}+3)$

18. _____

In Problems 19 – 25, perform the indicated operation and simplify. Assume all variables are positive or zero.

19. $(\sqrt{6}-2\sqrt{2})(2\sqrt{6}+3\sqrt{2})$ 　　　　**20.** $(\sqrt{7}-\sqrt{3})^2$

19. _____

20. _____

21. $(\sqrt{5}-\sqrt{3})^2 - \sqrt{60}$ 　　　　**22.** $(4+\sqrt{2x+3})^2$

21. _____

22. _____

23. $(\sqrt{3a}-\sqrt{4b})(\sqrt{3a}+\sqrt{4b})+4\sqrt{b^2}$ 　　　　**24.** $(\sqrt{2}-\sqrt{7})^2 - \sqrt{56}$

23. _____

24. _____

25. $\dfrac{4}{5}\cdot\left(-\dfrac{\sqrt{5}}{5}\right)+\left(-\dfrac{3}{5}\right)\cdot\left(-\dfrac{2\sqrt{5}}{5}\right)$

25. _____

Name:
Instructor:

Date:
Section:

Five-Minute Warm-Up 9.6
Rationalizing Radical Expressions

1. By what would you multiply 75 so that the product is a perfect square? There is more than one right answer so choose the smallest possible factor that yields a perfect square.

   1. _____

2. Simplify: $\sqrt{121a^2}$, $a > 0$

   2. _____

3. Multiply: $\dfrac{\sqrt{3}}{\sqrt{2}} \cdot \dfrac{\sqrt{6}}{\sqrt{2}}$

   3. _____

4. Multiply: $(2 + \sqrt{3})(2 - \sqrt{3})$

   4. _____

5. Multiply: $(2\sqrt{5} + \sqrt{3})(3\sqrt{5} - 2\sqrt{3})$

   5. _____

Sullivan/Struve/Mazzarella, *Elementary & Intermediate Algebra*, 3e
Copyright © 2014 Pearson Education, Inc.

375

Name:
Instructor:

Date:
Section:

Guided Practice 9.6
Rationalizing Radical Expressions

Objective 1: Rationalize a Denominator Containing One Term

1. In your own words, what does it mean to rationalize the denominator of a rational expression?

2. To rationalize a denominator containing a single square root, we multiply the numerator and denominator of the quotient by a square root so that the radicand in the denominator becomes _____.

3. Determine what to multiply each quotient by so that the denominator contains a radicand which is a perfect square. *(See textbook Example 1)*

   (a) $\dfrac{3}{\sqrt{3}}$ (b) $\dfrac{4}{\sqrt{20}}$ (c) $\dfrac{1}{\sqrt{8a}}$

   3a. _____

   3b. _____

   3c. _____

4. Determine what to multiply each quotient by so that the denominator contains a radicand which is a perfect power. *(See textbook Example 2)*

   (a) $\dfrac{6}{\sqrt[3]{5}}$ (b) $\sqrt[3]{\dfrac{2}{12}}$ (c) $\dfrac{4x}{\sqrt[4]{27x^2 y}}$

   4a. _____

   4b. _____

   4c. _____

Sullivan/Struve/Mazzarella, *Elementary & Intermediate Algebra*, 3e
Copyright © 2014 Pearson Education, Inc.

Guided Practice 9.6

Objective 2: Rationalize a Denominator Containing Two Terms

5. To rationalize a denominator containing two terms, we multiply both numerator and denominator by the _____ of the denominator.

6. Identify the conjugate of each expression. Then multiply the expression by its conjugate.
(See textbook Example 3)

(a) $3 + \sqrt{5}$ (b) $2\sqrt{3} - 5\sqrt{2}$

6a. _____

6b. _____

7. Determine what to multiply the quotient by to rationalize the denominator $\dfrac{\sqrt{8}}{\sqrt{2} - 3}$.

7. _____

(See textbook Example 3)

8. Rationalize the denominator: $\dfrac{12}{\sqrt{2} - \sqrt{5}}$

8. _____

Do the Math Exercises 9.6
Rationalizing Radical Expressions

In Problems 1 – 14, rationalize each denominator. Assume all variables are positive.

1. $\dfrac{2}{\sqrt{3}}$

2. $-\dfrac{3}{2\sqrt{3}}$

3. $\dfrac{5}{\sqrt{20}}$

4. $\dfrac{\sqrt{3}}{\sqrt{11}}$

5. $\sqrt{\dfrac{5}{z}}$

6. $\dfrac{\sqrt{32}}{\sqrt{a^5}}$

7. $-\sqrt[3]{\dfrac{4}{p}}$

8. $\sqrt[3]{\dfrac{-5}{72}}$

9. $\dfrac{8}{\sqrt[3]{36z^2}}$

10. $\dfrac{6}{\sqrt[4]{9b^2}}$

11. $\dfrac{6}{\sqrt{7}-2}$

12. $\dfrac{10}{\sqrt{10}+3}$

13. $\dfrac{\sqrt{3}}{\sqrt{15}-\sqrt{6}}$

14. $\dfrac{2\sqrt{3}+3}{\sqrt{12}-\sqrt{3}}$

1. _____
2. _____
3. _____
4. _____
5. _____
6. _____
7. _____
8. _____
9. _____
10. _____
11. _____
12. _____
13. _____
14. _____

Do the Math Exercises 9.6

In Problems 15 – 18, perform the indicated operation and simplify.

15. $\sqrt{5} - \dfrac{1}{\sqrt{5}}$

16. $\dfrac{\sqrt{5}}{2} + \dfrac{3}{\sqrt{5}}$

15. _____

16. _____

17. $\sqrt{\dfrac{2}{5}} + \sqrt{20} - \sqrt{45}$

18. $\sqrt{\dfrac{4}{3}} + \dfrac{4}{\sqrt{48}}$

17. _____

18. _____

In Problems 19 – 23, simplify each expression so that the denominator does not contain a radical.

19. $\dfrac{\sqrt{2}}{\sqrt{18}}$

20. $\sqrt{\dfrac{9}{5}}$

19. _____

20. _____

21. $\dfrac{\sqrt{2} - 5}{\sqrt{2} + 5}$

22. $\dfrac{5}{\sqrt{6} + 4}$

21. _____

22. _____

23. $\dfrac{\sqrt{75}}{\sqrt{3}}$

23. _____

In Problems 24 and 25, rationalize the numerator.

24. $\dfrac{\sqrt{3} + 2}{2}$

25. $\dfrac{\sqrt{a} - \sqrt{b}}{\sqrt{2}}$

24. _____

25. _____

Five-Minute Warm-Up 9.7
Functions Involving Radicals

In Problems 1 and 2, simplify each expression.

1. $\sqrt{144}$

2. $\sqrt{n^2}$, $n > 0$

1. _____

2. _____

3. Solve: $-3x - 9 \geq 0$

3. _____

4. Given $f(x) = -2x^2 + 16x$, find $f(-2)$.

4. _____

5. Graph $f(x) = x^2 + 2$ using point plotting.

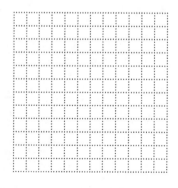

Name:
Instructor:

Date:
Section:

Guided Practice 9.7
Functions Involving Radicals

Objective 1: Evaluate Functions Involving Radicals

1. For the functions $f(x) = 2\sqrt{3x-1}$, $g(x) = \sqrt[3]{\dfrac{3x}{x+4}}$, and $h(x) = \sqrt{\dfrac{x+2}{x-2}}$, find each of the following.

   (See textbook Example 1)

   (a) $f(7)$ (b) $g(-3)$ (c) $h(6)$

   1a. _____

   1b. _____

   1c. _____

Objective 2: Find the Domain of a Function Involving a Radical

2. If the index on a radical is even, then the radicand must be _____.

3. If the index on a radical is odd, then the radicand can be _____.

4. Find the domain of each of the following functions. *(See textbook Example 2)*

   (a) $f(x) = \sqrt{2x-3}$ (b) $g(x) = \sqrt[3]{6x-9}$ (c) $h(t) = \sqrt[4]{14-7t}$

   4a. _____

   4b. _____

   4c. _____

Objective 3: Graph Functions Involving Square Roots

5. Given the function $f(x) = \sqrt{x+3}$, *(See textbook Example 3)*

   (a) find the domain.

   5a. _____

   (b) graph the function using point-plotting.

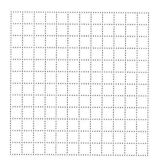

   (c) Based on the graph, determine the range.

   5c. _____

Sullivan/Struve/Mazzarella, *Elementary & Intermediate Algebra*, 3e
Copyright © 2014 Pearson Education, Inc.

Guided Practice 9.7

Objective 4: Graph Functions Involving Cube Roots

6. Given the function $g(x) = \sqrt[3]{x} - 3$, *(See textbook Example 4)*

   (a) find the domain. 6a. _____

   (b) graph the function using point-plotting.

   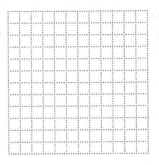

   (c) Based on the graph, determine the range. 6c. _____

Name:
Instructor:

Date:
Section:

Do the Math Exercises 9.7
Functions Involving Radicals

In Problems 1 – 3, evaluate each radical function at the indicated values.

1. $f(x) = \sqrt{x+10}$

   (a) $f(6)$

   (b) $f(2)$

   (c) $f(-6)$

2. $G(t) = \sqrt[3]{t-6}$

   (a) $G(7)$

   (b) $G(-21)$

   (c) $G(22)$

1a. _____

1b. _____

1c. _____

2a. _____

2b. _____

2c. _____

3. $f(x) = \sqrt{\dfrac{x-4}{x+4}}$

   (a) $f(5)$

   (b) $f(8)$

   (c) $f(12)$

3a. _____

3b. _____

3c. _____

In Problems 4 – 8, find the domain of the radical function.

4. $f(x) = \sqrt{x+4}$

5. $G(x) = \sqrt{5-2x}$

4. _____

5. _____

6. $G(z) = \sqrt[3]{5z-3}$

7. $C(y) = \sqrt[4]{3y-2}$

6. _____

7. _____

8. $f(x) = \sqrt{\dfrac{3}{x-3}}$

8. _____

Sullivan/Struve/Mazzarella, *Elementary & Intermediate Algebra*, 3e
Copyright © 2014 Pearson Education, Inc.

Do the Math Exercises 9.7

In Problems 9 – 12, (a) determine the domain of the function; (b) determine the range of the function; (c) graph the function using point-plotting.

9. $f(x) = \sqrt{x-1}$

10. $F(x) = \sqrt{4-x}$

9a. _____

9b. _____

10a. _____

10b. _____

11. $f(x) = \sqrt{x} + 1$

12. $g(x) = \sqrt[3]{x-4}$

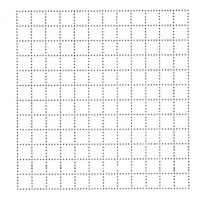
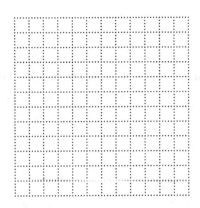

11a. _____

11b. _____

12a. _____

12b. _____

Five-Minute Warm-Up 9.8
Radical Equations and Their Applications

1. Solve: $4x + 12 = 0$

    1. _____

2. Solve: $-4x^2 + 12x - 8 = 0$

    2. _____

3. Simplify: $\left(\sqrt{2x}\right)^2$, $x > 0$

    3. _____

4. Multiply: $\left(\sqrt{2x} + 5\right)^2$; $x > 0$

    4. _____

5. Simplify: $\left[(3x+8)^{\frac{3}{2}}\right]^{\frac{2}{3}}$; $x > 0$

    5. _____

6. Evaluate: $\sqrt{2(-3) + (-3)(-18)}$

    6. _____

Sullivan/Struve/Mazzarella, *Elementary & Intermediate Algebra*, 3e
Copyright © 2014 Pearson Education, Inc.

Guided Practice 9.8
Radical Equations and Their Applications

Objective 1: Solve Radical Equations Containing One Radical

1. Solve: $\sqrt{4x+1} - 2 = 3$ *(See textbook Example 1)*

Step 1: Isolate the radical.

$\sqrt{4x+1} - 2 = 3$

Add 2 to both sides: (a) _____

Step 2: Raise both sides to the power of the index.

The index is 2, so we square both sides: (b) _____

Simplify: (c) _____

Step 3: Solve the equation that results.

(d) _____

Step 4: Check

Substitute the value you found in (d) into the original equation. Does this yield a true statement? (e) _____

State the solution set: (f) _____

2. What is meant by *extraneous solutions*?

3. When solving radical equations, what is the key first step?

4. Solve: $\sqrt{7x+2} + 5 = 2$ *(See textbook Example 2)*

(a) What value did you find for *x*? 4a. _____

(b) Does this satisfy the original equation? 4b. _____

(c) State the solution set. 4c. _____

(d) Without solving, how can you tell that this equation has no real solution? 4d. _____

Guided Practice 9.8

5. Solve: $\sqrt{x+5}+1=x$ *(See textbook Example 3)*

(a) What value(s) did you find for x? 5a. _____

(b) Does this value (or values) satisfy the original equation? 5b. _____

(c) State the solution set. 5c. _____

6. Solve: $(3x+1)^{\frac{3}{2}}-2=6$ *(See textbook Example 5)*

(a) Isolate the expression containing the rational exponent. 6a. _____

(b) To what power do we need to raise both sides in order to eliminate the exponent on the variable expression? 6b. _____

(c) Solve the resulting equation. What value(s) did you find for x? 6c. _____

(d) Check and then state the solution set. 6d. _____

Objective 2: Solve Radical Equations Containing Two Radicals

7. Solve: $\sqrt{2x^2-5x-20}=\sqrt{x^2-3x+15}$ *(See textbook Example 6)*

$$\sqrt{2x^2-5x-20}=\sqrt{x^2-3x+15}$$

Step 1: Isolate one of the radicals.	The radical on the left side of the equation is isolated:	(a)	_____
Step 2: Raise both sides to the power of the index.	The index is 2, so we square both sides:	(b)	_____
Step 3: Because there is no radical, we solve the equation that results.	Solve:	(c)	_____
Step 4: Check.	Substitute the value(s) you found in (b) into the original equation. Does this yield a true statement?	(d)	_____
	State the solution set:	(e)	_____

Name:
Instructor:

Date:
Section:

Do the Math Exercises 9.8
Radical Equations and Their Applications

In Problems 1 – 18, solve each equation.

1. $\sqrt{y-5} = 3$

2. $\sqrt{6p-5} = -5$

1. _____

2. _____

3. $\sqrt[3]{9w} = 3$

4. $\sqrt{q} - 5 = 2$

3. _____

4. _____

5. $\sqrt{x-4} + 4 = 7$

6. $4\sqrt{t} - 2 = 10$

5. _____

6. _____

7. $\sqrt{w} = 6 - w$

8. $\sqrt{1-4x} - 5 = x$

7. _____

8. _____

9. $\sqrt{3x+1} = \sqrt{2x+7}$

10. $\sqrt{2x^2 + 7x - 10} = \sqrt{x^2 + 4x + 8}$

9. _____

10. _____

11. $\sqrt{2x-1} - \sqrt{x-1} = 1$

12. $\sqrt{4x+1} - \sqrt{2x+1} = 2$

11. _____

12. _____

13. $(6p+3)^{1/5} = (4p-9)^{1/5}$

14. $(2x+3)^{1/2} = 3$

13. _____

14. _____

Sullivan/Struve/Mazzarella, *Elementary & Intermediate Algebra*, 3e
Copyright © 2014 Pearson Education, Inc.

Do the Math Exercises 9.8

15. $\sqrt{x+20} = x$

16. $\sqrt[5]{x+23} = 2$

17. $\sqrt{x-3} + \sqrt{x+4} = 7$

18. $\sqrt{a} - 5 = -2$

15. _____

16. _____

17. _____

18. _____

In Problems 19 – 21, solve for the indicated variable.

19. Solve $v = \sqrt{ar}$ for a.

20. Solve $r = \sqrt{\dfrac{S}{4\pi}}$ for S.

19. _____

20. _____

21. Solve $V = \sqrt{\dfrac{2U}{C}}$ for U.

21. _____

22. **Money** The annual rate of interest r (expressed as a decimal) required to have A dollars after t years from an initial deposit of P dollars can be calculated with the following formula:

$$r = \sqrt[t]{\dfrac{A}{P}} - 1$$

Suppose that you deposit $1,000 in an account that pays 5% annual interest so that $r = 0.05$. How much will you have after $t = 2$ years?

22. _____

Name:
Instructor:

Date:
Section:

Five-Minute Warm-Up 9.9
The Complex Number System

1. List the numbers in the set $\left\{8, \dfrac{-6}{3}, |-5|, 0.\overline{3}, \dfrac{15}{0}, -\dfrac{2}{5}, \pi, \dfrac{0}{-12}\right\}$ which are:

   (a) Natural numbers

   1a. _____

   (b) Whole numbers

   1b. _____

   (c) Integers

   1c. _____

   (d) Rational numbers

   1d. _____

   (e) Irrational numbers

   1e. _____

   (f) Real numbers

   1f. _____

In Problems 2 – 5, perform the indicated operations and simplify.

2. $(x^2 - 1) + (2x^2 - 1)$

   2. _____

3. $4p^5(-3p + 9)$

   3. _____

4. $(2x - 7)(-3x - 4)$

   4. _____

5. $(x^2 + 4)(x^2 - 4)$

   5. _____

Name: _____ Date: _____
Instructor: _____ Section: _____

Guided Practice 9.9
The Complex Number System

1. The *imaginary unit*, denoted by *i*, is the number whose square is −1. $i^2 = $ _____ or $i = $ _____.

2. *Complex numbers* are numbers of the form _____ where *a* and *b* are _____. When a number is in a form such as $6 - 2i$, we say that the number is in _____ form. The *real part* of the complex number is _____ and the *imaginary part* is _____.

Objective 1: Evaluate the Square Root of Negative Real Numbers

3. Write $\sqrt{-81}$ as a pure imaginary number. *(See textbook Example 1)*

    3. _____

4. Write $6 + \sqrt{-4}$ in standard form. *(See textbook Example 2)*

    4. _____

Objective 2: Add or Subtract Complex Numbers

5. Before beginning any operations with complex numbers, you must write the number in _____ form.

6. In your own words, explain how to add complex numbers _____
_____

Objective 3: Multiply Complex Numbers

7. Use the Distributive Property to multiply $\frac{3}{2}i(6 - 8i)$. *(See textbook Example 5)*

    7. _____

8. Multiply $\sqrt{-4} \cdot \sqrt{-9}$. *(See textbook Example 6)*
   (a) Explain why the Product Property of Radicals cannot be used to multiply these radicals.

    (a)_____

   (b) Express the radicals as pure imaginary numbers.

    8b. _____

   (c) Multiply.

    8c. _____

Sullivan/Struve/Mazzarella, *Elementary & Intermediate Algebra*, 3e
Copyright © 2014 Pearson Education, Inc.

Guided Practice 9.9

Objective 4: Divide Complex Numbers

9. Divide: $\dfrac{3+6i}{4+4i}$. *(See textbook Example 8)*

Step 1: Write the numerator and denominator in standard form, $a + bi$.	The numerator and denominator are already in standard form.	
Step 2: Multiply the numerator and denominator by the complex conjugate of the denominator.	Identify the conjugate of the denominator:	(a) _____
	Multiply the quotient by 1, written with the conjugate:	(b) _____
Step 3: Simplify by writing the quotient in standard form, $a + bi$.	Multiply the numerator; the denominator is of the form $(a+bi)(a-bi) = a^2 + b^2$:	(c) _____
	Combine like terms; $i^2 = -1$:	(d) _____
	Divide the denominator into each term of the numerator to write in standard form:	(e) _____
	Write each fraction in lowest terms:	(f) _____

Objective 5: Evaluate the Powers of i

10. The powers of *i* are a cyclic function, meaning that the values cycle through a set list. Complete the table to see the only values for the powers of *i*.

$i^1 = i$

(a) $i^2 = $ _____

(b) $i^3 = $ _____

(c) $i^4 = $ _____

(d) $i^5 = $ _____

(e) $i^6 = $ _____

(f) $i^7 = $ _____

(g) $i^8 = $ _____

Do the Math Exercises 9.9
The Complex Number System

In Problems 1 and 2, write each expression as a pure imaginary number.

1. $-\sqrt{-100}$

2. $\sqrt{-162}$

1. _____

2. _____

In Problems 3 – 5, write each expression as a complex number in standard form.

3. $10 + \sqrt{-32}$

4. $\dfrac{10 - \sqrt{-25}}{5}$

3. _____

4. _____

5. $\dfrac{15 - \sqrt{-50}}{5}$

5. _____

In Problems 6 – 9, add or subtract as indicated.

6. $(-6 + 2i) + (3 + 12i)$

7. $(-7 + 3i) - (-3 + 2i)$

6. _____

7. _____

8. $\left(-4 + \sqrt{-25}\right) + \left(1 - \sqrt{-16}\right)$

9. $\left(-10 + \sqrt{-20}\right) - \left(-6 + \sqrt{-45}\right)$

8. _____

9. _____

In Problems 10 – 17, multiply.

10. $3i(-2 - 6i)$

11. $(3 - i)(1 + 2i)$

10. _____

11. _____

12. $(2 + 8i)(-3 - i)$

13. $\left(-\dfrac{2}{3} + \dfrac{4}{3}i\right)\left(\dfrac{1}{2} - \dfrac{3}{2}i\right)$

12. _____

13. _____

Do the Math Exercises 9.9

14. $(2+5i)^2$

15. $(2-7i)^2$

14. _____

15. _____

16. $\sqrt{-36} \cdot \sqrt{-4}$

17. $\left(1-\sqrt{-64}\right)\left(-2+\sqrt{-49}\right)$

16. _____

17. _____

In Problems 18 – 21, divide.

18. $\dfrac{2-i}{2i}$

19. $\dfrac{2}{4+i}$

18. _____

19. _____

20. $\dfrac{-4}{-5-3i}$

21. $\dfrac{2+5i}{5-2i}$

20. _____

21. _____

In Problems 22 – 24, simplify.

22. i^{72}

23. i^{110}

22. _____

23. _____

24. i^{131}

24. _____

Five-Minute Warm-Up 10.1
Solving Quadratic Equations by Completing the Square

1. Multiply: $(3x-1)^2$

2. Factor: $x^2 - 4x + 4$

In Problems 3 and 4, solve each polynomial equation.

3. $x^2 - \dfrac{2}{3}x + \dfrac{1}{9} = 0$

4. $25n^2 - 49 = 0$

In Problems 5 and 6, simplify each expression.

5. $\sqrt{\dfrac{81}{25}}$

6. $\sqrt{(8x-3)^2}$

In Problems 7 and 8, simplify each expression using complex numbers.

7. $\sqrt{-12}$

8. $\dfrac{8 + \sqrt{-4}}{4}$

9. Find the complex conjugate of $-15 + 7i$.

Name: Date:
Instructor: Section:

Guided Practice 10.1
Solving Quadratic Equations by Completing the Square

Objective 1: Solve Quadratic Equations Using the Square Root Property

1. State the Square Root Property: If $x^2 = p$, where x is any variable expression and p is a real number,

then _____.

2. If the solution to a quadratic equation is $n = -1 \pm 2\sqrt{3}$, write the solution set. _____

3. *True or False* $\sqrt{x^2 - 16} = \sqrt{81}$ simplifies to $x - 4 = \pm 9$. _____

4. *True or False* $y^2 = -9$ has no solution. _____

5. Solve: $n^2 - 144 = 0$ *(See textbook Example 1)*

Step 1: Isolate the expression containing the square term. $n^2 - 144 = 0$

Add 144 to both sides: (a) _____

Step 2: Use the Square Root Property. Don't forget the $\pm$ symbol.

Take the square root of both sides of the equation: (b) _____

Simplify the radical: (c) _____

Step 3: Isolate the variable, if necessary.

The variable is already isolated.

Step 4: Verify your solution(s).

State the solution set: (d) _____

6. There is no reason that the solution to a quadratic equation must be real. List the possible nature of the solution(s) of a quadratic equation.

(a) Real, which includes: _____, _____, _____ and

(b) _____ in form _____

Objective 2: Complete the Square in One Variable

7. If a polynomial is of the form $x^2 + bx + c$, c must be equal to _____ in order to be a perfect square trinomial.

Guided Practice 10.1

8. Determine the number that must be added to the expression to make it a perfect square trinomial. Then factor the expression. *(See textbook Example 5)*

(a) $p^2 - 14p$ (b) $n^2 + 9n$ (c) $z^2 + \dfrac{4}{3}z$

8a. _____

8b. _____

8c. _____

Objective 3: Solve Quadratic Equations by Completing the Square

9. Solve: $p^2 - 6p - 18 = 0$ *(See textbook Example 6)*

Step 1: Rewrite $x^2 + bx + c = 0$ as $x^2 + bx = -c$ by adding or subtracting the constant from both sides of the equation.

$p^2 - 6p - 18 = 0$

Add 18 to both sides: (a) _____

Step 2: Complete the square in the expression $x^2 + bx$ by making it a perfect square trinomial.

What value must be added to both sides to make the expression on the left a perfect square trinomial? (b) _____

Add this number to both sides of the equation and simplify: (c) _____

Step 3: Factor the perfect square trinomial on the left side of the equation.

Use: $A^2 - 2AB + B^2 = (A-B)^2$ (d) _____

Step 4: Solve the equation using the Square Root Property.

Take the square root of both sides of the equation: (e) _____

Simplify the square root: (f) _____

Add 3 to both sides: (g) _____

Step 5: Verify your solutions(s).

State the solution set: (h) _____

10. To solve the equation $3x^2 - 9x + 12 = 0$ by the completing the square, the first step is_____.

Objective 4: Solve Problems Using the Pythagorean Theorem

11. State the Pythagorean Theorem in words. If x and y are the lengths of the legs and z is the length of the hypotenuse, write an equation which uses these variables to state the Pythagorean Theorem.

Name:
Instructor:

Date:
Section:

Do the Math Exercises 10.1
Solving Quadratic Equations by Completing the Square

In Problems 1 – 6, solve each equation using the Square Root Property.

1. $z^2 = 48$

2. $w^2 - 6 = 14$

3. $(y-2)^2 = 9$

4. $(2p+3)^2 = 16$

5. $\left(y + \dfrac{3}{2}\right)^2 = \dfrac{3}{4}$

6. $q^2 - 6q + 9 = 16$

1. _____

2. _____

3. _____

4. _____

5. _____

6. _____

In Problems 7 – 9, find the value to complete the square in each expression. Then factor the perfect square trinomial.

7. $p^2 - 4p$

8. $z^2 - \dfrac{1}{3}z$

9. $m^2 + \dfrac{5}{2}m$

7. _____
   _____

8. _____
   _____

9. _____
   _____

In Problems 10 – 15, solve each quadratic equation by completing the square.

10. $y^2 + 3y - 18 = 0$

11. $q^2 + 7q + 7 = 0$

12. $x^2 - 5x - 3 = 0$

13. $n^2 = 10n + 5$

14. $3a^2 - 4a - 4 = 0$

15. $2z^2 + 6z + 5 = 0$

10. _____

11. _____

12. _____

13. _____

14. _____

15. _____

Sullivan/Struve/Mazzarella, *Elementary & Intermediate Algebra*, 3e
Copyright © 2014 Pearson Education, Inc.

Do the Math Exercises 10.1

In Problems 16 and 17, the lengths of the legs of a right triangle are given. Find the length of the hypotenuse. Give the exact answers and decimal approximations rounded to two decimal places.

16. $a = 7, b = 24$ 17. $a = 2, b = \sqrt{5}$

16. _____

17. _____

18. **Right Triangle** A right triangle has a leg of length 2 and hypotenuse of length 10. Find the length of the missing leg. Give the exact answer and a decimal approximation, rounded to 2 decimal places.

18. _____

19. Given that $h(x) = (x+1)^2$, find all values of x such that $h(x) = 32$.

19. _____

20. **Fire Truck Ladder** A fire truck has a 75-foot ladder. If the truck can safely park 20 feet from a building, how far up the building can the ladder reach assuming that the top of the base of the ladder is resting on top of the truck and the truck is 10 feet tall? Give the decimal approximation, rounded to 3 decimal places.

20. _____

21. **The converse of the Pythagorean Theorem is also true.** That is, in a triangle, if the square of the length of one side equals the sum of the squares of the lengths of the other two sides, then the triangle is a right triangle. The 90° angle is opposite the longest side.

The lengths of the sides of a triangle are: 20, 48, and 52. Determine whether the triangle is a right triangle. If it is, identify the hypotenuse.

21. _____

Five-Minute Warm-Up 10.2
Solving Quadratic Equations by the Quadratic Formula

In Problems 1 and 2, simplify each expression.

1. $\sqrt{125}$

2. $\dfrac{15 + \sqrt{72}}{6}$

In Problems 3 and 4, simplify each expression using complex numbers.

3. $\sqrt{-28}$

4. $\dfrac{6 - 2\sqrt{-4}}{6}$

5. Divide: $\dfrac{9x^2 - 27x + 3}{3}$

6. Evaluate the expression $\sqrt{b^2 - 4ac}$ if $a = 2$, $b = -4$, $c = -3$.

1. _____

2. _____

3. _____

4. _____

5. _____

6. _____

Name:
Instructor:

Date:
Section:

Guided Practice 10.2
Solving Quadratic Equations by the Quadratic Formula

Objective 1: Solve Quadratic Equations Using the Quadratic Formula

1. If $ax^2 + bx + c = 0$, then $x = $ _____.

2. When using the quadratic formula, the first step is to write the quadratic equation in _____.

3. Write the quadratic equation in standard form and then identify the values assigned to a, b, and c.
   Do not solve the equation.
   (a) $x - 2x^2 = -4$ (b) $2 - 3x^2 = 8$ (c) $3x^2 = 6x$

   3a. _____

   3b. _____

   3c. _____

4. Solve: $8n^2 - 2n - 3 = 0$ *(See textbook Example 1)*

Step 1: Write the equation in standard form $ax^2 + bx + c = 0$ and identify the values of a, b, and c.

Since the equation is already in standard form, identify the values for a, b, and c.

$8n^2 - 2n - 3 = 0$

(a) $a = $ _____; $b = $ _____; $c = $ _____

Step 2: Substitute the values of a, b, and c into the quadratic formula.

Write the quadratic formula: (b) _____

Substitute the values for a, b, and c. (c) _____

Step 3: Simplify the expression found in Step 2.

What is the value of the radicand? (d) _____

Simplify the expression in (c): (e) _____

Write the two expressions using $a \pm b$ means $a - b$ or $a + b$: (f) _____

Step 4: Check

State the solution set: (g) _____

Sullivan/Struve/Mazzarella, *Elementary & Intermediate Algebra*, 3e
Copyright © 2014 Pearson Education, Inc.

Guided Practice 10.2

Objective 2: Use the Discriminant to Determine the Nature of Solutions of a Quadratic Equation

5. In the quadratic equation $ax^2 + bx + c = 0$, the *discriminant* is used to determine the nature and number of solutions. To find the discriminant, substitute the identified values for a, b, and c into part of the quadratic formula, _____.

6. Determine the discriminant of each quadratic equation. Use the value of the discriminant to determine whether the quadratic equation has two rational solutions, two irrational solutions, one repeated real solution, or two complex solutions that are not real. *(See textbook Example 5)*

(a) $4x^2 + 5x - 9 = 0$ (b) $x^2 + 4x + 9 = 0$ (c) $4x^2 = 4x - 1$

6a. _____

_____

6b. _____

_____

6c. _____

_____

Objective 3: Model and Solve Problems Involving Quadratic Equations

7. **Projectile Motion** The height s of a toy rocket after t seconds, when fired straight up with an initial speed of 150 feet per second from an initial height of 2 feet, can be modeled by the function
$$s(t) = -16t^2 + 150t + 2$$
When will the height of the rocket be 200 feet? Round your answer to the nearest tenth of a second. *(See textbook Examples 6 and 7)*

(a) **Step 1: Identify** Here we want to know the value t when $s = $? _____

   Step 2: Name The variables are named in the problem. t is the time the rocket has traveled and s is height of the rocket.

(b) **Step 3: Translate** Use the information from (a) and the given formula to write a model for this problem.

_____

(c) **Step 4: Solve** Solve the equation in (b). What values did you find for t? _____

   Step 5: Check

(d) **Step 6: Answer the Question** _____

(e) Will the rocket ever reach a height of 500 feet? _____

(f) When will the rocket hit the ground? _____

Do the Math Exercises 10.2
Solving Quadratic Equations by the Quadratic Formula

In Problems 1 – 6, solve each equation using the quadratic formula.

1. $p^2 - 4p - 32 = 0$

2. $10x^2 + x - 2 = 0$

3. $2q^2 - 4q + 1 = 0$

4. $x + \dfrac{1}{x} = 3$

5. $2z^2 + 7 = 4z$

6. $1 = 5w^2 + 6w$

1. _____
2. _____
3. _____
4. _____
5. _____
6. _____

In Problems 7 – 9, determine the discriminant of each quadratic equation. Use the value of the discriminant to determine whether the quadratic equation has two rational solutions, two irrational solutions, one repeated real solution, or two complex solutions that are not real.

7. $p^2 + 4p - 2 = 0$

8. $16x^2 + 24x + 9 = 0$

9. $6x^2 - x = -4$

7. _____
8. _____
9. _____

In Problems 10 – 15, solve each equation using any method you wish.

10. $q^2 - 7q + 7 = 0$

11. $3x^2 + 5x = 2$

10. _____
11. _____

Sullivan/Struve/Mazzarella, *Elementary & Intermediate Algebra*, 3e
Copyright © 2014 Pearson Education, Inc.

Do the Math Exercises 10.2

12. $5m - 4 = \dfrac{5}{m}$

13. $8p^2 - 40p + 50 = 0$

12. _____

13. _____

14. $(a-3)(a+1) = 2$

15. $\dfrac{x-1}{x^2+4} = 1$

14. _____

15. _____

In Problem 16, suppose that $f(x) = x^2 + 2x - 8$.

16a. Solve for x, if $f(x) = 0$

16b. Solve for x, if $f(x) = -8$

16a. _____

16b. _____

17. **Area** The area of a rectangle is 60 square inches. The length of the rectangle is 6 inches more than the width. What are the dimensions of the rectangle?

17. _____

18. **Area** The area of a triangle is 35 square inches. The height of the rectangle is 2 inches less than the base. What are the base and height of the triangle?

18. _____

19. **Roundtrip** A Cessna aircraft flies 200 miles due west into the jet stream and flies back home on the same route. The total time of the trip (excluding the time on the ground) takes 4 hours. The Cessna aircraft can fly 120 miles per hour in still air. What is the net effect of the jet stream on the aircraft?

19. _____

20. Given the quadratic equation $ax^2 + bx + c = 0$, show that the sum of the solutions to any quadratic equation in this form is $-\dfrac{b}{a}$. Show that the product of the solutions to this quadratic equation is $\dfrac{c}{a}$.

Five-Minute Warm-Up 10.3
Solving Equations Quadratic in Form

In Problems 1 and 2, factor completely.

1. $a^4 - 7a^2 - 18$

2. $6(x-1)^2 - 13(x-1) + 6$

1. _____

2. _____

In Problems 3 and 4, simplify each expression.

3. $\left(5x^{-1}\right)^2$

4. $\left(-\dfrac{2}{3}x^3\right)^2$

3. _____

4. _____

5. Solve: $2x^2 + x - 6 = 0$

5. _____

Name: Date:
Instructor: Section:

Guided Practice 10.3
Solving Equations Quadratic in Form

Objective 1: Solve Equations That Are Quadratic in Form

1. In general, if a substitution u transforms an equation into one of the form $au^2 + bu + c = 0$, then the original equation is called an **equation quadratic in form.** Here u represents any variable expression and we solve the equation by substituting u for the variable expression written with the coefficient b. For these equations quadratic in form, identify the substitution for u and then write a new equation using your substitution.

(a) $2y - 11\sqrt{y} + 15 = 0$ $u = $ _____ _____

(b) $v^4 + 10v^2 + 1 = 0$ $u = $ _____ _____

(c) $(x-1)^2 - 5(x-1) + 6 = 0$ $u = $ _____ _____

(d) $x^{\frac{2}{3}} - x^{\frac{1}{3}} - 3 = 0$ $u = $ _____ _____

(e) $4x^{-2} + x^{-1} - 3 = 0$ $u = $ _____ _____

2. Solve: $x^4 - 6x^2 - 16 = 0$ *(See textbook Example 1)*

Step 1: Determine the appropriate substitution and write the equation in the form $au^2 + bu + c = 0$.

Here we let $u = ?$ (a) _____

$x^4 - 6x^2 - 16 = 0$

Rewrite the equation using your substitution from (a): (b) _____

Step 2: Solve the equation $au^2 + bu + c = 0$.

Factor: (c) _____

Set each factor to 0 and solve:

(d) _____

Step 3: Solve for the variable in the original equation using the value of u from (a).

Substitute from (a) into your solution from (d): (e) _____

Solve the equations. In this case we use the Square Root Property: (f) _____

Simplify the radicals: (g) _____

Step 4: Verify your solution(s).

State the solution set: (h) _____

Sullivan/Struve/Mazzarella, *Elementary & Intermediate Algebra*, 3e

Guided Practice 10.3

3. Solve: $(x^2 + 3)^2 - 6(x^2 + 3) + 8 = 0$ (See textbook Example 2)

(a) Here we let $u =$ _____.

(b) In this problem, $u = 4$ or $u = 2$. Replace u and solve for x. State the solution set. _____

4. Solve: $2x + \sqrt{x} - 10 = 0$ (See textbook Example 3)

(a) Here we let $u =$ _____.

(b) In this problem, $u = -\dfrac{5}{2}$ or $u = 2$. Replace u and solve for x. State the solution set. _____

5. Solve: $n^{\frac{2}{3}} - 2n^{\frac{1}{3}} - 3 = 0$ (See textbook Example 5)

(a) Here we let $u =$ _____.

(b) In this problem, $u = 3$ or $u = -1$. Replace u and solve for x. State the solution set. _____

6. As always, it is important to check for extraneous solutions. List two cases when extraneous solutions might appear.

(a) _____ **(b)** _____

Do the Math Exercises 10.3
Solving Equations Quadratic in Form

In Problems 1 – 14, solve each equation.

1. $x^4 - 10x^2 + 9 = 0$

2. $4b^4 - 5b^2 + 1 = 0$

3. $(x + 2)^2 - 3(x + 2) - 10 = 0$

4. $x - 5\sqrt{x} - 6 = 0$

5. $z + 7\sqrt{z} + 6 = 0$

6. $q^{-2} + 2q^{-1} = 15$

7. $10a^{-2} + 23a^{-1} = 5$

8. $y^{\frac{2}{3}} - 2y^{\frac{1}{3}} - 3 = 0$

9. $\dfrac{1}{x^2} - \dfrac{7}{x} + 12 = 0$

10. $y^6 - 7y^3 - 8 = 0$

1. _____
2. _____
3. _____
4. _____
5. _____
6. _____
7. _____
8. _____
9. _____
10. _____

Do the Math Exercises 10.3

11. $6b^{-2} - b^{-1} = 1$

12. $x^4 + 3x^2 = 4$

11. _____

12. _____

13. $c^{\frac{1}{2}} + c^{\frac{1}{4}} - 12 = 0$

14. $\left(\dfrac{1}{x-1}\right)^2 + \dfrac{7}{x-1} = 8$

13. _____

14. _____

In Problems 15 and 16, suppose that $f(x) = x^4 + 5x^2 + 3$. Find the values of x such that

15. $f(x) = 3$

16. $f(x) = 17$

15. _____

16. _____

In Problems 17 and 18, suppose that $h(x) = 3x^4 - 9x^2 - 8$. Find the values of x such that

17. $h(x) = -8$

18. $h(x) = 22$

17. _____

18. _____

In Problems 19 and 20, find the zeros of the function.

19. $f(x) = x^4 - 13x^2 + 42$

20. $h(p) = 8p - 18\sqrt{p} - 35$

19. _____

20. _____

Five-Minute Warm-Up 10.4
Graphing Quadratic Functions Using Transformations

In Problems 1 and 2, graph using the point-plotting method.

1. $y = -x^2$

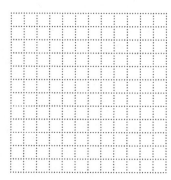

2. $y = x^2 + 2$

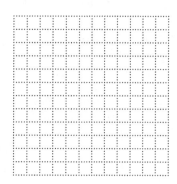

In Problems 3 and 4, find the function value.

3. $f(x) = -x^2 + 6x - 2;\ f(-3)$

4. $f(x) = 2x^2 + x + 5;\ f(-4)$

3._____

4._____

5. What is the domain of $f(x) = \dfrac{1}{2}x^2 + \dfrac{4}{3}x - 6$?

5._____

Name:
Instructor:
Date:
Section:

Guided Practice 10.4
Graphing Quadratic Functions Using Transformations

Objective 1: Graph Quadratic Functions of the Form $f(x) = x^2 + k$

In Problems 1 and 2, match the function to either graph (a) or graph (b).

1. $f(x) = x^2$ **(a)**   **(b)**   1. _____

2. $f(x) = -x^2$ 2. _____

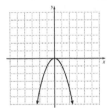 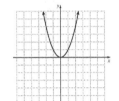

3. To obtain the graph of $f(x) = x^2 - 2$, shift the graph of $f(x) = x^2$

(a) horizontally or vertically? *(See textbook Examples 1 and 2)*   3a. _____

(b) How many units will the graph shift?   3b. _____

(c) Does the graph shift right, left, up, or down?   3c. _____

Objective 2: Graph Quadratic Functions of the Form $f(x) = (x - h)^2$

4. To obtain the graph of $f(x) = -(x - 1)^2$, shift the graph of $f(x) = -x^2$

(a) horizontally or vertically? *(See textbook Examples 3 and 4)*   4a. _____

(b) How many units will the graph shift?   4b. _____

(c) Does the graph shift right, left, up, or down?   4c. _____

Objective 3: Graph Quadratic Functions of the Form $f(x) = ax^2$

5. The graph of $f(x) = ax^2 + bx + c$ is a parabola that opens either up or down. This is determined by the coefficient a. *(See textbook Example 6)*
(a) If $a > 0$, the parabola opens   5a. _____

(b) If $a < 0$, the parabola opens   5b. _____

6. The value of a will also determine the breadth of the parabola.

(a) If $|a|$ _____ we say the graph is vertically stretched (taller, thinner, steeper).

(b) If _____ $|a|$ _____ we say the graph is vertically compressed (shorter, fatter, flatter).

Sullivan/Struve/Mazzarella, *Elementary & Intermediate Algebra*, 3e
Copyright © 2014 Pearson Education, Inc.

Guided Practice 10.4

Objective 4: Graph Quadratic Functions of the Form $f(x) = ax^2 + bx + c$

7. Graph $f(x) = 3x^2 - 12x + 7$. *(See textbook Examples 7 and 8)*

Step 1: Write the function $f(x) = ax^2 + bx + c$ as $f(x) = a(x - h)^2 + k$ by completing the square.

$f(x) = 3x^2 - 12x + 7$

Group the terms involving x: (a) _____

Factor out the coefficient of the square term, 3, from the parentheses: (b) _____

Identify the number required to complete the square: (c) _____

When this added to the right side of the equation, what must be done to maintain the equality? (d) _____

Write the amended equation: (e) _____

Factor the perfect square trinomial: (f) _____

Step 2: Graph the function $f(x) = a(x - h)^2 + k$ using transformations.

Identify the vertex: (g) _____

Opening direction: (h) _____

Axis of symmetry: (i) _____

Do the Math Exercises 10.4
Graphing Quadratic Functions Using Transformations

In Problems 1 – 6, use the graph of $y = x^2$ to graph the quadratic function.

1. $f(x) = x^2 - 1$

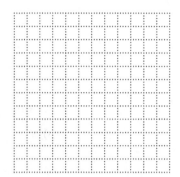

2. $f(x) = (x + 4)^2$

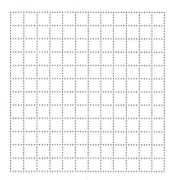

3. $G(x) = 5x^2$

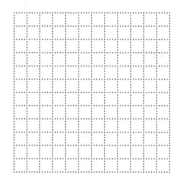

4. $P(x) = -3x^2$

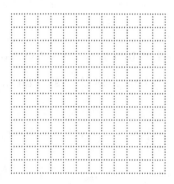

5. $g(x) = (x + 2)^2 - 1$

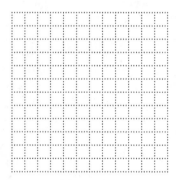

6. $G(x) = (x - 4)^2 + 2$

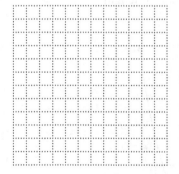

Do the Math Exercises 10.4

In Problems 7 – 10, write each function in the form $f(x) = a(x-h)^2 + k$. Then determine the vertex and the axis of symmetry.

7. $h(x) = x^2 - 7x + 10$

8. $f(x) = 3x^2 + 18x + 25$

7. _____

8. _____

9. $g(x) = -x^2 - 8x - 14$

10. $h(x) = -4x^2 + 4x$

9. _____

10. _____

11. Write a quadratic function in the form $f(x) = a(x-h)^2 + k$ with the properties: opens up; vertically compressed by a factor of $\dfrac{1}{2}$; vertex at $(-5, 0)$.

11. _____

12. Determine the quadratic function whose graph is shown below. Each tick mark represents one unit of length.

12. _____

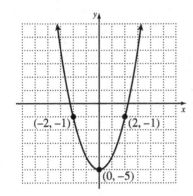

Sullivan/Struve/Mazzarella, *Elementary & Intermediate Algebra*, 3e
Copyright © 2014 Pearson Education, Inc.

Five-Minute Warm-Up 10.5
Graphing Quadratic Functions Using Properties

1. Find the intercepts of the graph of $3x - 4y = -12$.

    1. _____

       _____

2. Solve: $2x^2 - 13x - 24 = 0$

    2. _____

3. Find the zeros of $f(x) = -2x^2 + 10x + 12$

    3. _____

4. If $f(x) = -x^2 - 3x - 2$, find $f(-5)$.

    4. _____

Name:
Instructor:

Date:
Section:

Guided Practice 10.5
Graphing Quadratic Functions Using Properties

Objective 1: Graph Quadratic Functions of the Form $f(x) = ax^2 + bx + c$

1. The vertex is the turning point of the parabola. If $f(x) = ax^2 + bx + c$, the x-coordinate of the vertex is at

   $x = $ _____.

2. The x-intercepts of the parabola can be found by _____.

3. Graph $f(x) = x^2 + 2x - 8$ using its properties. *(See textbook Example 1)*

Step 1: Determine whether the parabola opens up or down.

Determine the values using $f(x) = ax^2 + bx + c$:
(a) $a = $ _____; $b = $ _____; $c = $ _____

Does the parabola open up or down?
(b) _____

Step 2: Determine the vertex and axis of symmetry.

Calculate the x-coordinate of the vertex $\left(x = -\dfrac{b}{2a}\right)$:
(c) _____

Calculate the y-coordinate of the vertex:
(d) _____

Identify the vertex:
(e) _____

Identify the axis of symmetry:
(f) _____

Step 3: Determine the y-intercept.

Evaluate $f(0)$:
(g) _____

Step 4: Find the discriminant, $b^2 - 4ac$, to determine the number of the x-intercepts. Then determine the x-intercepts, if any.

Evaluate the discriminant:
(h) _____

Number of x-intercepts:
(i) _____

Factor $0 = x^2 + 2x - 8$ and use the Zero-Product Property to identify the x-intercepts:
(j) _____

Sullivan/Struve/Mazzarella, *Elementary & Intermediate Algebra*, 3e

Guided Practice 10.5

Step 5: Plot the vertex, y-intercept, and x-intercepts. Use the axis of symmetry to find an additional point. Draw the graph of the quadratic function.

Graph: (k)

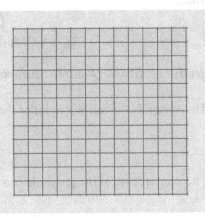

Objective 2: Find the Maximum or Minimum Value of a Quadratic Function

4. The maximum or minimum value of a quadratic function is the _____ of the vertex.

In Problems 5 and 6, determine whether the function has a maximum or a minimum value.
(See textbook Example 5)

5. $f(x) = -2x^2 + 4x + 3$ 6. $f(x) = 3x^2 + 12x - 1$

5. _____

6. _____

Objective 3: Model and Solve Optimization Problems Involving Quadratic Functions

7. Martin is making a dog run in which he will keep his show dogs and decides to enclose an area of his backyard. He uses one side of his house for the pen and encloses the other three sides with fencing. If he has 50 feet of fencing he plans to use, what are the dimensions of the pen that encloses the most area?
(See textbook Example 7)

  Step 1: Identify We wish to determine the dimensions of the rectangle that maximize the area.

  Step 2: Name We let x represent one side of the rectangle and y represent the other side.

  Step 3: Translate Since we are looking for the maximum area, we use the formula $A = lw$. Our goal is to have A written in terms of one variable. In this case we will choose x, but y will work as well.

(a) Our rectangle has sides x and y. Write an equation for the area using x and y. _____

(b) There is 50 feet of fencing for three sides. Write this equation: _____

(c) Solve for y: _____

(d) Substitute this expression into your equation from (a): _____

(e) **Step 4: Solve** _____

(f) Substitute your value for x into (b) to find the value for y: _____
    Step 5: Check

(g) **Step 6: Answer the Question** _____

(h) What is the maximum area Martin can enclose? _____

Do the Math Exercises 10.5
Graphing Quadratic Functions Using Properties

In Problems 1 – 3, use the discriminant to determine the number of x-intercepts the graph of each function will have. Then determine the x-intercepts.

1. $g(x) = 2x^2 - 7x - 4$ 2. $f(x) = x^2 - 6x + 9$

1. _____

2. _____

3. $P(x) = -2x^2 + 3x + 1$

3. _____

In Problems 4 – 9, graph each quadratic function.

4. $f(x) = x^2 - 2x - 8$ 5. $g(x) = -x^2 + 2x + 15$

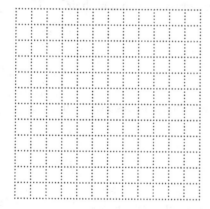 

6. $f(x) = x^2 - 4x + 7$ 7. $P(x) = -x^2 - 12x - 36$

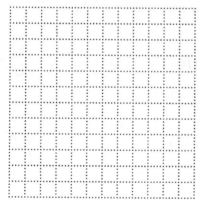 

Do the Math Exercises 10.5

8. $g(x) = (x+2)^2 - 1$

9. $G(x) = (x-4)^2 + 2$

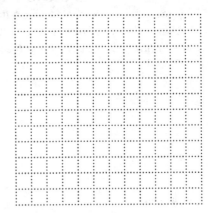

In Problems 10 and 11, determine whether the quadratic function has a maximum or minimum. Then find the maximum or minimum value.

10. $H(x) = -3x^2 + 12x - 1$

11. $G(x) = 5x^2 + 10x - 1$

10. _____

11. _____

12. **Fun with Numbers** The sum of two numbers is 50. Find the numbers such that their product is a maximum.

12. _____

13. **Fun with Numbers** The difference of two numbers is 10. Find the numbers such that their product is a minimum.

13. _____

14. **Punkin Chunkin** Suppose that catapult in the Punkin Chunkin contest releases a pumpkin 8 feet above the ground at an angle of 45° to the horizontal with an initial speed 220 feet per second. The model $s(t) = -16t^2 + 155t + 8$ can be used to estimate the height s of an object after t seconds.

   (a) Determine the time at which the pumpkin is at a maximum height.

14a. _____

   (b) Determine the maximum height of the pumpkin.

14b. _____

   (c) After how long will the pumpkin strike the ground?

14c. _____

Name:
Instructor:

Date:
Section:

Five-Minute Warm-Up 10.6
Polynomial Inequalities

In Problems 1 – 4, write in interval notation.

1. $-4 \leq x < -2$

2. $-3 < x \leq 1$

1. _____

2. _____

3. $x > -2$

4. $x \leq 5$

3. _____

4. _____

5. Solve: $6x + 3 \leq 11x - 7$

5. _____

6. Solve: $(x-3)(x+1) = 2$

6. _____

Name:
Instructor:

Date:
Section:

Guided Practice 10.6
Polynomial Inequalities

Objective 1: Solve Quadratic Inequalities

1. A quadratic inequality is an inequality written in one of the following forms:
   $ax^2 + bx + c > 0$; $ax^2 + bx + c \geq 0$; $ax^2 + bx + c < 0$; $ax^2 + bx + c \leq 0$.

   (a) If $f(x)$ is a quadratic function and $f(x) > 0$, we are interested in finding x values for which the function is _____ (above or below) the x-axis.

   1a. _____

   (b) If $f(x)$ is a quadratic function and $f(x) < 0$, we are interested in finding x values for which the function is _____ (above or below) the x-axis.

   1b. _____

2. We will present two methods for solving quadratic inequalities. These methods are:

   2a. _____

   2b. _____

3. Solve $x^2 + 2x - 3 \leq 0$ using the graphical method. *(See textbook Example 1)*

   Step 1: Write the inequality so that $ax^2 + bx + c$ is on one side of the inequality and 0 is on the other.

   The inequality is already in this form: $x^2 + 2x - 3 \leq 0$

   Step 2: Graph the function $f(x) = ax^2 + bx + c$. Be sure to label the x-intercepts on the graph.

   Function to be graphed: **(a)** _____

   x-intercepts: **(b)** _____

   Vertex: **(c)** _____

   (d) Graph:

   Step 3: From the graph, determine where the function is positive and where the function is negative. Use the graph to determine the solution set to the inequality.

   Are we looking for positive or negative function values in this problem? **(e)** _____

   State the solution set: **(f)** _____

Sullivan/Struve/Mazzarella, *Elementary & Intermediate Algebra*, 3e

Guided Practice 10.6

4a. If $f(x)$ is a quadratic function in factored form and $f(x) > 0$, we are interested in finding the product of the factors for which the function is _____ (positive or negative).

4a. _____

4b. If $f(x)$ is a quadratic function in factored form and $f(x) < 0$, we are interested in finding the product of the factors for which the function is _____ (positive or negative).

4b. _____

5. Solve $x^2 + 2x - 35 > 0$ using the algebraic method. *(See textbook Example 2)*

Step 1: Write the inequality so that $ax^2 + bx + c$ is on one side of the inequality and 0 is on the other.

The inequality is already in this form: $x^2 + 2x - 35 > 0$

Step 2: Determine the solutions to the equation $ax^2 + bx + c = 0$.

Factor: **(a)** _____

Use Zero-Product Property: **(b)** _____

Step 3: Use the solutions to the equation solved in Step 2 to separate the real number line into intervals.

List the intervals: **(c)** _____

Step 4: Write the expression in factored form. Within each interval formed in Step 3, choose a test point and determine the sign of each factor. Then determine the sign of the product. Also determine the value of the expression at each solution found in Step 2.

(d) Complete the chart below:

Interval	$(-\infty, \_\_\_)$	$x = \_\_\_$	$(\_\_\_, \_\_\_)$	$x = \_\_\_$	$(\_\_\_, \infty)$
Test Point					
Sign of ___					
Sign of ___					
Sign of Product					
Conclusion					

To find the solution, select the intervals where the product is (positive or negative)? **(e)** _____

Is 0 included in the solution set? **(f)** _____

State the solution set: **(g)** _____

432   Sullivan/Struve/Mazzarella, *Elementary & Intermediate Algebra*, 3e
Copyright © 2014 Pearson Education, Inc.

Name:
Instructor:

Date:
Section:

Do the Math Exercises 10.6
Polynomial Inequalities

In Problems 1 – 10, solve each inequality. Write the solution set in interval notation.

1. $(x-8)(x+1) \leq 0$

2. $(x-4)(x-10) > 0$

1. _____

2. _____

3. $p^2 + 5p + 4 < 0$

4. $2b^2 + 5b < 7$

3. _____

4. _____

5. $x + 6 < x^2$

6. $x^2 - 3x - 5 \geq 0$

5. _____

6. _____

7. $-3p^2 < 3p - 5$

8. $y^2 + 3y + 5 \geq 0$

7. _____

8. _____

9. $3w^2 + w < -2$

10. $p^2 - 8p + 16 \leq 0$

9. _____

10. _____

Sullivan/Struve/Mazzarella, *Elementary & Intermediate Algebra*, 3e
Copyright © 2014 Pearson Education, Inc.

Do the Math Exercises 10.6

In Problems 11 and 12, for each function find the values of x that satisfy the given condition.

11. Solve $f(x) > 0$ if $f(x) = x^2 + 4x$ 12. Solve $f(x) \leq 0$ if $f(x) = x^2 + 2x - 48$

11. _____

12. _____

In Problems 13 and 14, find the domain of the given function.

13. $f(x) = \sqrt{x^2 - 5x}$ 14. $G(x) = \sqrt{x^2 + 2x - 63}$

13. _____

14. _____

15. **Revenue Function** Suppose that the marketing department of Samsung has found that, when a certain model of cellular telephone is sold a price of p dollars, the daily revenue R (in dollars) as a function of the price p is $R(p) = -5p^2 + 600p$. Determine the prices for which revenue will exceed $17,500. That is, solve $R(p) > 17,500$.

15. _____

In Problems 16 and 17, solve each polynomial inequality.

16. $(3x+4)(x-2)(x-6) \geq 0$ 17. $3x^3 + 5x^2 12x - 20 < 0$

16. _____

17. _____

Sullivan/Struve/Mazzarella, *Elementary & Intermediate Algebra*, 3e
Copyright © 2014 Pearson Education, Inc.

Five-Minute Warm-Up 10.7
Rational Inequalities

1. Write in interval notation: $-3 \leq x < 2$

   1. _____

2. Solve and graph the solution set: $-5x + 1 > 9 + 3(2x + 1)$

   2. _____

   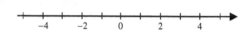

3. Solve and write the solution set in interval notation: $\dfrac{3z-1}{4} - 1 \leq \dfrac{6z+5}{2}$

   3. _____

4. Determine whether $x = -2$ satisfies the inequality $5x + 17 \geq 7$.

   4. _____

Name:
Instructor:

Date:
Section:

Guided Practice 10.7
Rational Inequalities

Objective 1: Solve a Rational Inequality

1. Solving rational inequalities depends on finding intervals where the polynomials in the rational expression are positive or negative. This means that the inequality must be of the form:

   (a) $\dfrac{p(x)}{q(x)} > \underline{}$ or $\dfrac{p(x)}{q(x)} \geq \underline{}$ when the quotient is positive and

   (b) $\dfrac{p(x)}{q(x)} < \underline{}$ or $\dfrac{p(x)}{q(x)} \leq \underline{}$ when the quotient is negative.

2. The quotient is positive when both $p(x)$ and $q(x)$ are _____ or when both $p(x)$ and $q(x)$ are _____.

3. The quotient is negative when $p(x)$ is _____ and $q(x)$ is _____ or when $p(x)$ is _____ and $q(x)$ is _____.

4. We always exclude the values that cause _____.

5. Solve $\dfrac{x+2}{x-4} \leq 0$. Graph the solution set. *(See textbook Example 1)*

Step 1: Write the inequality so that a rational expression is on one side of the inequality and zero is on the other. Be sure to write the rational expression as a single quotient.

$\dfrac{x+2}{x-4} \leq 0$ is already in the required form.

Step 2: Determine the numbers for which the rational expression equals 0 or is undefined.

Value(s) when the rational expression = 0: **(a)** _____

Value(s) when the rational expression is undefined: **(b)** _____

Step 3: Use the numbers found in Step 2 to separate the real number line into intervals.

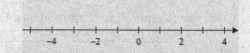

Step 4: Choose a test point within each interval formed in Step 3 to determine the sign of $x+2$ and $x-4$. Then determine the sign of the quotient.

Select a test point in the interval $(-\infty, -2)$: **(c)** _____

Sign of $x+2$: **(d)** _____

Sign of $x-4$: **(e)** _____

Sign of the quotient: **(f)** _____

continued next page

Guided Practice 10.7

Select a test point in the interval $(-2, 4)$: (g) _____

Sign of $x + 2$: (h) _____

Sign of $x - 4$: (i) _____

Sign of the quotient: (j) _____

Select a test point in the interval $(4, \infty)$: (k) _____

Sign of $x + 2$: (l) _____

Sign of $x - 4$: (m) _____

Sign of the quotient: (n) _____

In this problem, are we looking for the quotient to be positive or negative? (o) _____

What value must be excluded? (p) _____

Graph the solution set: (q)

6. Rewrite the rational expression $\dfrac{5}{x-1} - \dfrac{7}{x+1} > 0$ as single rational expression which is positive.

6. _____

7. Rewrite the rational expression $\dfrac{3x-7}{x+2} < 2$ as single rational expression which is negative.

7. _____

8. Complete the following chart used to solve $\dfrac{(x+4)(x-3)}{x-2} \geq 0$. *(See textbook Example 2)*

Interval	$(-\infty, -4)$	-4	$(-4, 2)$	2	$(2, 3)$	3	$(3, \infty)$
Test Point							
Sign of $x + 4$							
Sign of $x - 3$							
Sign of $x - 2$							
Sign of quotient							
Conclusion							

9. Graph the solution set to Problem 8.

Do the Math Exercises 10.7
Rational Inequalities

In Problems 1 – 10, solve each rational inequality. Express your answer in interval notation.

1. $\dfrac{x+8}{x+2} > 0$

2. $\dfrac{x+12}{x-2} \geq 0$

1. _____

2. _____

3. $\dfrac{x-10}{x+5} \leq 0$

4. $\dfrac{(5x-2)(x+4)}{x-5} < 0$

3. _____

4. _____

5. $\dfrac{(3x-2)(x-6)}{x+1} \geq 0$

6. $\dfrac{x+3}{x-4} > 1$

5. _____

6. _____

7. $\dfrac{3x-7}{x+2} \leq 2$

8. $\dfrac{3x+20}{x+6} < 5$

7. _____

8. _____

9. $\dfrac{2}{x+3} + \dfrac{2}{x} \leq 0$

10. $\dfrac{1}{x-4} \geq \dfrac{3}{2x+1}$

9. _____

10. _____

Sullivan/Struve/Mazzarella, *Elementary & Intermediate Algebra*, 3e
Copyright © 2014 Pearson Education, Inc.

Do the Math Exercises 10.7

In Problems 11 and 12, for each function find the values of x that satisfy the given condition.

11. Solve $R(x) \geq 0$ if $R(x) = \dfrac{x+3}{x-8}$

12. Solve $R(x) < 0$ if $R(x) = \dfrac{3x+2}{x-4}$

11. _____

12. _____

In Problems 13 and 14, find the x-intercept of the graph of each function.

13. $h(x) = -3x^2 - 7x + 20$

14. $R(x) = \dfrac{x^2 + 5x + 6}{x+2}$

13. _____

14. _____

15. **Average Cost** Suppose that the daily cost C of manufacturing x bicycles is given by $C(x) = 90x + 5000$. Then the average daily cost $\overline{C}$ is given by $\overline{C}(x) = \dfrac{90x + 5000}{x}$. How many bicycles must be produced each day in order for the average cost to be no more than $130?

15. _____

Name:
Instructor:

Date:
Section:

Five-Minute Warm-Up 11.1
Composite Functions and Inverse Functions

1. Determine the domain: $R(x) = \dfrac{4}{x^2 - 3x + 2}$.

   1. _____

2. Use the function $f(x) = -3x^2 + 1$ to find the following.

   (a) $f(-4)$ (b) $f(a-2)$

   2a. _____

   2b. _____

   (c) $f(x+h)$

   2c. _____

In Problem 3, use the Vertical Line Test to determine whether the following is the graph of a function.

3a. 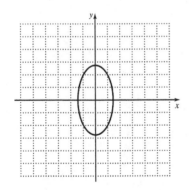   3b.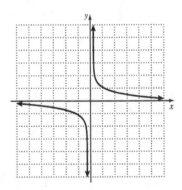

3a. _____

3b. _____

Sullivan/Struve/Mazzarella, *Elementary & Intermediate Algebra*, 3e
Copyright © 2014 Pearson Education, Inc.

Name:
Instructor:

Date:
Section:

Guided Practice 11.1
Composite Functions and Inverse Functions

Objective 1: Form the Composite Function

1. Function composition is the process of evaluating one function and then carrying the result forward to be evaluated in a second function. The notation $f(g(x))$ means evaluate the inner function, $g(x)$, and then evaluate the result in the outer function, $f(x)$. This function composition is written with the notation

   _____. This notation is read "f composed with g". The mathematical notation "$\circ$" is called small circle, so we can also say "f small circle g of x".

In Problems 2 and 3, suppose that $f(x) = 2x^2 + 7$ and $g(x) = x - 1$.
(See textbook Example 1)

2. Find $(f \circ g)(-2)$.

   (a) Evaluate $g(-2)$.

   (b) Use this result to find $f(g(-2))$.

   (c) What is $(f \circ g)(-2)$?

3. Find $(g \circ f)(4)$.

   (a) Evaluate $f(4)$.

   (b) Use this result to find $g(f(4))$.

   (c) What is $(g \circ f)(4)$?

2a. _____

2b. _____

2c. _____

3a. _____

3b. _____

3c. _____

Objective 2: Determine Whether a Function is One-to-One

4. In your own words, what does it mean for a function to be one-to-one?

5. Determine whether or not the function is one-to-one. *(See textbook Example 3)*
   (a) $\{(1, -3), (-3, 1), (2, -4), (-2, 4)\}$
   (b) $\{(0, 1), (1, 2), (2, 3), (3, 1)\}$

5a. _____

5b. _____

6. Graphically, how can you test whether a function is one-to-one?

6. _____

Objective 3: Find the Inverse of a Function Defined by a Map or Set of Ordered Pairs

7. In order for the inverse of a function to also be a function, the function must be _____.

Sullivan/Struve/Mazzarella, Elementary & Intermediate Algebra, 3e
Copyright © 2014 Pearson Education, Inc.

Guided Practice 11.1

8. If $f(x)$ is a one-to-one function, then its inverse function is written with the notation _____. This means that if the ordered pair (a, b) satisfies $f(x)$, then the ordered pair _____ satisfies the inverse function.

9. Given the function $\{(-3, 15), (-1, -5), (0, 0), (2, 10)\}$, find
   (a) the inverse function *(See textbook Example 6)*
   (b) the domain of the function
   (c) the range of the function
   (d) the domain of the inverse function
   (e) the range of the inverse function

9a. _____
9b. _____
9c. _____
9d. _____
9e. _____

Objective 4: Obtain the Graph of the Inverse Function from the Graph of the Function

10. Graphs can be transformed in several different ways: translation (slide), reflection (flip), and rotation (turn). A graph is symmetric about a line if one part of the graph is a mirror image of the other. The line on which the graph is flipped (or folded) is called the axis of symmetry. For instance, $y = x^2$ is symmetric about the y-axis. The graph of a function and its inverse are symmetric about the line _____.

Objective 5: Find the Inverse of a Function Defined by an Equation

11. Find the inverse of $f(x) = 4x - 8$. *(See textbook Example 8)*

$f(x) = 4x - 8$

Step 1: Replace f(x) with y in the equation for f(x).

Replace $f(x)$ with y.   (a) _____

Step 2: Interchange the variables to write in inverse.

Rewrite equation (a) exchanging the variables x and y.   (b) _____

Step 3: Solve the equation found in Step 2 for y in terms of x.

Add 8 to both sides: (c) _____

Divide both sides by 4:   (d) _____

Step 4: Replace y with $f^{-1}(x)$.   (e) _____

Step 5: Verify your result by showing that
$f^{-1}(f(x)) = x$ and
$f(f^{-1}(x)) = x$.

Name:
Instructor:

Date:
Section:

Do the Math Exercises 11.1
Composite Functions and Inverse Functions

In Problems 1 – 4, use the functions $f(x) = x^2 - 3$ and $g(x) = 5x + 1$ to find each of the following.

1. $(f \circ g)(3)$
2. $(g \circ f)(-2)$

1. _____

2. _____

3. $(f \circ f)(1)$
4. $(g \circ g)(-4)$

3. _____

4. _____

5. Given the functions $f(x) = x - 3$ and $g(x) = 4x$, find $(f \circ g)(x)$.

5. _____

6. Given the functions $f(x) = \sqrt{x+2}$ and $g(x) = x - 2$, find $(g \circ f)(x)$.

6. _____

7. Given the functions $f(x) = \dfrac{2}{x-1}$ and $g(x) = \dfrac{4}{x}$, find $(f \circ f)(x)$.

7. _____

In Problems 8 – 10, determine whether the function is one-to-one.

8. $\{(-2, -8), (-1, -1), (0, 0), (1, 1), (2, 8)\}$

8. _____

9. $\{(-3, 0), (-2, 3), (-1, 0), (0, -3)\}$

9. _____

10.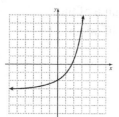

10. _____

Do the Math Exercises 11.1

11. Find the inverse of the function $\{(-10, 1), (-5, 4), (0, 3), (-5, 2)\}$.

11. _____

In Problems 12 and 13, verify that the functions f and g are inverses of each other.

12. $f(x) = 10x$; $g(x) = \dfrac{x}{10}$

13. $f(x) = \dfrac{2}{x+4}$; $g(x) = \dfrac{2}{x} - 4$

12. _____

13. _____

In Problems 14 – 19, find the inverse function of the given one-to-one function.

14. $g(x) = x + 6$

15. $H(x) = 3x + 8$

14. _____

15. _____

16. $f(x) = x^3 - 2$

17. $G(x) = \dfrac{2}{3-x}$

16. _____

17. _____

18. $R(x) = \dfrac{2x}{x+4}$

19. $g(x) = \sqrt[3]{x+2} - 3$

18. _____

19. _____

20. Volume of a Balloon The volume V of a hot-air balloon (in cubic meters) as a function of its radius r is given by $V(r) = \dfrac{4}{3}\pi r^3$. If the radius r of the balloon is increasing as a function of time t (in minutes) according to $r(t) = 3\sqrt[3]{t}$, for $t \geq 0$, find the volume of the balloon as a function of time t. What will be the volume of the balloon after 30 minutes?

20. _____

Name:
Instructor:

Date:
Section:

Five-Minute Warm-Up 11.2
Exponential Functions

In Problems 1 – 4, evaluate each expression.

1. 2^4

2. 2^{-2}

3. 3^0

4. 10^{-1}

1. _____

2. _____

3. _____

4. _____

5. Write 6.023455 as a decimal
   (a) rounded to 4 decimal places
   (b) truncated to 4 decimal places.

5a. _____

5b. _____

In Problems 6 – 8, simplify each expression.

6. $5x^3 \cdot 2x^{-5}$

7. $\dfrac{4a^2}{10a^{-4}}$

8. $\left(4p^5\right)^3$

6. _____

7. _____

8. _____

9. Solve: $6x^2 = 7x + 5$

9. _____

Sullivan/Struve/Mazzarella, *Elementary & Intermediate Algebra*, 3e
Copyright © 2014 Pearson Education, Inc.

Name:
Instructor:

Date:
Section:

Guided Practice 11.2
Exponential Functions

Objective 1: Evaluate Exponential Expressions

1. An exponential function is a function of the form $f(x) = a^x$ where a is _____ and $a \neq$ _____.

In Problems 2 and 3, evaluate each expression to 5 decimal places. (See textbook Example 1)

2. $2^{1.4}$ 3. $2^{\sqrt{2}}$

2. _____

3. _____

Objective 2: Graph Exponential Functions

In Problems 4 and 5, match the function to the graph.
(See textbook Examples 2 and 3)

4. $f(x) = a^x; a > 1$

5. $f(x) = a^x; 0 < a < 1$

(a)

(b)

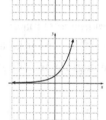

4. _____

5. _____

Objective 3: Define the Number e

6. What is an approximate value of e to the nearest thousandth?

6. _____

7. Evaluate each of the following to two decimal places.

(a) e^3 (b) e^{-2}

7a. _____

7b. _____

Objective 4: Solve Exponential Equations

8. The Property for Solving Exponential Equations states that if two exponential functions have the same base and the exponential functions are equal, then it must be true that _____.

Sullivan/Struve/Mazzarella, *Elementary & Intermediate Algebra*, 3e
Copyright © 2014 Pearson Education, Inc.

Guided Practice 11.2

9. Solve: $3^{2x+1} = 27$ *(See textbook Example 5)*

Step 1: Use the Laws of Exponents to write both sides of the equation with the same base.

$3^{2x+1} = 27$

Prime factor 27 and write in exponential form:

(a) _____

Step 2: Set the exponents on each side of the equation equal to each other.

(b) _____

Step 3: Solve the equation resulting from Step 2.

Subtract 1 from both sides: (c) _____

Divide both sides by 2: (d) _____

Step 4: Verify your solution(s). Substitute your value into the original equation.

We leave it to you to verify your solution.

State the solution set: (e) _____

Objective 5: Use Exponential Models that Describe Our World

10. Strontium 90 is a radioactive material that decays according to the function $A(t) = A_0 e^{-0.087t}$, where A_0 is the initial amount present and A is the amount present at time t (in days). If you begin with 100 grams of Strontium 90, how much will be present after 5 days?

10. _____

11. What is the compound interest formula? Identify each of the variables in the formula.

12. Suppose that you deposit $100 into a savings account that earns 7% compounded quarterly for a period of 5 years.

(a) Identify the value of each variable in the compound interest formula.

12a. _____

(b) How much is in the account 5 years after the $100 deposit?

12b. _____

(c) You also deposit $100 in another bank because you believe you are getting a better deal when you receive 7% compounded daily. Assuming there are 360 days per business year, identify the value of each variable in the compound interest formula.

12c. _____

(d) How much is in the account after 5 years?

12d. _____

Name:
Instructor:

Date:
Section:

Do the Math Exercises 11.2
Exponential Functions

In Problem 1 and 2, approximate each number using a calculator. Round your answer to three decimal places.

1a. $5^{1.4}$

1b. $5^{1.41}$

1a. _____

1b. _____

1c. $5^{1.414}$

1d. $5^{1.4142}$

1c. _____

1d. _____

1e. $5^{\sqrt{2}}$

1e. _____

2a. e^3

2b. $e^{1.5}$

2a. _____

2b. _____

In Problems 3 – 5, graph each function.

3. $g(x) = 10^x$

4. $H(x) = 2^{x-2}$

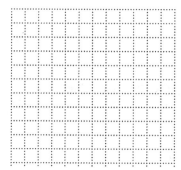
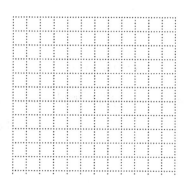

5. $f(x) = e^x - 1$

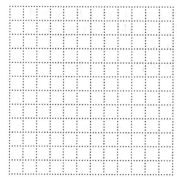

Sullivan/Struve/Mazzarella, *Elementary & Intermediate Algebra*, 3e
Copyright © 2014 Pearson Education, Inc.

451

Do the Math Exercises 11.2

In Problems 6–11, solve each equation.

6. $3^x = 3^{-2}$

7. $2^{x+3} = 128$

8. $3^{-x+4} = 27^x$

9. $9^{2x} \cdot 27^{x^2} = 3^{-1}$

10. $\left(\dfrac{1}{6}\right)^x - 36 = 0$

11. $e^{3x} = e^2$

6. _____

7. _____

8. _____

9. _____

10. _____

11. _____

In Problem 12 and 13, suppose that $f(x) = 3^x$.

12. What is $f(2)$? What point is on the graph of f?

13. If $f(x) = \dfrac{1}{81}$, what is x? What point is on the graph of f?

12. _____

13. _____

14. **A Population Model** According to the U.S. Census Bureau, the population of the world in 2012 was 7018 million people. In addition, the population of the world was growing at a rate of 1.26% per year. Assuming that this growth rate continues, the model $P(t) = 7018(1.0126)^{t-2012}$ represents the population P (in millions of people) in year t.

(a) According to this model, what will be the population of the world in 2015?

(b) According to this model, what will be the population of the world in 2025?

14a. _____

14b. _____

Five-Minute Warm-Up 11.3
Logarithmic Functions

1. Solve: $4x + 6 > 0$

    1. _____

2. Solve: $\sqrt{2x + 3} = x$

    2. _____

3. Solve: $x^2 = -7x - 12$

    3. _____

In Problems 4 and 5, evaluate each expression.

4. 4^{-2} 5. $\left(\dfrac{1}{2}\right)^{-3}$

    4. _____

    5. _____

In Problems 6 and 7, find the function value.

6. $f(x) = x^2;\ f(-2)$ 7. $f(x) = 2^x;\ f(-2)$

    6. _____

    7. _____

Guided Practice 11.3
Logarithmic Functions

1. The logarithmic function to the base a, where $a > 0$ and $a \neq 1$, is denoted by $y = \log_a x$ and is read as "y is the logarithm to the base a of x". This is defined as

$y = \log_a x$ is equivalent to _____

Objective 1: Change Exponential Equations to Logarithmic Equations

In Problems 2 and 3, rewrite each exponential expression as an equivalent expression involving a logarithm. (See textbook Example 1)

2. $2^{-3} = \dfrac{1}{8}$ **3.** $a^4 = 3$

2. _____

3. _____

Objective 2: Change Logarithmic Equations to Exponential Equations

In Problems 4 and 5, rewrite each logarithmic expression as an equivalent expression involving an exponent. (See textbook Example 2)

4. $p = \log_4 30$ **5.** $\log_n 3 = -5$

4. _____

5. _____

Objective 4: Determine the Domain of a Logarithmic Function
(See textbook Example 5)

6a. The domain of the logarithmic function is

6b. The range of the logarithmic function is

6c. Determine the domain of the function $f(x) = \log_3 (10 - 2x)$.

6a. _____

6b. _____

6c. _____

Guided Practice 11.3

Objective 5: Graph Logarithmic Functions

7. Because exponential functions and logarithmic functions are inverses of each other, we know that the graph of $y = \log_5 x$ is a reflection of _____ across the line $y = x$. Graph both of these functions in the same coordinate plane. *(See textbook Example 6)*

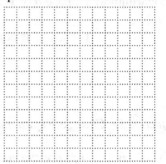

Objective 6: Work with Natural and Common Logarithms

8. The natural logarithm: $y = \ln x$ is written in exponential form as

8. _____

9. The common logarithm: $y = \log x$ is written in exponential form as:

9. _____

Objective 7: Solve Logarithmic Equations

In Problems 10 and 11, solve the logarithmic equation. Be sure to check your answers as extraneous solutions may appear. *(See textbook Examples 9 and 10)*

10. $\log_3 (5 - 2x) = 2$ **11.** $\ln x = -2$

10. _____

11. _____

Objective 8: Use Logarithmic Models That Describe Our World

12. List 3 applications of logarithmic functions that are used in the world today.

12a. _____

12b. _____

12c. _____

Do the Math Exercises 11.3
Logarithmic Functions

In Problems 1 – 3, change each exponential expression to an equivalent expression involving a logarithm.

1. $64 = 4^3$

2. $b^4 = 23$

3. $10^{-3} = z$

1. _____

2. _____

3. _____

In Problems 4 – 6, change each logarithmic expression to an equivalent expression involving an exponent.

4. $\log_3 81 = 4$

5. $\log_6 x = -4$

6. $\log_a 16 = 2$

4. _____

5. _____

6. _____

In Problems 7 – 8, find the exact value of each logarithm without using a calculator.

7. $\log_4 16$

8. $\log_{\sqrt{3}} 3$

7. _____

8. _____

In Problems 9 –10, find the domain of each function.

9. $f(x) = \log_3(x - 2)$

10. $G(x) = \log_4(3 - 5x)$

9. _____

10. _____

Do the Math Exercises 11.3

In Problems 11 and 12, use a calculator to evaluate each expression. Round your answer to three decimal places.

11. log 0.78

12. $\ln \frac{1}{2}$

11. _____

12. _____

In Problems 13 – 18, solve each logarithmic equation.

13. $\log_3(5x - 3) = 3$

14. $\log_4(8x + 10) = 3$

13. _____

14. _____

15. $\log_a 81 = 2$

16. $\log(2x + 3) = 1$

15. _____

16. _____

17. $\ln e^{2x} = 8$

18. $\log_3(x^2 + 1) = 2$

17. _____

18. _____

19. Alaska, 1964 According to the United States Geological Survey, an earthquake on March 28, 1964 in Prince William Sound, Alaska resulted in a seismographic reading of 1,584,893 millimeters 100 kilometers from its epicenter. What was the magnitude of this earthquake? This earthquake was the second largest ever recorded, with the largest being the Great Chilean Earthquake of 1960, whose magnitude was 9.5 on the Richter scale.

19. _____

Name:
Instructor:

Date:
Section:

Five-Minute Warm-Up 11.4
Properties of Logarithms

1. Write 1.13985 as a decimal
   (a) rounded to 3 decimals places
   (b) truncated to 3 decimal places.

   1a. _____

   1b. _____

In Problems 2 and 3, write each expression with a rational exponent.

2. $\sqrt{x}$

3. $\sqrt[4]{a^3}$

2. _____

3. _____

In Problems 4 and 5, write in exponential form using prime numbers in the base.

4. $\dfrac{1}{16}$

5. $\dfrac{27}{8}$

4. _____

5. _____

In Problems 6 and 7, simplify each expression.

6. $x^0, x \neq 0$

7. $6a^0, a \neq 0$

6. _____

7. _____

8. Find the inverse function of $y = \log_3 x$. Write the equation in exponential form.

8. _____

Sullivan/Struve/Mazzarella, *Elementary & Intermediate Algebra, 3e*
Copyright © 2014 Pearson Education, Inc.

Name:
Instructor:

Date:
Section:

Guided Practice 11.4
Properties of Logarithms

Objective 1: Understand the Properties of Logarithms

1. There are four useful properties that can be quickly derived from the definition of a logarithm. While it is not essential, it will make your work with logarithms faster and easier if you memorize them.

For any real number a, for which the logarithm is defined:

$$\log_a 1 = 0 \qquad \log_a a = 1 \qquad \log_a a^r = r \qquad a^{\log_a M} = M$$

Use these properties to simplify the following. *(See textbook Examples 1 and 2)*

(a) $3^{\log_3 \sqrt{2}}$ (b) $\log_4 1$ (c) $\ln e$ (d) $\log_x x^7$

1a. _____

1b. _____

1c. _____

1d. _____

Objective 2: Write a Logarithmic Expression as a Sum or Difference of Logarithms

2. There are three important rules that you will need as well. These should be easy to remember because they are very similar to rules you already know, Laws of Exponents. For each of these, M, N, and a are positive real numbers with $a \neq 1$.

(a) The Product Rule: $\log_a (M \cdot N) =$ _____

(b) The Quotient Rule: $\log_a \left(\dfrac{M}{N} \right) =$ _____

(c) The Power Rule: $\log_a M^r =$ _____

3. Use the Product Rule of Logarithms to write $\log(9x)$ as the sum of logarithms. *(See textbook Example 4)*

3. _____

4. Use the Quotient Rule of Logarithms to write $\ln \left(\dfrac{2}{x} \right)$ as the difference of logarithms. *(See textbook Example 5)*

4. _____

Guided Practice 11.4

Objective 3: Write a Logarithmic Expression as a Single Logarithm

5. Now we will reverse the process and write an expanded logarithmic expression as a single logarithm. This will be an important skill in the next section.

Use the Quotient Rule of Logarithms to write $\log_4(x+1) - \log_4(x^2-1)$ as a single logarithm.
(See textbook Example 9)

5. _____

Objective 4: Evaluate a Logarithm Whose Base is Neither 10 Nor e

6. Calculators only have keys for the common logarithm, log, and the natural logarithm, ln. When we need to find the logarithmic value of an expression that uses a base other than 10 or e, we use the Change-of-Base Formula. If M, a and b are positive real numbers with $a \neq 1$, $b \neq 1$, then

Change-of-Base Formula: $\log_a M = $ _____

7. Approximate $\log_2 9$ using the Change-of-Base Formula. Round you answer to three decimal places.
(See textbook Example 12)

7. _____

8. Write an examples to illustrate: $\log_2(x+y) \neq \log_2 x + \log_2 y$

Name:
Instructor:

Date:
Section:

Do the Math Exercises 11.4
Properties of Logarithms

In Problem 1, use properties of logarithms to find the exact value of each expression. Do not use a calculator.

1a. $\log_5 5^{-3}$

1b. $5^{\log_5 \sqrt{2}}$

1a. _____

1b. _____

1c. $e^{\ln 10}$

1c. _____

In Problem 2, suppose that $\ln 2 = a$ and $\ln 3 = b$. Use properties of logarithms to write each logarithm in terms of a and b.

2a. $\ln 4$

2b. $\ln 18$

2a. _____

2b. _____

In Problems 3 – 10, write each expression as a sum and/or difference of logarithms. Express exponents as factors.

3. $\log_4 \left(\dfrac{a}{b} \right)$

4. $\log_3 (a^3 b)$

3. _____

4. _____

5. $\log_2 (8z)$

6. $\log_2 \left(\dfrac{16}{p} \right)$

5. _____

6. _____

7. $\log_2 \left(32 \sqrt[4]{z} \right)$

8. $\ln \left(\dfrac{\sqrt[5]{x}}{(x+2)^2} \right)$

7. _____

8. _____

Do the Math Exercises 11.4

9. $\log_6 \sqrt[3]{\dfrac{x-2}{x+1}}$

10. $\log_4 \left[\dfrac{x^3(x-3)}{\sqrt[3]{x+1}} \right]$

9. _____

10. _____

In Problems 11 – 19, write each expression as a single logarithm.

11. $\log_4 32 + \log_4 2$

12. $\log_2 48 - \log_2 3$

11. _____

12. _____

13. $8 \log_2 z$

14. $4\log_2 a + 2\log_2 b$

13. _____

14. _____

15. $\dfrac{1}{3}\log_4 z + 2\log_4(2z+1)$

16. $\log_7 x^4 - 2\log_7 x$

15. _____

16. _____

17. $\dfrac{1}{3}[\ln(x-1) + \ln(x+1)]$

18. $\log_5(x^2 + 3x + 2) - \log_5(x+2)$

17. _____

18. _____

19. $10\log_4 \sqrt[5]{x} + 4\log_4 \sqrt{x} - \log_4 16$

19. _____

In Problem 20, use the Change-of-Base Formula and a calculator to evaluate each logarithm. Round your answer to three decimal places.

20a. $\log_7 5$

20b. $\log_{\sqrt{3}} \sqrt{6}$

20a. _____

20b. _____

Five-Minute Warm-Up 11.5
Exponential and Logarithmic Equations

In Problems 1 – 4, solve each equation.

1. $\dfrac{2}{3}x - 7 = -1$

2. $x^2 + 5x = 24$

3. $4n^2 = 2 - 7n$

4. $(x-1)^2 - 5(x-1) - 6 = 0$

5. Find the domain: $f(x) = \log_2(-2x + 6)$

1. _____

2. _____

3. _____

4. _____

5. _____

Name: Date:
Instructor: Section:

Guided Practice 11.5
Exponential and Logarithmic Equations

Objective 1: Solve Logarithmic Equations Using the Properties of Logarithms

1. We use the following property where M, N, and a are positive real numbers with $a \neq 1$ to solve logarithmic equations where the log function appears on both sides of the equation.

$$\text{If } \log_a M = \log_a N, \text{ then } M = N.$$

M and N are called arguments so we say, "if there is equality between two logarithmic expressions which have the same base, set the arguments equal." Note that each log function has to be completely simplified to a single logarithm. Also, be careful to check for extraneous solutions.

Use this property to solve the following equations. *(See textbook Examples 1 and 2)*

(a) $\log_2 x = \log_2 (3x - 5)$ **(b)** $\log_4 (x + 3) - \log_4 x = \log_4 10$ 1a. _____

1b. _____

(c) $\dfrac{1}{2} \ln x = 3 \ln 2$

1c. _____

Objective 2: Solve Exponential Equations

2. In Section 11.2, we were able to solve exponential equations of the form $a^u = a^v$ by using the Property for Solving Exponential Equations. This states that if two exponential functions have the same base and the exponential functions are equal, it must be true that the exponents are equal.

Now we will encounter exponential equations which cannot be written on a common base. For this circumstance, we use the definition of a logarithm to convert from exponential form to logarithmic form. Use this approach to solve each of the following. Give both the exact and approximate solution. *(See textbook Examples 3 and 4)*

(a) $3^x = 12$ **(b)** $\dfrac{1}{2} e^{3x} = 9$ 2a. _____

2b. _____

Guided Practice 11.5

Objective 3: Solve Equations Involving Exponential Models

3. Radioactive Decay The half-life of carbon-10 is 19.255 seconds. Suppose that a researcher possesses a 200-gram sample of carbon-10. The amount A (in grams) of carbon-10 after t seconds is given by

$$A(t) = 100 \cdot \left(\frac{1}{2}\right)^{\frac{t}{19.255}}$$

(See textbook Example 5)

(a) Write a model to find when there will be 90 grams of carbon-10 left in the sample. 3a. _____

(b) Use logarithms to solve the equation from part (a). 3b. _____

4. The population of a small town is growing at a rate of 3% per year. The population of the town can be calculated by the exponential function $P(t) = 2500e^{0.03t}$ where t is the number of years after 1950. *(See textbook Examples 5 and 6)*

(a) What was the population in 1965? 4a. _____

(b) Write a model to find when the population reached 7500 people. 4b. _____

(c) Use logarithms to solve the equation from (b) and find when the population reached 7500 people. 4c. _____

(d) How long will it take for the population to double? (That is, for the town to have a population of 5000 people.) 4d. _____

Do the Math Exercises 11.5
Exponential and Logarithmic Equations

In Problems 1 – 18, solve each equation. Express irrational solutions in exact form and as a decimal rounded to 3 decimal places.

1. $\log_5 x = \log_5 13$

2. $\dfrac{1}{2}\log_2 x = 2\log_2 2$

1. _____

2. _____

3. $\log_2 (x - 7) + \log_2 x = 3$

4. $\log_3 (x + 5) - \log_3 x = 2$

3. _____

4. _____

5. $\log_5 (x + 3) + \log_5 (x - 4) = \log_5 8$

6. $3^x = 8$

5. _____

6. _____

7. $4^x = 20$

8. $e^x = 3$

7. _____

8. _____

9. $10^x = 0.2$

10. $2^{2x} = 5$

9. _____

10. _____

11. $3 \cdot 4^x = 15$

12. $\log_6 x + \log_6 (x + 5) = 2$

11. _____

12. _____

Do the Math Exercises 11.5

13. $3 \log_2 x = \log_2 8$

14. $5 \log_4 x = \log_4 32$

13. _____

14. _____

15. $-4e^x = -16$

16. $9^x = 27^{x-4}$

15. _____

16. _____

17. $\log_7 x^2 = \log_7 8$

18. $\log_3 (x-5) + \log_3 (x+1) = \log_3 7$

17. _____

18. _____

19. A Population Model According to the *United States Census Bureau*, the population of the world in 2012 was 7018 million people. In addition, the population of the world was growing at a rate of 1.26% per year. Assuming that this growth rate continues, the model $P(t) = 7018(1.0126)^{t-2012}$ represents the population P (in millions of people) in year t. According to this model, when will the population of the world be 11.58 billion people?

19. _____

20. Depreciation Based on data obtained from the *Kelley Blue Book*, the value V of a Chevy Malibu that is t years old can be modeled by $V(t) = 25{,}258(0.84)^t$. According to the model, when will the car be worth $15,000?

20. _____

Name:
Instructor:

Date:
Section:

Five-Minute Warm-Up 12.1
Distance and Midpoint Formulas

In Problems 1 – 3, simplify each expression.

1. $\sqrt{40}$
2. $\sqrt{108}$
3. $\sqrt{32p^4}$

1. _____

2. _____

3. _____

4. Simplify the expression: $\sqrt{(2x-5)^2}$

4. _____

In Problems 5 and 6 simplify each expression.

5. $-2\sqrt{25}$
6. $\sqrt{(-5-2)^2 + (16-(-8))^2}$

5. _____

6. _____

7. Find the area of a triangle whose base has length of $\sqrt{18}$ cm and whose height has length of $\sqrt{8}$ cm.

7. _____

8. _____

8. Find the length of the hypotenuse of a right triangle whose legs are 6 and $6\sqrt{3}$.

Name:
Instructor:
Date:
Section:

Guided Practice 12.1
Distance and Midpoint Formulas

Objective 1: Use the Distance Formula

1. We can find the distance between two points in the Cartesian plane using the Pythagorean Theorem. This can be accomplished by plotting the points, drawing a right triangle, and applying $a^2 + b^2 = c^2$ where c is the length of the hypotenuse or, in this case, the distance between the points. If you get stuck or forget the distance formula, you can always use this approach.

We use the Distance Formula to find the length of a line segment quickly and easily. If the two points in the Cartesian plane are $P_1 = (x_1, y_1)$ and $P_2 = (x_2, y_2)$, the distance between P_1 and P_2, denoted $d(P_1, P_2)$, is

_____

2. Be sure to use the Rule for Order of Operations when finding the distance between two points. If we want to find the distance between (a, b) and (c, d), the steps are:

(a) subtract _____ ; **(b)** subtract _____ ; **(c)** square the value from part _____ ;

(d) square the value from part _____ ; **(e)** _____ ; **(f)** _____ ; **(g)** _____ .

3. Find the distance between $(9, 3)$ and $(1, -1)$. Find both the exact value and the approximate distance to two decimal places. *(See textbook Example 1)*

3. _____

Guided Practice 12.1

Objective 2: Use the Midpoint Formula

4. A *midpoint* is a point (in this case an ordered pair) which divides a line segment into _____.

5. To find the coordinates of the midpoint, we use the Midpoint Formula. This states that if a line segment has endpoints at $P_1 = (x_1, y_1)$ and $P_2 = (x_2, y_2)$, the midpoint, M, is an ordered pair such that

$$M = \left(\frac{}{2}, \frac{}{2} \right).$$

You can see that the Midpoint Formula averages the x values to find a coordinate in the middle and averages the y values to find a coordinate in the middle. You can verify that the midpoint divides the segment into two segments of equal length by finding the distance from P_1 to M and the finding the distance from M to P_2. These distances will be equal. This step is not necessary, but it is a good check.

6. Find the midpoint of the line segment joining $P_1 = (-3, 2)$ and $P_2(-5, -8)$. *(See textbook Example 4)*

6. _____

Name:
Instructor:

Date:
Section:

Do the Math Exercises 12.1
Distance and Midpoint Formulas

In Problems 1 – 6, find the distance $d(P_1, P_2)$ between points P_1 and P_2.

1. $P_1 = (1, 3)$; $P_2 = (4, 7)$

2. $P_1 = (-10, -3)$; $P_2 = (14, 4)$

1. _____

2. _____

3. $P_1 = (-1, 2)$; $P_2 = (-1, 0)$

4. $P_1 = (5, 0)$; $P_2 = (-1, -4)$

3. _____

4. _____

5. $P_1 = (\sqrt{6}, -2\sqrt{2})$; $P_2 = (3\sqrt{6}, 10\sqrt{2})$

6. $P_1 = (-1.7, 1.3)$; $P_2 = (0.3, 2.6)$

5. _____

6. _____

In Problems 7 – 12, find the midpoint of the line segment formed by joining points P_1 and P_2.

7. $P_1 = (1, 3)$; $P_2 = (5, 7)$

8. $P_1 = (-10, -3)$; $P_2 = (14, 7)$

7. _____

8. _____

9. $P_1 = (-1, 2)$; $P_2 = (3, 9)$

10. $P_1 = (5, 0)$; $P_2 = (-1, -4)$

9. _____

10. _____

11. $P_1 = (\sqrt{6}, -2\sqrt{2})$; $P_2 = (3\sqrt{6}, 10\sqrt{2})$

12. $P_1 = (-1.7, 1.3)$; $P_2 = (0.3, 2.6)$

11. _____

12. _____

Do the Math Exercises 12.1

13. Consider the three points $A = (-2, 3)$, $B = (2, 0)$, and $C = (5, 6)$.

   (a) Plot each point in the Cartesian plane and form the triangle ABC.

   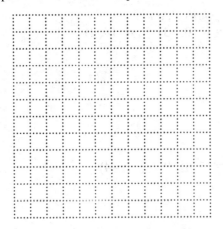

   (b) Find the length of each side of the triangle.

   13b. _____

14. Find all points having a y-coordinate of -3 whose distance from the point $(-4, 2)$ is 13.   14. _____

Name:
Instructor:

Date:
Section:

Five-Minute Warm-Up 12.2
Circles

In Problems 1 and 2, complete the square. Determine the number that must be added to the expression to make it a perfect square trinomial. Then factor the expression.

1. $x^2 - 14x$

2. $y^2 - 5y$

1. _____

2. _____

In Problems 3 and 4, factor completely.

3. $y^2 + 16y + 64$

4. $2x^2 - 12x + 18$

3. _____

4. _____

5. Find **(a)** the area and **(b)** the circumference of a circle whose diameter is 15 inches. Give both the exact answer and then the answer rounded to 2 decimal places.

5a. _____
   _____

5b. _____
   _____

Sullivan/Struve/Mazzarella, *Elementary & Intermediate Algebra*, 3e
Copyright © 2014 Pearson Education, Inc.

Name:
Instructor:

Date:
Section:

Guided Practice 12.2
Circles

Objective 1: Write the Standard Form of the Equation of a Circle

1. A *circle* is the set of all points in the Cartesian plane that are a fixed distance r from a fixed point (h, k).

   (a) The point (h, k) is called the _____.

   (b) The fixed distance r is called the _____.

   (c) We also know that if d is the length of the diameter of the circle, then _____.

2. The *standard form of an equation of a circle* with radius r and center (h, k) is _____.

3. Write the standard form of the equation of the circle with radius 5 and center $(-1, 3)$.
   (See textbook Example 1)

   3. _____

Objective 2: Graph a Circle

4. Graph the equation: $(x - 2)^2 + (x + 3)^2 = 4$ *(See textbook Example 2)*

   (a) Identify the center: _____   (b) Length of radius: _____   (c) Graph:

Guided Practice 12.2

Objective 3: Find the Center and Radius of a Circle Given an Equation in General Form

5. General form expands (multiplies) the binomials, regroups like terms, and has all the terms on one side of the equation and zero on the other side of the equation. What is the general form of the equation of a circle?

_____

6. Graph the equation: $x^2 + y^2 + 6x - 2y + 1 = 0$ (See textbook Example 3)

(a) Complete the square for the x terms and complete the square for the y terms:

$$x^2 + 6x + \underline{} + y^2 - 2y + \underline{} = -1 + \underline{} + \underline{}$$

(b) Factor: $(x + \underline{})^2 + (y - \underline{})^2 = \underline{}$

(c) Identify the center: _____ and the length of radius: _____

(d) Graph:

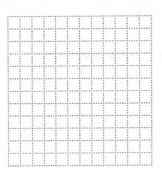

7. Are circles functions? Why or why not?

8. Is $3x^2 - 12x + 3y^2 - 15 = 0$ The equation of a circle? Why or why not? If so, what is the center and radius?

8. _____

Name:
Instructor:

Date:
Section:

Do the Math Exercises 12.2
Circles

In Problems 1 – 4, write the standard form of the equation of each circle whose radius is r and center is (h, k). Graph each circle.

1. $r = 5$; $(h, k) = (0, 0)$

2. $r = 2$; $(h, k) = (1, 0)$

1. _____

2. _____

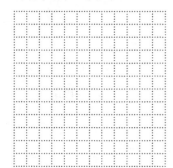
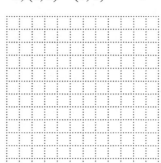

3. $r = 4$; $(h, k) = (-4, 4)$

4. $r = \sqrt{7}$; $(h, k) = (5, 2)$

3. _____

4. _____

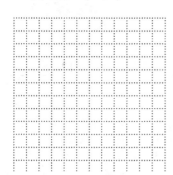
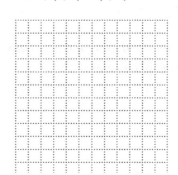

In Problems 5 – 8, identify the center (h, k) and radius r of each circle. Graph each circle.

5. $x^2 + y^2 = 25$

6. $(x - 5)^2 + (y + 2)^2 = 49$

5. _____

6. _____

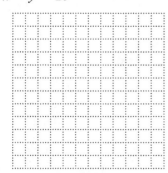

7. $(x - 6)^2 + y^2 = 36$

8. $(x - 2)^2 + (y + 2)^2 = \dfrac{1}{4}$

7. _____

8. _____

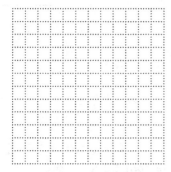
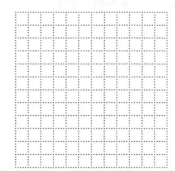

Do the Math Exercises 12.2

In Problems 9 – 11, find the center (h, k) and radius r of each circle.

9. $x^2 + y^2 + 2x - 8y + 8 = 0$ 10. $x^2 + y^2 + 4x - 12y + 36 = 0$

9. _____

10. _____

11. $2x^2 + 2y^2 - 28x + 20y + 20 = 0$

11. _____

In Problems 12 – 14, find the standard form of the equation of each circle.

12. Center at (0, 3) and containing the point (3, 7).

12. _____

13. Center at (2, –3) and tangent to the x-axis.

13. _____

14. With endpoints of a diameter at (–5, –3) and (7, 2).

14. _____

15. Find the area and circumference of the circle $(x - 1)^2 + (y - 4)^2 = 49$.

15. _____

Five-Minute Warm-Up 12.3
Parabolas

In Problems 1 – 3, use the function $f(x) = 2(x-2)^2 + 1$.

1. Identify the vertex.

2. Does the parabola open up or down?

3. Name the axis of symmetry.

1. _____

2. _____

3. _____

In Problems 4 and 5, complete the square. Determine the number that must be added to the expression to make it a perfect square trinomial. Then factor the expression.

4. $x^2 + 10x$

5. $-3x^2 + 12x$

4. _____

5. _____

6. Solve: $(x-4)^2 = 16$

6. _____

Name:
Instructor:
Date:
Section:

Guided Practice 12.3
Parabolas

1. A *parabola* is the set of all points P in the plane that are the same distance from a fixed point F as they are from a fixed line D. In other words, a parabola is the set points P such that $d(F, P) = d(P, D)$.

   (a) The point F is called the _____ of the parabola.

   (b) The line D is its _____.

   (c) The turning point of the parabola is its _____.

   (d) The line through the point F and perpendicular to the line D is called the _____.

Objective 1: Graph Parabolas Whose Vertex Is the Origin

2. In this course, the axis of symmetry is parallel to either the *x*-axis or *y*-axis. This means that the parabola opens up, down, left, or right. In the equation of the parabola, the coefficient of the linear variable $(x^1$ or $y^1)$ will determine the direction that the parabola opens. If k is a real number (in the text $k = 4a$ as this coefficient can be used to determine the breadth of the parabola):

   (a) $y^2 = kx$ opens either   2a. _____

   (b) If $k > 0$, the parabola opens   2b. _____

   (c) If $k < 0$, the parabola opens   2c. _____

   Notice that in (b), the positive *x*-axis has an arrow which points to the right and the parabola opens right.

   (d) $x^2 = ky$ opens either   2d. _____

   (e) If $k > 0$, the parabola opens   2e. _____

   (f) If $k < 0$, the parabola opens   2f. _____

   Notice that in (e), the positive *y*-axis has an arrow which points up and the parabola opens up.

3. Determine which direction each parabola opens. *(See textbook Examples 1 and 2)*

   (a) $x^2 = -4y$ (b) $y^2 = 4x$ 3a. _____

   3b. _____

Guided Practice 12.3

Objective 2: Find the Equation of a Parabola

4. Find the equation of a parabola with vertex at $(0, 0)$ if its axis of symmetry is the x-axis and its graph contains the point $(-2, -1)$. *(See textbook Examples 3 and 4)*

(a) Which direction does the parabola open?

4a. _____

(b) Review the equations on page 702 of your text. Which matches the given information?

4b. _____

(c) Substitute the given information to find the equation for the parabola.

4c. _____

Objective 3: Graph a Parabola Whose Vertex Is Not the Origin

5. Graph the parabola $x^2 - 8x + 4y + 20 = 0$. *(See textbook Example 5)*

(a) Isolate the terms involving the second-degree variable:

5a. _____

(b) Complete the square:

5b. _____

(c) Simplify:

5c. _____

(d) Factor:

5d. _____

(e) Which direction does the parabola open?

5e. _____

(f) Identify the vertex.

5f. _____

(g) Find two more points on the parabola and then graph.

Do the Math Exercises 12.3
Parabolas

In Problems 1 – 5, find the equation of the parabola described.

1. vertex at (0, 0); focus at (0, 5)

    1. _____

2. vertex at (0, 0); focus at (–8, 0)

    2. _____

3. vertex at (0, 0); contains the point (2, 2); axis of symmetry the *x*-axis

    3. _____

4. vertex at (0, 0); directrix $x = -4$

    4. _____

5. focus at (0, –2); directrix $y = 2$

    5. _____

In Problems 6 – 13, graph the parabola. Find (a) the vertex, (b) the focus, and (c) the directrix.

6. $x^2 = 28y$ 7. $y^2 = 10x$

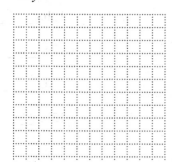   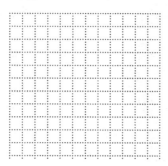

6a. _____

6b. _____

6c. _____

7a. _____

7b. _____

7c. _____

Do the Math Exercises 12.3

8. $x^2 = -16y$

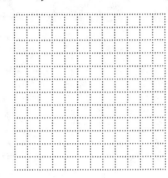

9. $(x+4)^2 = -4(y-1)$

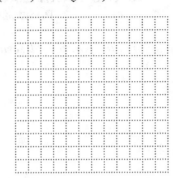

8a. _____
8b. _____
8c. _____
9a. _____
9b. _____
9c. _____

10. $(y-2)^2 = 12(x+5)$

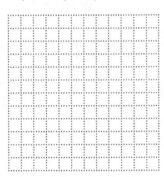

11. $x^2 + 2x - 8y + 25 = 0$

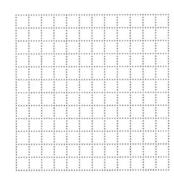

10a. _____
10b. _____
10c. _____
11a. _____
11b. _____
11c. _____

12. $y^2 - 8y + 16x - 16 = 0$

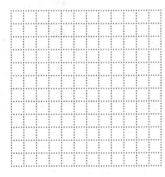

13. $x^2 - 4x + 10y + 4 = 0$

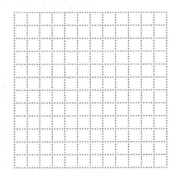

12a. _____
12b. _____
12c. _____
13a. _____
13b. _____
13c. _____

14. **Suspension Bridge** The cables of a suspension bridge are in the shape of a parabola. The towers supporting the cable are 400 feet apart and 80 feet high. If the cables touch the road surface midway between the towers, what is the height of the cable at a point 100 feet from the center of the bridge?

14. _____

Name:
Instructor:

Date:
Section:

Five-Minute Warm-Up 12.4
Ellipses

In Problems 1 – 3, complete the square. Determine the number that must be added to the expression to make it a perfect square trinomial. Then factor the expression.

1. $y^2 - 9y$

2. $x^2 + 12x$

1. _____

2. _____

3. $4y^2 - 2y$

3. _____

In Problems 4 – 7, use the function $y = 9x^2 + 54x - 87$.

4. Write the equation in standard form by completing the square.

5. Identify the vertex.

4. _____

5. _____

6. _____

6. Does the parabola open up or down?

7. Name the axis of symmetry.

7. _____

Sullivan/Struve/Mazzarella, *Elementary & Intermediate Algebra*, 3e
Copyright © 2014 Pearson Education, Inc.

Name: Date:
Instructor: Section:

Guided Practice 12.4
Ellipses

Objective 1: Graph an Ellipse Whose Center Is the Origin

1. An *ellipse* is the set of all points in the plane such that the sum of the distances from two fixed points is a constant.

   (a) The fixed points are called the _____.

   (b) The long axis contains the fixed points and is called the _____.

   (c) The other axis is perpendicular to long axis and is called the _____.

   (d) The point where the two axes intersect is the _____ of the ellipse.

   (e) The major axis contains turning points of the ellipse called _____.

2. The standard form of an ellipse is either $\dfrac{x^2}{a^2} + \dfrac{y^2}{b^2} = 1$ or $\dfrac{x^2}{b^2} + \dfrac{y^2}{a^2} = 1$.
   (See textbook Examples 1 and 2)

   (a) In either of these equations, the larger of the two denominators is _____.

   (b) The distance from the center to the vertex is _____ units of length.

   (c) Therefore, the length of the major axis is _____.

   (d) The length of the minor axis is _____.

   (e) The term with larger denominator tell us which axis is the _____ axis.

   (f) The foci are c units from the center, on the major axis, where $c^2 =$ _____.

   (g) For these two equations, the center of the ellipse is _____.

3. Find the intercepts to graph each ellipse: *(See textbook Examples 1 and 2)*

   (a) $\dfrac{x^2}{9} + \dfrac{y^2}{25} = 1$ (b) $4x^2 + 16y^2 = 64$

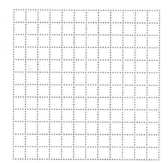 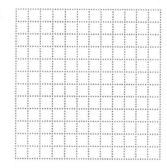

Guided Practice 12.4

Objective 2: Find the Equation of an Ellipse Whose Center Is the Origin

4. Find the equation, in standard form, of the ellipse whose center is at (0, 0), one focus is at (0, 5) and one vertex is at (0, 13). *(See textbook Example 3)*

4. _____

Objective 3: Graph an Ellipse Whose Center Is Not the Origin

In Problems 5 – 7, **(a)** write the equation in the form $\dfrac{(x-h)^2}{a^2} + \dfrac{(y-k)^2}{b^2} = 1$ or

$\dfrac{(x-h)^2}{b^2} + \dfrac{(y-k)^2}{a^2} = 1$, then identify **(b)** the center, **(c)** the vertices and **(d)** the foci.
(See textbook Example 4)

5. $\dfrac{(x+5)^2}{9} + \dfrac{(y-2)^2}{25} = 1$

6. $4(x-1)^2 + 9(y-3)^2 = 36$

7. $16x^2 + 9y^2 - 128x + 54y - 239 = 0$

5a. _____

5b. _____

5c. _____

5d. _____

6a. _____

6b. _____

6c. _____

6d. _____

7a. _____

7b. _____

7c. _____

7d. _____

Do the Math Exercises 12.4
Ellipses

In Problems 1 – 4, graph the ellipse. Find (a) the vertices and (b) the foci of each ellipse.

1. $\dfrac{x^2}{25} + \dfrac{y^2}{4} = 1$

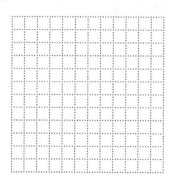

2. $\dfrac{x^2}{16} + \dfrac{y^2}{36} = 1$

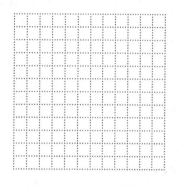

1a. _____

1b. _____

2a. _____

2b. _____

3. $\dfrac{x^2}{64} + y^2 = 1$

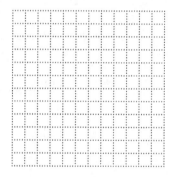

4. $9x^2 + y^2 = 81$

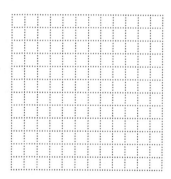

3a. _____

3b. _____

4a. _____

4b. _____

In Problems 5 – 8, find an equation for each ellipse.

5. center at (0, 0); focus at (2, 0); vertex at (5, 0)

5. _____

6. center at (0, 0); focus at (0, −1); vertex at (0, 5)

6. _____

7. foci at (0, ±2); vertices at (0, ±7)

7. _____

8. foci at (±6, 0); length of the major axis is 20

8. _____

Do the Math Exercises 12.4

In Problems 9 and 10, graph each ellipse.

9. $\dfrac{(x+8)^2}{81} + (y-3)^2 = 1$

10. $9(x-3)^2 + (y-4)^2 = 81$

11. Consider the graph of the ellipse: $25x^2 + 150x + 9y^2 - 72y + 144 = 0$.

   (a) Write the equation of the ellipse in standard form. 11a. _____

   (b) Find the center. 11b. _____

   (c) Find the vertices. 11c. _____

   (d) Find the foci. 11d. _____

12. **London Bridge** An arch in the shape of the upper half of an ellipse is used to support London Bridge. The main span is 45.6 meters wide. Suppose that the center of the arch is 15 meters above the center of the river.

   (a) Write the equation for the ellipse in which the x-axis coincides with the water and the y-axis passes though the center of the arch. 12a. _____

   (b) Can a rectangular barge that is 20 meters wide and sits 12 meters above the surface of the water fit through the opening of the bridge? 12b. _____

Name:
Instructor:

Date:
Section:

Five-Minute Warm-Up 12.5
Hyperbolas

In Problems 1 and 2, solve the equation.

1. $y^2 = 16$

2. $(y+3)^2 = 4$

1. _____

2. _____

In Problems 3 – 4, complete the square. Determine the number that must be added to the expression to make it a perfect square trinomial. Then factor the expression.

3. $x^2 + 8x$

4. $y^2 - \dfrac{4}{3}y$

3. _____

4. _____

5. Graph $y = \pm \dfrac{2}{3}x$. That is, graph $y = \dfrac{2}{3}x$ and $y = -\dfrac{2}{3}x$ on the same coordinate plane.

5. _____

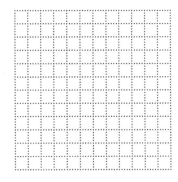

Sullivan/Struve/Mazzarella, *Elementary & Intermediate Algebra*, 3e

Name: Date:
Instructor: Section:

Guided Practice 12.5
Hyperbolas

Objective 1: Graph a Hyperbola Whose Center Is the Origin

1. A *hyperbola* is the collection all points in the plane the difference of whose distances from two fixed points is a positive constant.

    (a) The two fixed points are called _____.

    (b) The line containing these points, as well as the center and the vertices, is called the _____.

    (c) The midpoints of the line segment joining the foci is the _____.

    (d) The line through the center, perpendicular to the transverse axis, is called the _____.

    (e) The branches of the hyperbola have turning points called _____.

2. The standard form of a hyperbola is either $\dfrac{x^2}{a^2} - \dfrac{y^2}{b^2} = 1$ or $\dfrac{y^2}{a^2} - \dfrac{x^2}{b^2} = 1$.
(See textbook Examples 1 and 2)

    (a) In either of these equations, the first denominator is _____, whether it is larger or smaller.

    (b) The distance from the center to the vertex is _____ units of length.

    (c) Therefore, the length of the transverse axis is _____.

    (d) The length of the conjugate axis is _____.

    (e) If the first, or positive, term is $\dfrac{x^2}{a^2}$, the hyperbola opens _____.

    (f) If the first, or positive, term is $\dfrac{y^2}{a^2}$, the hyperbola opens _____.

    (g) The foci are c units from the center, on the transverse axis, where $c^2 =$ _____.

    (h) For these two equations, the center of the hyperbola is _____.

3. Graph the hyperbola $\dfrac{x^2}{16} - \dfrac{y^2}{9} = 1$. *(See textbook Example 1)*

    (a) Find the center. 3a. _____

    (b) The equation is of the form $\dfrac{x^2}{a^2} - \dfrac{y^2}{b^2} = 1$, where $c^2 = a^2 + b^2$. Find the values of a and b. 3b. _____

Guided Practice 12.5

(c) Find the value of c. 3c. _____

(d) Find the vertices and foci. 3d. _____

(e) Let $x = \pm c$ to locate points above and below the foci. 3e. _____

(f) Plot the vertices, foci, and the four points found in (e) to graph the hyperbola.

Objective 3: Find the Asymptotes of a Hyperbola Whose Center Is the Origin

4. In your own words, what is an *asymptote*?

5. The equations of the two asymptotes of the hyperbola can be found using the values of a and b as determined from the standard form. The equations for the asymptotes changes when the direction the hyperbola opening changes. Determine the equations of the asymptotes for each hyperbola:

Hyperbola	Equation of Asymptotes
$\dfrac{x^2}{a^2} - \dfrac{y^2}{b^2} = 1$	(a) $y = \pm$
$\dfrac{y^2}{a^2} - \dfrac{x^2}{b^2} = 1$	(b) $y = \pm$

Name:
Instructor:

Date:
Section:

Do the Math Exercises 12.5
Hyperbolas

In Problems 1 – 4, graph each hyperbola. Find (a) the vertices and (b) the foci.

1. $\dfrac{x^2}{9} - \dfrac{y^2}{16} = 1$

2. $\dfrac{y^2}{81} - \dfrac{x^2}{9} = 1$

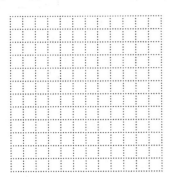
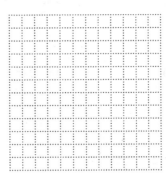

1a. _____

1b. _____

2a. _____

2b. _____

3. $x^2 - 9y^2 = 36$

4. $4y^2 - 9x^2 = 36$

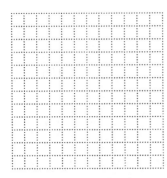
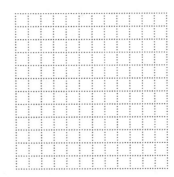

3a. _____

3b. _____

4a. _____

4b. _____

In Problems 5 – 8, find the equation for the hyperbola described.

5. center at (0, 0); focus at (–4, 0); and vertex (–1, 0)

5. _____

6. vertices at (0, 6) and (0, –6); focus at (0, 8)

6. _____

7. vertices at (0, –4) and (0, 4); asymptote the line $y = 2x$

7. _____

8. foci at (–9, 0) and (9, 0); asymptote the line $y = -3x$

8. _____

Sullivan/Struve/Mazzarella, *Elementary & Intermediate Algebra*, 3e
Copyright © 2014 Pearson Education, Inc.

Do the Math Exercises 12.5

In Problems 9 – 14, identify the graph of each equation as a circle, parabola, ellipse, or hyperbola.

9. $x^2 + 4y = 4$ 10. $3y^2 - x^2 = 9$ 9. _____

10. _____

11. $y^2 - 4y - 4x^2 + 8x = 4$ 12. $4x^2 + 8x + 4y^2 - 4y - 12 = 0$ 11. _____

12. _____

13. $4y^2 + 3x - 16y + 19 = 0$ 14. $9x^2 + 4y^2 - 18x + 8y - 23 = 0$ 13. _____

14. _____

15. Consider the hyperbola $\dfrac{y^2}{25} - \dfrac{x^2}{4} = 1$.

(a) Find the vertices.   15a. _____

(b) Find the foci.   15b. _____

16. Consider the hyperbola $\dfrac{x^2}{16} - \dfrac{y^2}{36} = 1$. Find the equation of the asymptotes.   16. _____

Five-Minute Warm-Up 12.6
Systems of Nonlinear Equations

In Problems 1 and 2, solve the system using substitution.

1. $\begin{cases} y = \dfrac{3}{4}x \\ x - 4y = -4 \end{cases}$

2. $\begin{cases} -x + 3y = 4 \\ 2x - 6y = -8 \end{cases}$

1. _____

2. _____

In Problems 3 and 4, solve the system using elimination.

3. $\begin{cases} 6x - 5y = 1 \\ 8x - 2y = -22 \end{cases}$

4. $\begin{cases} 6x - 4y = -5 \\ -12x + 8y = 2 \end{cases}$

3. _____

4. _____

Guided Practice 12.6
Systems of Nonlinear Equations

Objective 1: Solve a System of Nonlinear Equations Using Substitution

1. Solve the following system of equations using substitution: $\begin{cases} y = x - 4 & (1) \\ x^2 + y^2 = 16 & (2) \end{cases}$

(See textbook Example 1)

Step 1: Graph each equation in the system.

Graph of equation (1) is what? **(a)** _____

Graph of equation (2) is what? **(b)** _____

Graph each equation in the system: **(c)**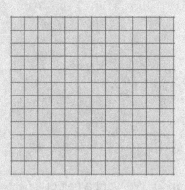

Based on (c), how many solutions will this system have? **(d)** _____

Step 2: Solve equation (1) for y.   This is already done.

Step 3: Substitute the expression for y into equation (2).

Equation (2): **(e)** _____

Substitute the expression for y into equation (2): **(f)** _____

Simplify: **(g)** _____

Step 4: Solve for x.

Factor: **(h)** _____

Zero-Product Property: **(i)** $x = $ _____ or $x = $ _____

Step 5: Use your values from (i) and equation (1) to determine the ordered pairs that satisfy the system.

Equation (1): **(j)** _____

Substitute your first value of x: **(k)** _____

Solve for y: **(l)** _____

Equation (1): **(m)** _____

Substitute your second value of x: **(n)** _____

Solve for y: **(o)** _____

Step 6: Check

State the solution set: **(p)** _____

Guided Practice 12.6

Objective 2: Solve a System of Nonlinear Equations Using Elimination

2. Solve the following system of equations by elimination: $\begin{cases} x^2 + y^2 = 4 & (1) \\ x^2 + 4y^2 = 16 & (2) \end{cases}$ *(See textbook Example 3)*

Step 1: Graph each equation in the system.

Graph of equation (1) is what? **(a)** _____

Graph of equation (2) is what? **(b)** _____

Graph each equation in the system: **(c)**

Based on (c), how many solutions will this system have? **(d)** _____

Step 2: We want a pair of coefficients to be additive inverses so that when we add equation (1) and equation (2), one of the variables will be eliminated.

Let's eliminate x^2. What will equation (1) need to be multiplied by in order to eliminate x^2? **(e)** _____

Multiply equation (1) by the value determined in (e) and write the system: **(f)** $\begin{cases} \underline{\hspace{2in}} & (1) \\ \underline{\hspace{2in}} & (2) \end{cases}$

Step 3: Add equations (1) and (2) to eliminate x^2. Solve the resulting equation for y.

Add: **(g)** _____

Divide by 3: **(h)** _____

Use the Square Root Property: **(i)** $y =$ _____

Step 4: Solve for x using either equation (1) or equation (2). We will use your values from (i) and equation (1) to determine the ordered pairs that satisfy the system.

Equation (1): **(j)** _____

Substitute your first value of y: **(k)** _____

Solve for x: **(l)** _____

Equation (1): **(m)** _____

Substitute your second value of y: **(n)** _____

Solve for x: **(o)** _____

Step 5: Check

State the solution set: **(p)** _____

Name:
Instructor:

Date:
Section:

Do the Math Exercises 12.6
Systems of Nonlinear Equations

In Problems 1 – 4, solve the system of nonlinear equations by substitution.

1. $\begin{cases} y = x^3 + 2 \\ y = x + 2 \end{cases}$

2. $\begin{cases} y = \sqrt{100 - x^2} \\ x + y = 14 \end{cases}$

1. _____

2. _____

3. $\begin{cases} x^2 + y^2 = 16 \\ y = x^2 - 4 \end{cases}$

4. $\begin{cases} xy = 1 \\ x^2 - y = 0 \end{cases}$

3. _____

4. _____

In Problems 5 – 8, solve the system of nonlinear equations by elimination.

5. $\begin{cases} x^2 + y^2 = 8 \\ x^2 + y^2 + 4y = 0 \end{cases}$

6. $\begin{cases} 4x^2 + 16y^2 = 16 \\ 2x^2 - 2y^2 = 8 \end{cases}$

5. _____

6. _____

7. $\begin{cases} 2x^2 + y^2 = 18 \\ x^2 - y^2 = 9 \end{cases}$

8. $\begin{cases} 2x^2 - 5x + y = 12 \\ 14x - 2y = -16 \end{cases}$

7. _____

8. _____

Sullivan/Struve/Mazzarella, *Elementary & Intermediate Algebra*, 3e

Do the Math Exercises 12.6

In Problems 9 – 12, solve the system of nonlinear equations by any method.

9. $\begin{cases} y = x^2 + 4x + 5 \\ x - y = 9 \end{cases}$

10. $\begin{cases} x^2 + y^2 = 25 \\ x^2 - y^2 = 25 \end{cases}$

9. _____

10. _____

11. $\begin{cases} 9x^2 + 4y^2 = 36 \\ x^2 + (y-7)^2 = 4 \end{cases}$

12. $\begin{cases} x^2 + y^2 = 65 \\ y = -x^2 + 9 \end{cases}$

11. _____

12. _____

13. **Fun with Numbers** The sum of two numbers is 8. The sum of their squares is 160. Find the numbers.

13. _____

14. **Perimeter and Area of a Rectangle** The perimeter of a rectangle is 64 meters. The area of the rectangle is 240 square feet. Find the dimensions of the rectangle.

14. _____

Name:
Instructor:

Date:
Section:

Five-Minute Warm-Up 13.1
Sequences

1. If $f(x) = -2x^2 - 3x$, find the function value.
   (a) $f(1)$ (b) $f(-3)$

   1a. _____
   1b. _____

2. Evaluate the expression $(-1)^n (2n - 3)$ for each of the following.

   (a) $n = 1$ (b) $n = 2$ (c) $n = 3$ (d) $n = 4$

   2a. _____
   2b. _____
   2c. _____
   2d. _____

3. Evaluate the expression $\left(-\dfrac{1}{3}\right)^{n+1}$ for each of the following.

   (a) $n = 1$ (b) $n = 2$ (c) $n = 3$

   3a. _____
   3b. _____
   3c. _____

4. If $f(x) = \dfrac{1}{2x}$, find $f(1) + f(2) + f(3) + f(4)$.

   4. _____

Sullivan/Struve/Mazzarella, *Elementary & Intermediate Algebra*, 3e
Copyright © 2014 Pearson Education, Inc.

Name:
Instructor:

Date:
Section:

Guided Practice 13.1
Sequences

1. A *sequence* is a function whose domain is the set of positive integers.

    (a) The numbers in the ordered list are called _____ of the sequence and we separate each entry on the list from the next entry by a comma.

    (b) Sequences can be either infinite or finite. If the list does not end, it is called *infinite* and we use three dots, called _____ , to indicate that the pattern continues indefinitely.

    (c) By contrast, a _____ sequence has a countable number of terms.

Objective 1: Write the First Few Terms of a Sequence

2. We use the notation a_6 to mean the sixth term of the sequence. The formula for the *n*th term, or *general term*, of the sequence is denoted

    2. _____

3. Write the first five terms of the sequence $\{a_n\} = \left\{\dfrac{2^n - 1}{n}\right\}$. *(See textbook Example 1)*

    3. _____

Sullivan/Struve/Mazzarella, *Elementary & Intermediate Algebra*, 3e
Copyright © 2014 Pearson Education, Inc.

Guided Practice 13.1

Objective 2: Find a Formula for the nth Term of a Sequence

4. Sometimes a sequence is indicated by an observed pattern and it is our job to find the pattern. Try subtracting successive terms to find a constant difference or dividing successive terms to find a constant ratio. Sometimes the pattern is neither of these, but rather something you can recognize such as perfect squares.

Find the formula for the *n*th term of the sequence: *(See textbook Example 3)*

(a) 3, 6, 9, 12, ...

(b) $\dfrac{1}{6}, -\dfrac{1}{36}, \dfrac{1}{216}, -\dfrac{1}{1296}, ...$

4a. _____

4b. _____

Objective 3: Use Summation Notation

5. We use *summation notation* to indicate that the terms of the sequence should be added. The *index* of summation can be any variable, but typically we use *i*. This tells you where to start the sum and where to end. When there are a finite number of terms to be added, the sum is called a *partial sum*.

Consider the partial sum: $\sum_{i=1}^{4}\left(i^2 + 3\right)$. We substitute the values $i = $ ___, $i = $ ___, $i = $ ___, $i = $ ___ into the formula $i^2 + 3$ to get the terms ___, ___, ___, ___. The sum is ___. *(See textbook Example 4)*

6. Write out the sum and determine its value: $\sum_{k=0}^{5}(2k - 5)$ _____

7. Express the sum using summation notation: $0 + 2 + 4 + 6 + 8 + 10 + 12$ *(See textbook Example 5)*

7. _____

Name:
Instructor:

Date:
Section:

Do the Math Exercises 13.1
Sequences

In Problems 1 – 4, write down the first five terms of each sequence.

1. $\{n-4\}$

2. $\left\{\dfrac{n+4}{n}\right\}$

1._____

2._____

3. $\{3^n - 1\}$

4. $\left\{\dfrac{n^2}{2}\right\}$

3._____

4._____

In Problems 5 – 8, find the nth term of each sequence suggested by the pattern.

5. 5, 10, 15, 20, …

6. $\dfrac{1}{2}, 1, \dfrac{3}{2}, 2, \dfrac{5}{2}, \ldots$

5._____

6._____

7. 0, 7, 26, 63, …

8. $1, -\dfrac{1}{2}, \dfrac{1}{4}, -\dfrac{1}{8}, \ldots$

7._____

8._____

In Problems 9 - 14, determine the value of the sum.

9. $\displaystyle\sum_{i=1}^{5}(3i+2)$

10. $\displaystyle\sum_{i=1}^{4}\dfrac{i^3}{2}$

9._____

10._____

11. $\displaystyle\sum_{k=1}^{4} 3^k$

12. $\displaystyle\sum_{k=1}^{8}\left[(-1)^k \cdot k\right]$

11._____

12._____

13. $\displaystyle\sum_{j=1}^{8} 2$

14. $\displaystyle\sum_{j=5}^{10}(k+4)$

13._____

14._____

Sullivan/Struve/Mazzarella, *Elementary & Intermediate Algebra*, 3e
Copyright © 2014 Pearson Education, Inc.

Do the Math Exercises 13.1

In Problems 15 – 18, express each sum using summation notation.

15. $1 + 3 + 5 + \ldots + 17$

16. $1 + \dfrac{1}{2} + \dfrac{1}{4} + \ldots + \dfrac{1}{2^{15}}$

15. _____

16. _____

17. $\dfrac{2}{3} - \dfrac{4}{9} + \dfrac{8}{27} + \ldots + (-1)^{15+1}\left(\dfrac{2}{3}\right)^{15}$

18. $3 + 3 \cdot \dfrac{1}{2} + 3 \cdot \dfrac{1}{4} + \ldots + 3 \cdot \left(\dfrac{1}{2}\right)^{11}$

17. _____

18. _____

19. The Future Value of Money Suppose that you place $5,000 into a company 401(k) plan that pays 8% interest compounded monthly. The balance in the account after n months is given by $a_n = 5{,}000\left(1 + \dfrac{0.08}{12}\right)^n$.

(a) Find the value in the account after 1 month.

19a. _____

(b) Find the value in the account after 1 year.

19b. _____

(c) Find the value in the account after 10 years.

19c. _____

Name:
Instructor:

Date:
Section:

Five-Minute Warm-Up 13.2
Arithmetic Sequences

1. Determine the slope of $y = 4x + 5$.

   1. _____

2. If $g(x) = \dfrac{2}{3}x - 1$, find $g\left(-\dfrac{6}{5}\right)$.

   2. _____

3. Solve the following systems of linear equations:

   (a) $\begin{cases} 3x - 5y = 22 \\ x - y = 10 \end{cases}$ (b) $\begin{cases} x - 2y = 9 \\ 2x + y = -2 \end{cases}$

   3a. _____

   3b. _____

4. Evaluate the expression: $\dfrac{9}{2}\left(-\dfrac{4}{3} + \left(-\dfrac{20}{3}\right)\right)$

   4. _____

Sullivan/Struve/Mazzarella, *Elementary & Intermediate Algebra*, 3e
Copyright © 2014 Pearson Education, Inc.

Name:
Instructor:
Date:
Section:

Guided Practice 13.2
Arithmetic Sequences

Objective 1: Determine Whether a Sequence Is Arithmetic

1. If there is a constant difference between the successive terms, the sequence is called an *arithmetic sequence*.

    (a) In formulas for arithmetic sequences, we label the common difference _____.

    (b) We label the first term _____.

2. Determine if the sequence 5, 8, 11, 14, ... is arithmetic. If it is, determine the first term and the common difference. *(See textbook Example 1)*

    2. _____ _____

3. Show that the sequence $\{a_n\} = \{2 + n^2\}$ is not arithmetic by listing the first six terms and calculating the difference between successive terms. *(See textbook Example 3)*

    3. _____

Objective 2: Find a Formula for the nth Term of an Arithmetic Sequence

4. The *n*th term of an arithmetic sequence whose first term is a_1 and whose common difference is *d*, is determined by the formula: $a_n = a_1 + (n-1)d$.

(a) Find a formula for the *n*th term of the arithmetic sequence whose first term is −3 and whose common difference is 5. *(See textbook Example 4)*   4a. _____

(b) Find the 8th term of this sequence.   4b. _____

5. The 5th term of an arithmetic sequence is 24, and the 11th term is 42. *(See textbook Example 5)*

(a) Write and solve a system of linear equations to find the first term and the common difference.   5a. _____

(b) Give a formula for the *n*th term.   5b. _____

Sullivan/Struve/Mazzarella, *Elementary & Intermediate Algebra*, 3e
Copyright © 2014 Pearson Education, Inc.

Guided Practice 13.2

Objective 3: Find the Sum of an Arithmetic Sequence

6. Let $\{a_n\}$ be an arithmetic sequence with first term a_1 and common difference d. The sum S_n of the first n terms of $\{a_n\}$ is $S_n = \dfrac{n}{2}(a_1 + a_n)$.

   Find the sum of the first 10 terms of the arithmetic sequence 12, 16, 20, 24, ... *(See textbook Example 6)*

   (a) Write a formula for the nth term.   6a. _____

   (b) Use the formula to find the 10th term of the sequence.   6b. _____

   (c) Find the sum of the first 10 terms of the arithmetic sequence.   6c. _____

7. Find the sum of the first 10 terms of the arithmetic sequence $\{-10n + 13\}$. *(See textbook Example 7)*

   (a) Use the formula to find the 1st and 10th term of the sequence.   7a. _____

   (b) Find the sum of the first 10 terms of the arithmetic sequence.   7b. _____

Do the Math Exercises 13.2
Arithmetic Sequences

In Problems 1 – 2, find the common difference and write out the first four terms.

1. $\{10n + 1\}$
2. $\left\{\dfrac{1}{4}n + \dfrac{3}{4}\right\}$

1. _____

2. _____

In Problems 3 – 5, find a formula for the nth term of the arithmetic sequence whose first term is a and common difference d is given. What is the fifth term?

3. $a = 8, d = 3$
4. $a = 12, d = -3$

3. _____

4. _____

5. $a = -3; d = \dfrac{1}{2}$

5. _____

In Problems 6 – 8, write a formula for the nth term of each arithmetic sequence. Use the formula to find the 20th term of the sequence.

6. $-5, -1, 3, 7, \ldots$
7. $20, 14, 8, 2, \ldots$

6. _____

7. _____

8. $10, \dfrac{19}{2}, 9, \dfrac{17}{2}, \ldots$

8. _____

In Problems 9 – 12, find the formula for the nth term of an arithmetic sequence using the given information.

9. 5th term is 7; 9th term is 19
10. 2nd term is -9; 8th term is 15

9. _____

10. _____

11. 6th term is -8; 12th term is -38
12. 5th term is 5; 13th term is 7

11. _____

12. _____

Sullivan/Struve/Mazzarella, *Elementary & Intermediate Algebra*, 3e
Copyright © 2014 Pearson Education, Inc.

Do the Math Exercises 13.2

13. Find the sum of the first 40 terms of the sequence 1, 8, 15, 22, ...

13. _____

14. Find the sum of the first 75 terms of the sequence –9, –5, –1, 3 ...

14. _____

15. Find the sum of the first 50 terms of the sequence 12, 4, –4, –12, ...

15. _____

16. Find the sum of the first 80 terms of the arithmetic sequence $\{2n - 13\}$.

16. _____

17. Find the sum of the first 35 terms of the arithmetic sequence $\{-6n + 25\}$.

17. _____

18. Find the sum of the first 28 terms of the arithmetic sequence $\left\{7 - \dfrac{3}{2}n\right\}$.

18. _____

19. Find x so that $2x$, $3x + 2$, and $5x + 3$ are consecutive terms of an arithmetic sequence.

19. _____

20. The Theater An auditorium has 40 seats in the first row and 25 rows in all. Each successive row contains 2 additional seats. How many seats are in the auditorium?

20. _____

Name:
Instructor:

Date:
Section:

Five-Minute Warm-Up 13.3
Geometric Sequences and Series

1. If $f(x) = \left(\dfrac{2}{3}\right)^x$, find each of the following.

   (a) $f(1)$ (b) $f(2)$ (c) $f(3)$

   1a. _____

   1b. _____

   1c. _____

2. If $g(n) = 3n^2$, find each of the following.

   (a) $g(1)$ (b) $g(2)$ (c) $g(3)$

   2a. _____

   2b. _____

   2c. _____

3. Simplify each expression.

   (a) $\dfrac{24x^5}{15x^2}$ (b) $\left(-3r^4\right)^2$

   3a. _____

   3b. _____

4. Evaluate the expression: $\dfrac{\dfrac{3}{2}}{1 - \dfrac{1}{4}}$

   4. _____

Sullivan/Struve/Mazzarella, *Elementary & Intermediate Algebra*, 3e
Copyright © 2014 Pearson Education, Inc.

Name:
Instructor:
Date:
Section:

Guided Practice 13.3
Geometric Sequences and Series

Objective 1: Determine Whether a Sequence Is Geometric

1. If there is a constant ratio between the successive terms, the sequence is called a *geometric sequence*.

   (a) In formulas for geometric sequences, we label the common ratio _____.

   (b) We label the first term _____.

2. Determine if the sequence $36, 18, 9, \frac{9}{2}, \ldots$ is geometric. If it is, determine the first term and the common ratio. *See Example 1*

   2. _____

3. Show that the sequence $\{b_n\} = \left\{\left(\frac{2}{5}\right)^{n-1}\right\}$ is geometric by listing the first four terms and calculating the ratio between successive terms. *(See textbook Example 3)*

   3. _____

Objective 2: Find a Formula for the nth Term of a Geometric Sequence

4. The *n*th term of a geometric sequence whose first term is a and whose common ratio is r, is determined by the formula: $a_n = a_1 r^{n-1}; r \neq 0$.

   (a) Find a formula for the *n*th term of the geometric sequence: $2, -\frac{2}{3}, \frac{2}{9}, -\frac{2}{27}, \ldots$.
   (See textbook Example 4)

   4a. _____

   (b) Find the 8th term of this sequence.

   4b. _____

Sullivan/Struve/Mazzarella, *Elementary & Intermediate Algebra*, 3e

Guided Practice 13.3

Objective 3: Find the Sum of a Geometric Sequence

5. Let $\{a_n\}$ be a geometric sequence with first term a_1 and common ratio r, where $r \neq 0, r \neq 1$.

The sum S_n of the first n terms of $\{a_n\}$ is $S_n = a_1 \cdot \dfrac{1-r^n}{1-r}$; $r \neq 0, r \neq 1$.

(a) Find the first term and the common ratio, r, of the geometric sequence 6, 24, 96, 384, ...
(See textbook Example 5)

5a. _____

(b) Find the sum of the first 10 terms of this sequence.

5b. _____

Objective 4: Find the Sum of a Geometric Series

6. An infinite sum of the terms of a geometric sequence is called a *geometric series*. We can find the sum of the series with the formula: $\sum\limits_{n=1}^{\infty} a_1 r^{n-1} = \dfrac{a_1}{1-r}$ provided that $-1 < r < 1$.

(a) Find the first term and the common ratio, r, of the geometric series: $6 - 2 + \dfrac{2}{3} - \dfrac{2}{9} + ...$
(See textbook Example 7)

6a. _____

(b) Find the sum of this series.

6b. _____

Objective 5: Solve Annuity Problems

7. If P represents the deposit in dollars made at each payment period for an annuity at i percent interest per payment period, the amount A of the annuity after n payment periods is: $A = P \cdot \dfrac{(1+i)^n - 1}{i}$.

Retirement Raymond is planning on retiring in 15 years, so he contributes $1,500 into his IRA every 6 months (semiannually). What will be the value of the IRA when Raymond retires if earns 10% interest compounded semiannually? (See textbook Example 10)

7. _____

Do the Math Exercises 13.3
Geometric Sequences and Series

In Problems 1 – 3, find the common ratio and write out the first four terms of each geometric sequence.

1. $\{(-2)^n\}$
2. $\left\{\dfrac{2^n}{3}\right\}$

3. $\left\{\dfrac{3^{-n}}{2^{n-1}}\right\}$

1. _____

2. _____

3. _____

In Problems 4 – 7, determine whether the given sequence is arithmetic, geometric, or neither.

4. $\{8 - 3n\}$
5. $\{n^2 - 2\}$

6. $100, 20, 4, \dfrac{4}{5}, \ldots$
7. $5, -2, 3, -1, 2, \ldots$

4. _____

5. _____

6. _____

7. _____

In Problems 8 and 9, find a formula for the nth term of the geometric sequence whose first term and common ratio are given. Use the formula to find the 8th term.

8. $a_1 = 30, r = \dfrac{1}{3}$
9. $a_1 = 1, r = -4$

8. _____

9. _____

In Problems 10 – 12, find the indicated term of each geometric sequence.

10. 12th term of $1, 3, 9, 27, \ldots$
11. 8th term of $10, -20, 40, -80, \ldots$

10. _____

11. _____

12. 10th term of $0.4, 0.04, 0.004, 0.0004, \ldots$

12. _____

Do the Math Exercises 13.3

In Problems 13 and 14, find the sum of each geometric series.

13. $3 + 9 + 27 + \ldots + 3^{10}$

14. $\sum_{n=1}^{12} \left[5 \cdot 2^n \right]$

13. _____

14. _____

In Problems 15 – 17, find the sum of each infinite geometric series.

15. $1 + \dfrac{1}{3} + \dfrac{1}{9} + \ldots$

16. $12 - 3 + \dfrac{3}{4} - \dfrac{3}{16} + \ldots$

15. _____

16. _____

17. $\sum_{n=1}^{\infty} \left(10 \cdot \left(\dfrac{1}{3} \right)^n \right)$

17. _____

In Problems 18 and 19, express each repeating decimal as a fraction in lowest terms.

18. $0.\overline{3}$

19. $0.\overline{45}$

18. _____

19. _____

20. **Depreciation of a Car** Suppose that you have just purchased a Chevy Impala for $16,000. Historically, the car depreciates by 10% each year, so that next year the car is worth $16,000(0.9). What will the value of the car be after you have owned it for four years?

20. _____

Name:
Instructor:

Date:
Section:

Five-Minute Warm-Up 13.4
The Binomial Theorem

In Problems 1 – 4, find the product.

1. $(x+2)^2$

2. $(y-3)^2$

1. _____

2. _____

3. $(4x-5y)^2$

4. $\left(3n+\dfrac{2}{3}\right)^2$

3. _____

4. _____

5. Simplify: $\dfrac{7 \cdot 6 \cdot 5 \cdot 4 \cdot 3 \cdot 2}{3 \cdot 2 \cdot 4 \cdot 3 \cdot 2}$

5. _____

Sullivan/Struve/Mazzarella, *Elementary & Intermediate Algebra*, 3e
Copyright © 2014 Pearson Education, Inc.

Name:
Instructor:
Date:
Section:

Guided Practice 13.4
The Binomial Theorem

Objective 1: Compute Factorials

1. If $n \geq 0$ is an integer, the **factorial symbol $n!$** (read "n factorial") is defined as:
$$0! = 1;\ 1! = 1;\ n! = n(n-1)(n-2) \cdot \ldots \cdot 3 \cdot 2 \cdot 1 \text{ for } n \geq 2.$$

Evaluate each of the following. *(See textbook Example 1)*

(a) $\dfrac{10!}{4!}$

(b) $\dfrac{6!}{2!(6-2)!}$

1a. _____

1b. _____

Objective 2: Evaluate a Binomial Coefficient

2. If j and n are integers with $0 \leq j \leq n$, the symbol $\binom{n}{j}$ (read "n choose j") is defined as
$$\binom{n}{j} = \frac{n!}{j!(n-j)!}.$$

Evaluate each of the following. *(See textbook Example 2)*

(a) $\binom{5}{3}$

(b) $\binom{13}{7}$

2a. _____

2b. _____

Guided Practice 13.4

Objective 3: Expand a Binomial

3. Multiplying binomials can become unwieldy when the exponents become large. We now introduce techniques for binomial expansion. We can find the binomial coefficients in a binomial expansion using Pascal's Triangle or the Binominal Theorem, which states that for any positive integer n,

$$(x+a)^n = \binom{n}{0}x^n + \binom{n}{1}ax^{n-1} + \binom{n}{2}a^2 x^{n-2} + \ldots + \binom{n}{j}a^j x^{n-j} + \ldots + \binom{n}{n}a^n.$$

Expand $(p+2)^4$ using the Binomial Theorem. *(See textbook Example 3)*

Do the Math Exercises 13.4
The Binomial Theorem

In Problems 1 – 4, evaluate each expression.

1. $5!$

2. $\dfrac{6!}{2!}$

3. $\dfrac{10!}{2! \cdot 8!}$

4. $0!$

1. _____

2. _____

3. _____

4. _____

In Problems 5 – 8, evaluate each expression.

5. $\dbinom{5}{3}$

6. $\dbinom{7}{5}$

7. $\dbinom{50}{49}$

8. $\dbinom{1000}{1000}$

5. _____

6. _____

7. _____

8. _____

In Problems 9 – 16, expand each expression using the Binomial Theorem.

9. $(x-1)^4$

10. $(x+5)^5$

9. _____

10. _____

Do the Math Exercises 13.4

11. $(2q+3)^4$ **12.** $(3w-4)^4$

11. _____

12. _____

13. $(y^2-3)^4$ **14.** $(3b^2+2)^5$

13. _____

14. _____

15. $(p-3)^6$ **16.** $(3x^2+y^3)^4$

15. _____

16. _____

17. Use the Binomial Theorem to find the numerical value of $(1.001)^5$ correct to five decimal places. [Hint: $(1.001)^5 = (1+10^{-3})^5$].

17. _____

Name:
Instructor:

Date:
Section:

Five-Minute Warm-Up Appendix A
Synthetic Division

In Problems 1 and 2, identify the coefficient.

1. $-x^2$

2. $5y^4$

1. _____

2. _____

In Problems 3 – 6, simplify each expression.

3. $\dfrac{r^7}{r^2}$

4. $\dfrac{63x^8}{27x^4}$

3. _____

4. _____

5. $25z^0$

6. $\dfrac{8a^4b}{16ab^3}$

5. _____

6. _____

7. Divide using long division: $\dfrac{2x^2 + 7x - 10}{x + 1}$

7. _____

Name:
Instructor:

Date:
Section:

Guided Practice Appendix A
Synthetic Division

Objective 1: Divide Polynomials Using Synthetic Division

1. When using synthetic division to divide two polynomials, it is essential that the dividend be in

   _____ form and, if any of the powers of the variable are missing, fill in a _____ coefficient for that term.

2. Synthetic division can be used only when the divisor is of the form _____ or _____ .

3. In each of the following, determine the value of c, when dividing polynomials using synthetic division.

   (a) $\dfrac{2x^2+11x+12}{x+4}$

   (b) $\dfrac{x^2-19x-14}{x-6}$

   3a. _____

   3b. _____

4. Use synthetic division to find the quotient and remainder when $(18 + x^4 - 9x^2 + 3x^3)$ is divided by $(x-3)$. *(See textbook Example 1)*

Step 1: Write the dividend in descending powers of x. Then copy the coefficients of the dividend. Remember to insert a 0 for any missing power of x.

(a) _____

Step 2: Insert the division symbol. Rewrite the divisor in the form $x-c$ and insert the value of c to the left of the division symbol.

(b) _____

Step 3: Bring the 1 down 2 rows and enter it in Row 3.

(c) _____

Step 4: Multiply the latest entry in Row 3 by 3 and place the result in Row 2, one column to the right.

(d) _____

Step 5: Add the entry in Row 2 to the entry above it in Row 1. Enter the sum in Row 3.

(e) _____

Step 6: Repeat steps 4 and 5 until no more entries are available in Row 1.

(f) _____

Step 7: The final entry in Row 3, 99, is the remainder; the other entries in Row 3, 1, 6, 9, and 27, are the coefficients of the quotient, in descending order of degree. The quotient is the polynomial whose degree is one less than the degree of the dividend. State the quotient and the remainder.

(g) _____

Step 8: Check: (Quotient)(Divisor) + Remainder = Dividend. Write your final answer as quotient + $\dfrac{\text{remainder}}{\text{divisor}}$

(h) _____

Sullivan/Struve/Mazzarella, *Elementary & Intermediate Algebra*, 3e
Copyright © 2014 Pearson Education, Inc.

Guided Practice Appendix A

Objective 2: Use the Remainder and Factor Theorems

5. The Remainder Theorem Let f be a polynomial function. If $f(x)$ is divided by $x - c$, then the remainder is _____.

6. The Factor Theorem Let f be a polynomial function. Then $x - c$ is a factor of $f(x)$ if and only if _____.

7. Use the Remainder Theorem to find the remainder when $f(x) = x^2 + 4x - 5$ is divided by $x + 2$. *(See textbook Example 3)*

7. _____

8. Use the Factor Theorem to determine whether the function $f(x) = 4x^3 - 7x^2 - 5x + 6$ has the factor *(See textbook Example 4)*

    (a) $x - 1$ (b) $x + 1$ 8a. _____

8b. _____

Name:
Instructor:

Date:
Section:

Do the Math Exercises Appendix A
Synthetic Division

In Problems 1 – 7, divide using synthetic division.

1. $\dfrac{x^2 - 4x - 21}{x + 3}$

2. $\dfrac{x^2 + 2x - 17}{x - 4}$

1. _____

2. _____

3. $\dfrac{x^3 + x^2 - 22x - 40}{x + 2}$

4. $\dfrac{a^3 - 49a + 120}{a + 8}$

3. _____

4. _____

5. $\dfrac{x^3 - 13x - 17}{x + 3}$

6. $\dfrac{a^4 - 65a^2 + 55}{a - 8}$

5. _____

6. _____

7. $\dfrac{3x^3 + 13x^2 + 8x - 12}{x - \dfrac{2}{3}}$

7. _____

Sullivan/Struve/Mazzarella, *Elementary & Intermediate Algebra*, 3e
Copyright © 2014 Pearson Education, Inc.

Do the Math Exercises Appendix A

In Problem 8 and 9, use the Remainder Theorem to find the remainder.

8. $f(x) = x^3 + 3x^2 - x + 1$ is divided by $x - 3$ 8. _____

9. $f(x) = 3x^3 + 2x^2 - 5$ is divided by $x + 3$ 9. _____

In Problem 10, use the Factor Theorem to determine whether $x - c$ is a factor of the given function for the given value of c.

10. $f(x) = x^2 + 5x + 6$; $c = 3$ 10. _____

11. If $\dfrac{f(x)}{x+3} = 2x + 7$, find $f(x)$.   11. _____

12. If $\dfrac{f(x)}{x-3} = x^2 + 2 + \dfrac{7}{x-3}$, find $f(x)$.   12. _____

Name:
Instructor:

Date:
Section:

Five-Minute Warm-Up Appendix B
Geometry Review

1. *List for the formulas for each of the following geometric figures.*

(a) Square	(b) Rectangle
Area:_____ Perimeter:_____	Area:_____ Perimeter:_____
(c) Triangle	**(d) Trapezoid**
Area:_____ Perimeter:_____	Area:_____ Perimeter:_____
(e) Parallelogram	**(f) Circle**
Area:_____ Perimeter:_____	Area:_____ Circumference:_____

In Problems 2 – 6, (a) round and then (b) truncate each decimal to the indicated number of places.

2. 15.96145; 3 decimal places

2a. _____

2b. _____

3. −0.098; 2 decimal places

3a. _____

3b. _____

4. 9.55; nearest whole number

4a. _____

4b. _____

5. 100.73; nearest tenth

5a. _____

5b. _____

6. $5.76715; nearest cent

6a. _____

6b. _____

Sullivan/Struve/Mazzarella, *Elementary & Intermediate Algebra*, 3e
Copyright © 2014 Pearson Education, Inc.

Name: _____ Date: _____
Instructor: _____ Section: _____

Guided Practice Appendix B
Geometry Review

B.1 Lines and Angles

1. In your own words, describe each of the following:

   (a) point _____

   (b) line _____

   (c) ray _____

   (d) line segment _____

   (e) congruent _____

   (f) right angle _____

   (g) acute angle _____

   (h) obtuse angle _____

   (i) complementary angles _____

   (j) supplementary angles _____

   (k) parallel lines _____

   (l) perpendicular lines _____

2. Given that the lines in the figure are parallel, identify all of the following:

   (a) alternate interior angles _____

   (b) alternate exterior angles _____

   (c) corresponding angles _____

B.2 Polygons

3. The sum of the measures of the interior angles of a triangle is 180°. The measure of the first angle is 15° less than the second. The measure of the third angle is 45° more than half of the second. Find the measure of each interior angle of the triangle.

(a) **Step 1: Identify** What formula is needed to solve this problem? _____

Sullivan/Struve/Mazzarella, *Elementary & Intermediate Algebra*, 3e
Copyright © 2014 Pearson Education, Inc.

Guided Practice Appendix B

(b) Step 2: Name
If x represents the value of the second angle, what expression represents the value of the first angle?

_____

What expression represents the value of the third angle? _____

(c) Step 3: Translate Substitute the values from Step 2 into the formula from Step 1.

$$x° + y° + z° = 180°$$

_____

(d) Step 4: Solve

(e) Step 5: Check Substitute the value you found for x into your equation. Does the sum of the angles equal $180°$?

(f) Step 6: Answer the Question $m\angle x = $ _____ ; $m\angle y = $ _____ ; $m\angle z = $ _____
Note: $m\angle x$ means the measure of angle x.

4. In your own words, describe *congruent triangles*.

5. What 3 properties can be used to determine if two triangles are congruent?

6. What is the difference between *similar triangles* and *congruent triangles*?

B.4 Volume and Surface Area

7. If a solid figure has dimensions measured in feet, what units are used when determining the surface area?

7. _____

8. If a solid figure has dimensions measured in meters, what units are used when determining the volume?

8. _____

Name:
Instructor:

Date:
Section:

Do the Math Exercises Appendix B
Geometry Review

In Problems 1 – 4, find the complement of each angle.

1. 19°

2. 51°

3. $p°$

4. $(n-15)°$

1. _____

2. _____

3. _____

4. _____

In Problems 5 – 8, find the supplement of each angle.

5. 85°

6. 106°

7. $r°$

8. $(s+40)°$

5. _____

6. _____

7. _____

8. _____

In Problems 9 – 10, determine the missing angle of the triangle.

9. Two angles of the triangle are 32° and 65°.

10. A right triangle has an acute angle whose measure is 44°.

9. _____

10. _____

In Problems 11 – 15, determine (a) the perimeter and (b) the area.

11. A rectangle whose length is 12 miles and whose width is 8 miles.

11a. _____

11b. _____

12. A square which measures 13 yards on a side.

12a. _____

12b. _____

Do the Math Exercises Appendix B

13. A trapezoid with bases 10 m and 14m, height 5 m, and legs 6 m each.

13a. _____

13b. _____

14. A triangle whose base is 9 in., height 8 in., and the other two sides are 10 in. and 9 in.

14a. _____

14b. _____

15. A circle whose diameter is 3 miles.

15a. _____

15b. _____

16. Find (a) the volume and (b) the surface area of a rectangular box with length 2 meters, width 6 meters, and height 9 meters.

16a. _____

16b. _____

17. **Water for Horses** A trough for horses in the shape of a rectangular solid is 10 feet long, 2 feet wide, and 3 feet deep. How much water can the trough hold?

17. _____

18. Find (a) the volume and (b) the surface area of a right circular cylinder with radius 3 inches and height 6 inches.

18a. _____

18b. _____

19. **Water Cooler** The cups at the water cooler are cone-shaped. How much water can a cup hold if it has a 5-inch diameter and is 8 inches in height. Express your answer as a decimal rounded to the nearest hundredth.

19. _____

Name:
Instructor:

Date:
Section:

Five-Minute Warm-Up Appendix C.1
A Review of Systems of Linear Equations in Two Variables

1. Evaluate $5x - 2y$ for $x = 3$, $y = -1$.
   1. _____

2. Determine whether the point $\left(8, -\dfrac{4}{3}\right)$ is on the graph of the equation $x - 3y = 12$.
   2. _____

3. Graph the linear equation $y = -\dfrac{2}{3}x + 4$.

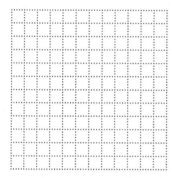

4. Find the equation of the line parallel to $x - y = 2$ containing the point $(-3, -2)$.
   4. _____

5. Determine the slope and y-intercept of $x - 3y = -9$.
   5.
   slope: _____

   y-intercept: ___

6. What is the additive inverse of 15?
   6. _____

7. Solve: $4x - 2(5x - 1) = -4$
   7. _____

Name:
Instructor:

Date:
Section:

Guided Practice Appendix C.1
A Review of Systems of Linear Equations in Two Variables

Objective 1: Determine Whether an Ordered Pair Is a Solution of a System of Linear Equations

1. Complete the following chart which describes the solutions to a system of linear equations in two variables.

Number of Solutions	Classification	Graph of the Two Lines
(a) no solution		
(b) infinitely many solutions		
(c) exactly one solution		

Objective 3: Solve a System of Two Linear Equations by Substitution *(See textbook Example 4)*

2. Solve the following system by substitution: $\begin{cases} 5x + 2y = -5 & (1) \\ 3x - y = -14 & (2) \end{cases}$

Step 1: Solve one of the equations for one of the unknowns. It is easiest to solve equation (2) for y since the coefficient of y is -1.

Equation (2): $\quad 3x - y = -14$

Subtract $3x$ from both sides: (a) _____

Multiply both sides by -1: (b) _____

Step 2: Substitute your expression for y in equation (1).

Equation (1): (c) _____

Substitute the expression from equation (2) into equation (1): (d) _____

Step 3: Solve the equation for x.

Distribute the 2: (e) _____

Combine like terms: (f) _____

Subtract 28 from both sides: (g) _____

Divide both sides by 11: (h) _____

Step 4: Substitute the value for x into the equation from Step 1(b) and then solve for y.

(i) $y =$ _____

Step 5: Check your answer in both of the original equations. If both equations yield a true statement, you have the correct answer.

Write the ordered pair that is the solution to the system. (j) _____

Sullivan/Struve/Mazzarella, *Elementary & Intermediate Algebra*, 3e
Copyright © 2014 Pearson Education, Inc.

Guided Practice C.1

Objective 4: Solve a System of Two Linear Equations Containing by Elimination *(See textbook Example 6)*

3. Solve the following system by elimination: $\begin{cases} 2x + y = -4 & (1) \\ 3x + 5y = 29 & (2) \end{cases}$

Step 1: Our first goal is to get the coefficients on one of the variables to be additive inverses. In looking at this system, we can make the coefficients of y be additive inverses by multiplying equation (1) by -5.

Multiply both sides of (1) by -5, use the Distributive Property, and then write the equivalent system of equations.

$\begin{cases} 2x + y = -4 \\ 3x + 5y = 29 \end{cases}$

(a) $\begin{cases} \underline{\hspace{3cm}} \quad (1) \\ \underline{\hspace{3cm}} \quad (2) \end{cases}$

Step 2: We now add equations (1) and (2) to eliminate the variable y and then solve for x.

Add (1) and (2): (b) _____

Divide both sides by -7: (c) _____

Step 3: Substitute your value for x into either equation (1) or equation (2). We will use equation (1) as it looks like less work.

Equation (1): $2x + y = -4$

Substitute your value for x. (d) _____

Solve for y. (e) _____

Step 4: Check your answer in both of the original equations. If both equations yield a true statement, you have the correct answer.

Write the ordered pair that is the solution to the system. (f) _____

Objective 5: Identify Inconsistent Systems

4. Algebraically, what occurs when you solve an inconsistent system of equations? _____

_____

Objective 6: Write the Solution of a System with Dependent Equations

5. Algebraically, what occurs when you solve a dependent system of equations? _____

_____

6. The following system is consistent and dependent. $\begin{cases} -x + 3y = 1 \\ 2x - 6y = -2 \end{cases}$

   Express the solution using set-builder notation.   6. _____

Name:
Instructor:

Date:
Section:

Do the Math Exercises Appendix C.1
A Review of Systems of Linear Equations in Two Variables

In Problems 1 and 2, determine whether the given ordered pairs are solutions of the system of linear equations.

1. $\begin{cases} x - 2y = -11 \\ 3x + 2y = -1 \end{cases}$

    (a) $(-5, 3)$ (b) $(-3, 4)$

2. $\begin{cases} -3x + y = 5 \\ 6x - 2y = 6 \end{cases}$

    (a) $(-2, -1)$ (b) $(2, 0)$

1a. _____

1b. _____

2a. _____

2b. _____

In Problems 3 and 4, solve the system of equations by graphing.

3. $\begin{cases} y = -2x + 4 \\ y = 2x - 4 \end{cases}$

4. $\begin{cases} -x + 2y = -9 \\ 2x + y = -2 \end{cases}$

3. _____

4. _____

In Problems 5 and 6, solve the system of equations using substitution.

5. $\begin{cases} y = \dfrac{1}{2}x \\ x - 4y = -4 \end{cases}$

6. $\begin{cases} 3x + 2y = 0 \\ 6x + 2y = 5 \end{cases}$

5. _____

6. _____

In Problems 7 and 8, solve the system of equations using elimination.

7. $\begin{cases} 6x - 4y = 6 \\ -3x + 2y = 3 \end{cases}$

8. $\begin{cases} x + 2y = -\dfrac{8}{3} \\ 3x - 3y = 5 \end{cases}$

7. _____

8. _____

Do the Math Exercises C.1

In Problems 9 – 11, solve the system of equations by any method.

9. $\begin{cases} 2x + y = -1 \\ -3x - 2y = 7 \end{cases}$

10. $\begin{cases} y = \dfrac{1}{2}x + 2 \\ x - 2y = -4 \end{cases}$

9. _____

10. _____

11. $\begin{cases} \dfrac{1}{3}x - \dfrac{1}{2}y = -5 \\ -\dfrac{4}{5}x + \dfrac{6}{5}y = 1 \end{cases}$

11. _____

In Problems 12 and 13, use slope-intercept form to determine the number of solutions the system has.

12. $\begin{cases} 4x - 2y = 8 \\ -10x + 5y = 5 \end{cases}$

13. $\begin{cases} 2x - y = -5 \\ -4x + 3y = 9 \end{cases}$

12. _____

13. _____

14. **Rhombus** A rhombus is a parallelogram whose adjacent sides are congruent. Consider the rhombus with vertices $(-1, 3)$, $(3, 6)$, $(3, 1)$, and $(-1, -2)$ to find the following.

(a) Find the equation of the line for the diagonal through the points $(-1, 3)$ and $(3, 1)$. 14a. _____

(b) Find the equation of the line for the diagonal through the points $(-1, -2)$ and $(3, 6)$. 14b. _____

(c) Find the point of intersection of the diagonals. 14c. _____

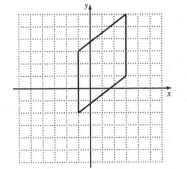

Name:
Instructor:

Date:
Section:

Five-Minute Warm-Up Appendix C.2
Systems of Linear Equations in Three Variables

1. Evaluate $2x - 5y - 8z$ for $x = -7,\ y = 4,\ z = -3$.

   1. _____

In Problems 2 – 4, solve the system of equations using elimination.

2. $\begin{cases} x - y = 10 \\ x + y = -20 \end{cases}$

   2. _____

3. $\begin{cases} 5x - y = 3 \\ -10x + 2y = 2 \end{cases}$

   3. _____

4. $\begin{cases} 4x + y = 3 \\ 8x + 2y = 6 \end{cases}$

   4. _____

Name: Date:
Instructor: Section:

Guided Practice Appendix C.2
Systems of Linear Equations in Three Variables

Objective 1: Solve Systems of Three Linear Equations

1. Geometrically, what does an equation in three variables represent? _____

2. A system of three linear equations containing three variables has one of the following possible solutions:

(a) **Exactly one solution** is a _____ system with _____ equations.

(b) **No solution** is an _____ system.

(c) **Infinitely many solutions** is a _____ system with _____ equations.

3. Use the method of elimination to solve the system: $\begin{cases} x + 3y + 3z = 9 & (1) \\ 3x + 5y + 4z = 8 & (2) \\ 5x + 3y + 7z = 9 & (3) \end{cases}$ *(See textbook Example 2)*

Step 1: Our goal is to eliminate the same variable from two of the equations. In looking at the system, we notice that we can use equation (1) to eliminate the variable x from equations (2) and (3). We can do this by multiplying equation (1) by -3 and adding the result to equation (2). The equation that results becomes equation (4). Why do we do this? Because the coefficients on x will be additive inverses and adding the equations eliminates the variable x. We also multiply equation (1) by -5 and add the result to equation (3). The equation that results becomes equation (5).

Multiply equation (1) by −3:   (a) $-3x - 9y - 9z = -27$

Equation (2):   (b) $3x + 5y + 4z = 8$

Add (a) and (b):   (c) $-4y - 5z = -19$ (4)

Multiply equation (1) by −5:   (d) $-5x - 15y - 15z = -45$

Equation (3):   (e) $5x + 3y + 7z = 9$

Add (d) and (e):   (f) $-12y - 8z = -36$ (5)

Step 2: We now concentrate on equations (4) and (5), treating them as a system of two equations containing two variables. It is easiest to eliminate the variable y by multiplying equation (4) by -3 and adding the result to equation (5). This results in an equation in one variable, equation (6).

Multiply equation (4) by −3:   (g) $12y + 15z = 57$

Equation (5):   (h) $-12y - 8z = -36$

Add (g) and (h):   (i) $7z = 21$ (6)

Step 3: We solve equation (6) for z.   On line (i), divide both sides by 7: (j) $z = 3$

Step 4: Back-substitute your value for z into either equation (4) or equation (5) to solve for y.   (k) $y = 1$

Step 5: Back-substitute your values for y and z into one of the equations (1), (2), or (3). Solve for x.   (l) $x = -3$

continued next page

Guided Practice C.2

Step 6: Check your answer in all three of the original equations. If all equations yield a true statement, you have the correct answer.

Write the ordered triple that is the solution to the system. (m) _____

Objective 2: Identify Inconsistent Systems

4. Whenever you solve a system of equations and end up with a false statement such as $0 = -7$, you have an _____ system. We say that the solution to the system is _____.

Objective 3: Write the Solution of a System with Dependent Equations

5. Whenever you solve a system of equations and end up with a true statement such as $3 = 3$ or $0 = 0$, you have a _____ system.

6. Typically, when writing the solution set, we express the values of x and y in terms of z, although this is not required. We know that $\begin{cases} 2x - y = 2 & (1) \\ -x + 5z = 3 & (2) \\ -y + 10z = 8 & (3) \end{cases}$ is a dependent system. *(See textbook Example 5)*

(a) Solve equation (2) for x: _____

(b) Solve equation (3) for y: _____

(c) Express the solution to the system: $\{(x, y, z) | x = $ _____, $y = $ _____, z is any real number$\}$

Objective 4: Model and Solve Problems Involving Three Linear Equations

7. **Theater Revenues** A theater has 600 seats, divided into orchestra, main floor, and balcony seating. Orchestra seats sell for $80, main floor seats for $60, and balcony seats for $25. If all the seats are sold, the total revenue to the theater is $33,500. One evening, all of the orchestra seats were sold, $\frac{3}{5}$ of the main seats were sold and $\frac{4}{5}$ of the balcony seats were sold. The total revenue collected was $24,640. How many are there of each kind of seat? *(See textbook Example 6)*

(a) Write an equation that expresses the total number of seats in the theater if a represents the number of orchestra, b represents the number of main floor, and c represents the number of balcony seats:

_____

(b) Write an equation that calculates the total revenue from all seats: _____

(c) Write an equation that calculates the revenue when a portion of the seats are sold:

_____

Do the Math Exercises Appendix C.2
Systems of Linear Equations in Three Variables

Determine whether the given ordered triples are solutions of the system of linear equations.

1. $\begin{cases} 2x+y-2z=6 \\ -2x+y+5z=1 \\ 2x+3y+z=13 \end{cases}$

   (a) $(3, 2, 1)$ **(b)** $(10, -4, 5)$

   1a. _____

   1b. _____

In Problems 2 – 6, solve each system of three linear equations containing three unknowns.

2. $\begin{cases} x+2y-z=4 \\ 2x-y+3z=8 \\ -2x+3y-2z=10 \end{cases}$

   2. _____

3. $\begin{cases} x-y+3z=2 \\ -2x+3y-8z=-1 \\ 2x-2y+4z=7 \end{cases}$

   3. _____

4. $\begin{cases} x-3z=-3 \\ 3y+4z=-5 \\ 3x-2y=6 \end{cases}$

   4. _____

Sullivan/Struve/Mazzarella, *Elementary & Intermediate Algebra*, 3e

Do the Math Exercises C.2

5. $\begin{cases} x - y + 2z = 3 \\ 2x + y - 2z = 1 \\ 4x - y + 2z = 0 \end{cases}$ 5. _____

6. $\begin{cases} x + y + z = 4 \\ 2x + 3y - z = 8 \\ x + y - z = 3 \end{cases}$ 6. _____

7. **Curve Fitting** The function $f(x) = ax^2 + bx + c$ is a quadratic function where a, b, and c are constants.

   (a) If $f(-1) = 6$, then $6 = a(-1)^2 + b(-1) + c$ or $a - b + c = 6$. Find two additional linear equations if $f(1) = 2$, and $f(2) = 9$. 7a. _____

   (b) Use the three linear equations found in part (a) to determine a, b, and c. 7b. _____

   (c) What is the quadratic function that contains the points $(-1, 6)$, $(1, 2)$, and $(2, 9)$? 7c. _____

8. **Nutrition** Antonio is on a special diet that requires he consume 1325 calories, 172 grams of carbohydrates, and 63 grams of protein for lunch. He wishes to have a Broccoli and Cheese Baked Potato, Chicken BLT Salad, and a medium Coke. Each Broccoli and Cheese Baked Potato has 480 calories, 80 g of carbohydrates, and 9 g of protein. Each Chicken BLT Salad has 310 calories, 10 g for carbohydrates, and 33 g for protein. Each Coke has 140 calories, 37 g of carbohydrates, and 0 g of protein. How many servings of each does Antonio need? 8. _____

Name:
Instructor:

Date:
Section:

Five-Minute Warm-Up Appendix C.3
Using Matrices to Solve Systems

1. Determine the coefficients of $-2x + y - 3z$.

   1. _____

2. Evaluate $-x - 5y + 11z$ for $x = 2$, $y = -5$, $z = -1$.

   2. _____

In Problems 3 and 4, solve for the indicated variable.

3. $7x - 5y = 10$ for y

4. $\dfrac{3}{2}x + \dfrac{2}{3}y = -2$ for x

   3. _____

   4. _____

In Problems 5 and 6, use the Distributive Property to remove the parentheses.

5. $-3(2x - 9y + z)$

6. $-\dfrac{5}{4}(8x - 4y + 12z)$

   5. _____

   6. _____

7. If $f(x) = -x^2 - 5x + 7$, find the value of each function.

   (a) $f(4)$

   (b) $f(-3)$

   7a. _____

   7b. _____

Name:
Instructor:
Date:
Section:

Guided Practice Appendix C.3
Using Matrices to Solve Systems

Objective 1: Write the Augmented Matrix of a System

1. A **matrix** is a rectangular array of numbers, meaning that the order of the numbers in the matrix is relevant. The size of the matrix, called the *dimension*, is denoted as the number of rows by the number of columns. If a matrix has 3 rows and 4 columns, we say the dimension of the matrix is 3×4.

    Find the dimension: $\begin{bmatrix} -1 & 8 & 3 & -7 \\ -5 & 0 & 1 & 2 \end{bmatrix}$

    1. _____

2. An **augmented matrix** can be used to represent a system of linear equations. Each row is created from one of the equations in the system and each column represents the coefficients of one of the variables. The vertical bar is the equal sign and the last column represents the constants. Be sure each equation is written in _____ form, filling in the coefficient of any missing variables with _____.

3. Write the system of equations as an augmented matrix. *(See textbook Example 1)*

    $\begin{cases} 2x + y + z = 3 \\ 4y - 7z = -1 \\ x + 3y = 0 \end{cases}$

    3. _____

Objective 2: Write the System from the Augmented Matrix

4. Write the system of linear equations corresponding to the augmented matrix. *(See textbook Example 2)*

    $\begin{bmatrix} 1 & 1 & | & 2 \\ -3 & 1 & | & 10 \end{bmatrix}$

    4. _____

Objective 3: Perform Row Operations on a Matrix

5. There are three basic row operations. These are similar to the types of operations that we used to solve systems of equations earlier in this chapter. List the row operations for matrices.

    (a) _____
    (b) _____
    (c) _____

6. The notation $-3r_1$ means multiply row 1 by -3. The notation $R_2 = r_1 + r_2$ means replace row 2 with the sum of row 1 plus row 2. Perform the following row operations and write the new augmented matrix. *(See textbook Example 3)*

    (a) $\begin{bmatrix} 2 & -1 & | & 5 \\ 1 & -7 & | & 4 \end{bmatrix}$ $R_2 = -2r_2 \rightarrow$

    (b) $\begin{bmatrix} 2 & -1 & | & 5 \\ 1 & -7 & | & 4 \end{bmatrix}$ $R_1 = r_2 + r_1 \rightarrow$

Objective 4: Solve Systems Using Matrices

7. When is a matrix in **row echelon form**? _____

Guided Practice C.3

8. Solve the following system using matrices: $\begin{cases} x + y + z = 1 \\ 2x + 2y = 6 \\ 3x + 4y - z = 13 \end{cases}$ (See textbook Example 6)

Step 1: Write the augmented matrix of the system. We will use the row operations from Objective 3 to solve the system.

(a) _____

Step 2: We want the entry in row 1, column 1 to be 1. This is already done.

Step 3: We want the entry in row 2, column 1 to be zero. We also want the entry in row 3, column 1 to be a zero. We use row operation #3 to accomplish this. The entries in row 1 remain unchanged.

$R_2 = -2r_1 + r_2 \rightarrow$

$R_3 = -3r_1 + r_3 \rightarrow$

(b) _____

Step 4: We want the entry in row 2, column 2 to be a 1. This is accomplished by interchanging rows 2 and 3, row operation #1.

(c) _____

Step 5: We want the entry in row 3, column 2 to be zero. This is already accomplished.

Step 6: We want the entry in row 3, column 3 to be a 1. We use row operation #2 to accomplish this.

$R_3 = -\dfrac{1}{2}r_3 \rightarrow$

(d) _____

Step 7: The augmented matrix is in row echelon form. Write the system of equations corresponding to the augmented matrix and solve. We know the value of z. Substitute into equation (2) and solve for y. Then use these values and substitute into equation (1) to solve for x.

State the solution as an ordered triple. (e) _____

Step 8: Check We leave it to you to verify the solution.

Objective 5: Solve Consistent Systems with Dependent Equations and Solve Inconsistent Systems

9. State the solution to the system represented by the augmented matrix: $\begin{bmatrix} 1 & 4 & | & 2 \\ 0 & 0 & | & 0 \end{bmatrix}$

9. _____

(See textbook Example 7)

10. State the solution to system represented by the augmented matrix: $\begin{bmatrix} 1 & 0 & -1 & | & -2 \\ 0 & 1 & 5 & | & -9 \\ 0 & 0 & 0 & | & -3 \end{bmatrix}$

10. _____

(See textbook Example 8)

Do the Math Exercises Appendix C.3
Using Matrices to Solve Systems

Write the augmented matrix of the given system of equations.

1. $\begin{cases} 6x + 4y + 2 = 0 \\ -x - y + 1 = 0 \end{cases}$

    1. _____

Perform each row operation on the given augmented matrix.

2. $\begin{bmatrix} 1 & -1 & 1 & | & 6 \\ -2 & 1 & -3 & | & 3 \\ 3 & 2 & -2 & | & -5 \end{bmatrix}$

    (a) $R_2 = 2r_1 + r_2$ followed by

    (b) $R_3 = -3r_1 + r_3$

    2a. _____

    2b. _____

In Problems 3 – 6, solve each system of equations using matrices. If a system has no solution, say that it is inconsistent.

3. $\begin{cases} 5x - 2y = 3 \\ -15x + 6y = -9 \end{cases}$

4. $\begin{cases} 3x + 3y = -1 \\ 2x + y = 1 \end{cases}$

    3. _____

    4. _____

Do the Math Exercises C.3

5. $\begin{cases} -x+2y+z=1 \\ 2x-y+3z=-3 \\ -x+5y+6z=2 \end{cases}$

5. _____

6. $\begin{cases} 2x-y+2z=13 \\ -x+2y-z=-14 \\ 3x+y-2z=-13 \end{cases}$

6. _____

7. **Finance** Marlon has $12,000 to invest. He decides to place some of the money into a savings account paying 2% annual interest, some in Treasury bonds paying 4% annual interest and some in a mutual fund paying 9% annual interest. Marlon would like to earn $440 per year in income. In addition, Marlon wants his investment in the savings account to be $4,000 more than the amount in Treasury bonds. How much should Marlon invest in each investment category?

7. _____

Name:
Instructor:

Date:
Section:

Five-Minute Warm-Up Appendix C.4
Determinants and Cramer's Rule

1. Evaluate: $-3 \cdot 5 - 2 \cdot (-7)$

   1. _____

2. Simplify each expression.

   (a) $\dfrac{13}{0}$

   (b) $\dfrac{0}{45}$

   2a. _____

   2b. _____

3. Simplify: $\dfrac{-12}{-10}$

   3. _____

4. Evaluate $x - 2y + z$ for $x = -\dfrac{3}{2}$, $y = \dfrac{2}{5}$, $z = \dfrac{3}{4}$.

   4. _____

In Problems 5 and 6, solve each equation.

5. $-8 - 3x = 1$

6. $\dfrac{9}{7}z - 4 = 32$

   5. _____

   6. _____

7. Solve: $\begin{cases} -6 - 2(3x - 6y) = 0 \\ 6 - 12(x - 2y) = 0 \end{cases}$

   7. _____

Sullivan/Struve/Mazzarella, *Elementary & Intermediate Algebra*, 3e
Copyright © 2014 Pearson Education, Inc.

Name:
Instructor:
Date:
Section:

Guided Practice Appendix C.4
Determinants and Cramer's Rule

Objective 1: Evaluate the Determinant of a 2×2 Matrix

1. The notation $\begin{vmatrix} a & b \\ c & d \end{vmatrix}$ denotes the **determinant** for the matrix $\begin{bmatrix} a & b \\ c & d \end{bmatrix}$. We use the definition $\begin{vmatrix} a & b \\ c & d \end{vmatrix} = ad - bc$ to evaluate the determinant and find a single number representing the array.

Evaluate each determinant. *(See textbook Example 1)*

(a) $\begin{vmatrix} 3 & 8 \\ -1 & 2 \end{vmatrix}$ (b) $\begin{vmatrix} -7 & 2 \\ 9 & 1 \end{vmatrix}$

1a. _____

1b. _____

Objective 2: Use Cramer's Rule to Solve a System of Two Equations

2. If the number of unknowns is the same as the number of equations in a system, we can use Cramer's Rule to solve the system. We first set up several different determinants.

(a) D is a determinant in which each entry is a _____ of the variables in the system.

(b) D_x is a determinant in which the first column (coefficients of x) is replaced by _____.

(c) D_y is a determinant in which the _____ column (coefficients of ___) is replaced by the constants on the right side of the equal sign.

3. Given the linear equations $-5y + 3x = 9$ and $2 - x - 2y = 0$, write the system with each equation expressed in standard form.

3. _____

4. Using your system above, determine each of the following. *(See textbook Example 2)*

(a) $D =$ (b) $D_x =$ (c) $D_y =$

5. According to Cramer's Rule, the solution to the system of equations can be found evaluating the determinants D, D_x, and D_y and simplifying the following ratios.

(a) $x = \dfrac{?}{?}$ (b) $y = \dfrac{?}{?}$

Objective 3: Evaluate the Determinant of a 3×3 Matrix

6. We cannot use the process from Objective 1 to find the value of a 3×3 determinant. Do you see why? Instead we use a process called **expansion by minors**. Although it is possible to expand by any row or column, typically we use the first row entries to be the coefficients of the corresponding minors. Be careful with the operations between the terms as they alternate in the expansion and can change when expanding by a different row or column. *(See textbook Example 3)*

Guided Practice C.4

Set up, but do not evaluate, the expansion by row 1: $\begin{vmatrix} 1 & -1 & 2 \\ 5 & -3 & 3 \\ 1 & 4 & -2 \end{vmatrix}$

_____

Objective 4: Use Cramer's Rule to Solve a System of Three Equations (See textbook Example 5)

7. Solve the following system using Cramer's Rule: $\begin{cases} x + 2y - z = -3 \\ 2x - 4y + z = -7 \\ -2x + 2y - 3z = 4 \end{cases}$

Step 1: Find the determinant of the coefficients of the variables, D.

Write the determinant, D: **(a)** _____

Evaluate D: **(b)** _____

Step 2: Because $D \neq 0$, we continue by writing and evaluating each of the determinants D_x, D_y, and D_z.

Write and evaluate D_x: **(c)** _____

Write and evaluate D_y: **(d)** _____

Write and evaluate D_z: **(e)** _____

Step 3: Solve for each of the variables.

(f) $x = \dfrac{D_x}{D} = \dfrac{?}{?}$ = _____

(g) $y = \dfrac{?}{?} = \dfrac{?}{?}$ = _____

(h) $z = \dfrac{?}{?} = \dfrac{?}{?}$ = _____

Step 4: Check you answer. We leave it to you to verify the solution.

State your solution as an ordered triple. **(i)** _____

8. If $D = 0$ Cramer's Rules does not apply. If at least one of the determinants D_x, D_y, or D_z is different from 0, then the system is _____ and the solution set is _____ .

9. If $D = 0$ Cramer's Rule does not apply. If all of the determinants D_x, D_y, and D_z equal 0, then the system is _____ and _____ and there are _____ solutions.

Do the Math Exercises Appendix C.4
Determinants and Cramer's Rule

In Problems 1 – 3, find the value of each determinant.

1. $\begin{vmatrix} 5 & 3 \\ 2 & 4 \end{vmatrix}$

2. $\begin{vmatrix} -8 & 5 \\ -4 & 3 \end{vmatrix}$

3. $\begin{vmatrix} -2 & 1 & 6 \\ -3 & 2 & 5 \\ 1 & 0 & -2 \end{vmatrix}$

1. _____

2. _____

3. _____

In Problems 4 – 7, solve each system of equations using Cramer's Rule, if possible.

4. $\begin{cases} 2x + 4y = -6 \\ 3x + 2y = 7 \end{cases}$

5. $\begin{cases} 3x - 6y - 2 = 0 \\ x + 2y - 4 = 0 \end{cases}$

4. _____

5. _____

6. $\begin{cases} x + y - z = 6 \\ x + 2y + z = 6 \\ -x - y + 2z = -7 \end{cases}$

6. _____

Do the Math Exercises C.4

7. $\begin{cases} x - 2y - z = 1 \\ 2x + 2y + z = 3 \\ 6x + 6y + 3z = 6 \end{cases}$

7. _____

8. *Solve for x:* $\begin{vmatrix} -2 & x \\ 3 & 4 \end{vmatrix} = 1$

8. _____

9. **Geometry: Area of a Triangle** Given the points $A = (-1, -1)$, $B = (3, 2)$, and $C = (0, 6)$, find the area of the triangle ABC.

9. _____

Chapter 1 Answers

Section 1.2

Five-Minute Warm-Up 1. Counting numbers such as 1, 2, 3... 2. Numbers whose only factors are 1 and itself. 3. Natural numbers which are not prime. 1 is neither prime nor composite. 4. The LCM of two or more numbers is the smallest number that is a multiple of each of the numbers. **5a.** 4, 9 **5b.** 36 **6.** 24, 48, 72,...
7. 1, 2, 4 **8.** 1, 2, 3, 6, 9, 18 **9.** $2 \cdot 3 \cdot 3$

Guided Practice **1a.** $2 \cdot 3 \cdot 5$ **1b.** $2 \cdot 2 \cdot 2 \cdot 3 \cdot 5$ **2a.** $2 \cdot 2 \cdot 2$ **2b.** $2 \cdot 2 \cdot 3$ **2c.** $2 \cdot 2$ **2d.** $2 \cdot 3$ **2e.** 24
3a. $\frac{3}{3}$ **3b.** $\frac{15}{24}$ **4a.** $2 \cdot 2 \cdot 3 \cdot 3$ **4b.** $2 \cdot 2 \cdot 2 \cdot 3$ **4c.** $2 \cdot 2 \cdot 2 \cdot 3 \cdot 3$ **4d.** 72 **4e.** $\frac{2}{2}$ **4f.** $\frac{3}{3}$
4g. $\frac{7}{36} = \frac{14}{72}; \frac{11}{24} = \frac{33}{72}$ **5a.** 94,000 **5b.** 94,200 **5c.** 94,205 **5d.** 94,204.6 **5e.** 94,200 **5f.** 94,204.64
6. 0.36 **7.** $\frac{4}{5}$ **8.** 0.75 **9.** 0.5%

Do the Math **1.** $2 \cdot 3 \cdot 3 \cdot 3$ **2.** $3 \cdot 3 \cdot 7$ **3.** 280 **4.** 180 **5.** $\frac{12}{15}$ **6.** $\frac{10}{28}$ **7.** $\frac{3}{36}$ and $\frac{10}{36}$ **8.** $\frac{25}{60}$ and $\frac{28}{60}$
9. $\frac{3}{5}$ **10.** $\frac{8}{9}$ **11.** tenths place **12.** ones place **13.** 7300 **14.** 37.4 **15.** $0.\overline{2}$ **16.** 0.34375 **17.** $\frac{2}{5}$ **18.** $\frac{179}{500}$
19. 0.59 **20.** 34.9% **21.** 180 laps

Section 1.3

Five-Minute Warm-Up **1.** 0 **2.** undefined **3.** 1 **4.** 3 **5a.** 3 **5b.** 7 **6.** $\frac{3}{6}$, 0.001, $\frac{0}{1}$, $\frac{4}{2}$ **7.** $\frac{3}{6}$, 0.001, $\frac{4}{2}$
8. $\frac{3}{6}$, 0.001 **9.** $\frac{0}{1}$ **10.** $0.8\overline{3}$ **11.** 0.8

Guided Practice **1a.** $B = \{0, 2, 4, 6, 8\}$ **1b.** $C = \{1, 3, 5, 7, 9\}$ **2a.** $\{14\}$ **2b.** $\left\{14, \frac{0}{3}\right\}$
2c. $\left\{14, \frac{0}{3}, \frac{-8}{2}, -10\right\}$ **2d.** $\left\{14, \frac{0}{3}, \frac{-8}{2}, -10, 2.\overline{6}\right\}$ **2e.** $\{-1.050050005...\}$
2f. $\left\{14, \frac{0}{3}, \frac{-8}{2}, -10, 2.\overline{6}, -1.050050005...\right\}$ **3a.** irrational, real **3b.** integer, rational, real **3c.** rational, real **3d.** natural, whole, integer, rational, real **4.** See Graphing Answer Section **5a.** > **5b.** = **5c.** > **6a.** True
6b. True **6c.** False **7a.** $x > 0$ **7b.** $x < 0$ **8.** positive **9a.** 0 **9b.** 4.2 **9c.** 120

Do the Math **1.** $B = \{1,2,3,4,...,24\}$ **2.** $C = \{-5,-4,-3,-2,-1,0,1,2,3\}$ **3.** $F = \{\ \}$ or $\varnothing$ **4.** 3, 0
5. $-4, 3, -\frac{13}{2}, 0$ **6.** All numbers listed **7.** See Graphing Answer Section **8.** False **9.** True **10.** > **11.** < **12.** =
13. 8 **14.** 7 **15.** $\frac{13}{9}$ **16.** False **17.** False **18.** True **19.** True

Section 1.4

Five-Minute Warm-Up **1.** division **2.** subtraction **3.** addition **4.** multiplication **5a.** $\frac{9}{18}$ **5b.** 9/18
5c. $9 \div 18$ **5d.** $18\overline{)9}$ **6a.** 13 **6b.** $\frac{15}{4}$ **6c.** 0 **7.** $\frac{5}{2}$

Guided Practice **1a.** positive **1b.** negative **2.** the same as the sign of the integer having the larger absolute value **3a.** 45 **3b.** 22 **3c.** 67 **3d.** negative **3e.** −67 **4a.** 73 **4b.** 81 **4c.** 8 **4d.** −81 **4e.** negative **4f.** negative
4g. −8 **5a.** $-\frac{9}{2}$ **5b.** 10.3 **6a.** $-32 + 61$ **6b.** 29 **7a.** positive **7b.** negative **7c.** zero **8a.** −72 **8b.** 60
8c. −117 **9a.** 36 **9b.** 12 **9c.** 3 **10a.** −72 **10b.** 72 **10c.** −1 **11.** $-\frac{1}{3}$

Do the Math 1. 8 2. –18 3. –213 4. 2 5. 34 6. –7 7. –7 8. –24 9. 71 10. 63 11. 110 12. –896
13. –192 14. $\dfrac{1}{10}$ 15. $-\dfrac{1}{3}$ 16. 4 17. –24 18. $\dfrac{20}{3}$ 19. $32 + (-64) = -32$ 20. $-40 \div 100$ or $\dfrac{-40}{100} = -\dfrac{2}{5}$
21. –3, –5 22. 6, –4

Section 1.5

Five-Minute Warm-Up 1. 90 2. –25 3. –78 4. –136 5. $\dfrac{1}{4}$ 6. 120 7. $\dfrac{25}{60}$

Guided Practice 1a. $-\dfrac{4}{5}$ 1b. –8 1c. $\dfrac{1}{5}$ 2. $-\dfrac{9}{40}$ 3a. $\dfrac{8}{15} \cdot \dfrac{45}{4}$ 3b. $\dfrac{2 \cdot 2 \cdot 2 \cdot 3 \cdot 3 \cdot 5}{2 \cdot 2 \cdot 3 \cdot 5}$

3c. $\dfrac{2 \cdot \cancel{2} \cdot \cancel{2} \cdot \cancel{3} \cdot 3 \cdot \cancel{5}}{\cancel{2} \cdot \cancel{2} \cdot \cancel{3} \cdot \cancel{5}}$ 3d. $\dfrac{6}{1} = 6$ 4a. $\dfrac{7-11}{12}$ 4b. $\dfrac{7+(-11)}{12}$ 4c. $\dfrac{-4}{12}$ 4d. $-\dfrac{1}{3}$ 5a. $2 \cdot 2 \cdot 3$

5b. $3 \cdot 3$ 5c. $2 \cdot 2 \cdot 3 \cdot 3$ 5d. 36 5e. $\dfrac{3}{3}$ 5f. $\dfrac{4}{4}$ 5g. $\dfrac{5}{12} = \dfrac{15}{36}$; $\dfrac{4}{9} = \dfrac{16}{36}$ 5h. $\dfrac{15+16}{36}$ 5i. $\dfrac{31}{36}$ 6a. $\dfrac{-3}{1}$

6b. $\dfrac{-24}{8} + \dfrac{5}{8}$ 6c. $\dfrac{-19}{8}$ 6d. $-\dfrac{19}{8}$ 7. at the right end of the integer 8. sum

Do the Math 1. $\dfrac{5}{12}$ 2. –27 3. $-\dfrac{8}{5}$ 4. 5 5. $\dfrac{4}{9}$ 6. $-\dfrac{1}{8}$ 7. 1 8. $-\dfrac{1}{16}$ 9. $-\dfrac{40}{27}$ 10. $\dfrac{1}{6}$ 11. 2 12. $\dfrac{25}{8}$
13. $-\dfrac{16}{15}$ 14. $-\dfrac{203}{60}$ 15. 43.2 16. 26.32 17. 33.79 18. 22.1 19. 42.1 20. –1600 21. 24 bags

Section 1.6

Five-Minute Warm-Up 1. –2 2. –12 3. $\dfrac{2}{3}$ 4. 45 5. 5

Guided Practice 1a. $\dfrac{4}{7}$ 1b. $-\dfrac{1}{4}$ 2a. $16\dfrac{1}{2}$ feet 2b. 900 seconds 3. order 4. division; subtraction
5a. $6 + (-12)$ 5b. $\dfrac{1}{2} \cdot (4 + 20)$ 6a. 4 6b. 90 7. grouping 8a. $-9 + (12 + 2)$ 8b. $\left(3 \cdot \dfrac{2}{3}\right) \cdot 15$ 9a. 20
9b. 50 10a. 0 10b. 0 10c. undefined

Do the Math 1. $43\dfrac{1}{3}$ yards = 43 yards, 1 foot 2. 59 meters 3. 14.5 gallons = 14 gallons, 2 quarts
4. 7.5 pounds = 7 pounds, 8 ounces 5. Commutative Property of Multiplication 6. Associative Property of Multiplication 7. $\dfrac{a}{0}$ is undefined 8. Multiplicative Inverse Property 9. 59 10. 28 11. –72 12. 13 13. 0
14. 2 15. 1 16. 30 17. $-\dfrac{21}{16}$ 18. $-6 - (4 + 10)$ 19. 88 feet per second

Section 1.7

Five-Minute Warm-Up 1. –81 2. –48 3. –32 4. –45 5. 70 6. 81
Guided Practice 1a. base 1b. exponent 1c. $2 \cdot 2 \cdot 2 \cdot 2 \cdot 2 = 32$ 2a. 9 2b. –9 3a. squared
3b. $(-6)^2$ 3c. $(-6)(-6) = 36$ 4a. cubed 4b. $(-5)^3$ 4c. $(-5)(-5)(-5) = -125$ 5a. Parentheses
5b. Exponents 5c. Multiply or Divide 5d. Add or Subtract 6. [ ], { }, | | 7. –31 8a. $\dfrac{4 \cdot 2^3 + (-16)}{3(-2)^2}$
8b. $\dfrac{4 \cdot 8 + (-16)}{3(4)}$ 8c. $\dfrac{32 + (-16)}{12}$ 8d. $\dfrac{16}{12}$ 8e. $\dfrac{4}{3}$

Do the Math 1. 4^5 2. $(-8)^3$ 3. 32 4. $\dfrac{625}{16}$ 5. 0.0016 6. –625 7. 1 8. $-\dfrac{243}{32}$ 9. 36 10. 40 11. 5 12. $\dfrac{4}{9}$
13. 23 14. 13.5 15. –26 16. $-\dfrac{49}{2}$ 17. $\dfrac{81}{64}$ 18. 24 19. $\dfrac{5}{24}$ 20. $3^3 \cdot 5^2$ 21. $4 \cdot (7-4^2)$

Section 1.8
Five-Minute Warm-Up 1. 23 2. –10 3. –45 4. 7 5a. 16 5b. –16 6. 225
Guided Practice 1. variable 2. constant 3. algebraic expression 4. evaluating 5. 21 6. term
7. coefficient 8. like terms 9. $x^2; -2xy; -y^2$ 10a. $\dfrac{1}{2}$ 10b. –1 10c. 14 10d. $-\dfrac{2}{5}$ 11a. unlike 11b. like
12a. $6x - 2$ 12b. $-2y - 1$ 13a. $4x$ 13b. $6x + 3$ 14. $9x + 12$

Do the Math 1. –1 2. –32 3. $3m^4, -m^3n^2, 4n, -1; 3, -1, 4, -1$ 4. $t^3, -\dfrac{t}{4}; 1, -\dfrac{1}{4}$ 5. like 6. unlike
7. $12s + 6$ 8. $-5k + 5n$ 9. $6x + 9y$ 10. $-5p^5$ 11. $6m - 9n + 8p$ 12. $9m - 6$ 13. $\dfrac{13}{10}y$ 14. $-21x + 21$ 15. $884

Chapter 2 Answers
Section 2.1
Five-Minute Warm-Up 1. $\dfrac{2}{3}$ 2. $\dfrac{1}{14}$ 3. 1 4. $-18x - 15$ 5. $11 - 6x$ 6. –26
Guided Practice 1. Substitute the value into the original equation. If this results in a true statement, then the value for the variable is a solution to the equation. 2a. Yes 2b. No 3. Whatever you add to one side of the equation, you must add to the other side of the equation to form an equivalent equation. 4. Yes. Since $a - b = a + (-b)$, subtraction is an alternate way of writing addition. Subtraction means to add the opposite.
5. isolating the variable 6a. $x + 13 - 13 = -11 - 13$ 6b. $x + 0 = -24$ 6c. $x = -24$ 6d. $\{-24\}$
7. When you multiply one side of an equation by a non-zero quantity, you must also multiply the other side by the same non-zero quantity to form an equivalent equation. Yes, this property applies to division. Division means to multiply by the reciprocal. 8a. $-\dfrac{1}{9} \cdot (-9x) = -\dfrac{1}{9} \cdot (324)$ 8b. $\left(-\dfrac{1}{9} \cdot (-9)\right)x = -\dfrac{1}{9} \cdot 324$
8c. $1 \cdot x = -36$ 8d. $x = -36$ 8e. $\{-36\}$ 9a. Divide both sides by 3; which is the same as multiply both sides by $\dfrac{1}{3}$. 9b. Multiply both sides by 15 9c. Multiply both sides by $-\dfrac{3}{2}$ 9d. Multiply both sides by $\dfrac{9}{4}$

Do the Math 1. no 2. no 3. yes 4. no 5. $\{19\}$ 6. $\{-15\}$ 7. $\left\{\dfrac{1}{2}\right\}$ 8. $\left\{\dfrac{13}{24}\right\}$ 9. $\left\{\dfrac{15}{2}\right\}$ 10. $\left\{-\dfrac{5}{2}\right\}$ 11. $\{12\}$
12. $\{30\}$ 13. $\{14\}$ 14. $\left\{-\dfrac{5}{9}\right\}$ 15. $799 16. An algebraic expression does not contain an equal sign and an equation does have an equal sign. We *solve* equations. We *simplify* or *evaluate* expressions. $x + 10 = 22$.

Section 2.2
Five-Minute Warm-Up 1. $-11x - 8$ 2. 93 3. x 4. $2x - 9$ 5. $\dfrac{3}{7}$ 6. 132.5

Guided Practice 1a. 9 1b. $6x - 9 + 9 = -27 + 9$ 1c. $6x = -18$ 1d. $6; \dfrac{1}{6}$ 1e. $\dfrac{6x}{6} = \dfrac{-18}{6}$ 1f. $x = -3$
1g. $\{-3\}$ 2a. $\dfrac{5}{4}n + 3 - 3 = -12 - 3$ 2b. $\dfrac{5}{4}n = -15$ 2c. $\dfrac{4}{5}\left(\dfrac{5}{4}n\right) = \dfrac{4}{5}(-15)$ 2d. $n = -12$ 2e. $\{-12\}$
3. combining like terms on each side of the equation. 4. Distributive 5a. $-2x + 5 = -4$
5b. $24n + 36 = -2n - 1$ 6. Add $5x$ to each side of the equation, and then subtract 4 from each side. Answers may vary. 7a. $3z + 15 - 16z = 2 - z - 3$ 7b. $-13z + 15 = -z - 1$ 7c. $-13z + z + 15 = -z + z - 1$
7d. $-12z + 15 = -1$ 7e. $-12z + 15 - 15 = -1 - 15$ 7f. $-12z = -16$ 7g. $\dfrac{-12z}{-12} = \dfrac{-16}{-12}$ 7h. $z = \dfrac{4}{3}$ 7i. $\left\{\dfrac{4}{3}\right\}$

Do the Math 1. $\{-3\}$ 2. $\left\{\dfrac{5}{3}\right\}$ 3. $\{8\}$ 4. $\{25\}$ 5. $\{-8\}$ 6. $\{-3\}$ 7. $\{-2\}$ 8. $\{-2\}$ 9. $\{2\}$ 10. $\left\{\dfrac{1}{3}\right\}$ 11. $\{2\}$
12. $\left\{\dfrac{8}{3}\right\}$ 13. 35 g 14. \$15 15. 68 inches, 70 inches, 72 inches 16. $d = -\dfrac{14}{5}$

Section 2.3

Five-Minute Warm-Up 1. 35 2. 225 3. $-3x + 4$ 4. $43x - 270$ 5. $x + 28$ 6. $4x - 8$ 7. $10x - 10$

Guided Practice 1a. 12 1b. $12 \cdot \left(\dfrac{3}{2}x - \dfrac{4}{3}x\right) = 12 \cdot \left(-\dfrac{9}{4}\right)$ 1c. $12 \cdot \left(\dfrac{3}{2}x\right) - 12 \cdot \left(\dfrac{4}{3}x\right) = 12 \cdot \left(-\dfrac{9}{4}\right)$

1d. $18x - 16x = -27$ 1e. $2x = -27$ 1f. $\dfrac{2x}{2} = \dfrac{-27}{2}$ 1g. $x = -\dfrac{27}{2}$ 1h. $\left\{-\dfrac{27}{2}\right\}$ 2a. 18

2b. $18\left(\dfrac{4x+3}{9} + 1\right) = 18\left(\dfrac{2x+1}{2}\right)$ 2c. $2(4x + 3) + 18 \cdot 1 = 9(2x + 1)$ 2d. $8x + 6 + 18 = 18x + 9$

2e. $8x + 24 = 18x + 9$ 3. $\{3000\}$ 4a. false 4b. $\emptyset$ 4c. contradiction 5a. 81 5b. yes; yes 5c. 46; 81

Do the Math 1. $\{3\}$ 2. $\left\{\dfrac{44}{9}\right\}$ 3. $\{8\}$ 4. $\left\{\dfrac{8}{3}\right\}$ 5. $\{20\}$ 6. $\{250\}$ 7. $\{8\}$ 8. $\{1\}$ 9. $\emptyset$ or $\{\ \}$; contradiction

10. all real numbers; identity 11. $\left\{\dfrac{3}{2}\right\}$; conditional equation 12. $\emptyset$ or $\{\ \}$; contradiction 13. \$48

14. 10 nickels 15. \$85,000

Section 2.4

Five-Minute Warm-Up 1. 28 2. 12.38 3. 0.0725 4. 78.5 5. $\{4\}$

Guided Practice 1. $68°F$ 2. I: interest earned; r: annual interest rate, as a decimal; P: Principal (amount of money); t time of deposit, typically in years 3a. $A = s^2$; $P = 4s$ 3b. $A = lw$; $P = 2l + 2w$

3c. $A = \dfrac{1}{2}bh$; $P = a + b + c$ 3d. $A = \dfrac{1}{2}h(B + b)$; $P = a + b + c + B$ 3e. $A = bh$; $P = 2a + 2b$

3f. $A = \pi r^2$; $C = 2\pi r = \pi d$ 3g. $V = s^3$; $S = 6s^2$ 3h. $V = lwh$; $S = 2lw + 2lh + 2wh$ 3i.

$V = \dfrac{4}{3}\pi r^3$; $S = 4\pi r^2$ 3j. $V = \pi r^2 h$; $S = 2\pi r^2 + 2\pi rh$ 3k. $V = \dfrac{1}{3}\pi r^2 h$ 4a. $P = 2l + 2w$ 4b. 30.5 m

4c. \$457.50 5a. $6x - 6x - 12y = -6x - 24$ 5b. $-12y = -6x - 24$ 5c. $\dfrac{-12y}{-12} = \dfrac{-6x - 24}{-12}$ 5d. $y = \dfrac{1}{2}x + 2$

6a. $C = R - P$ 6b. $C = \$2100$

Do the Math 1. 1066.8 meters 2. \$834 3. \$150 4a. 104 units 4b. 640 units^2 5a. 17.58 units

5b. 24.62 units^2 6. $m = \dfrac{F}{v^2}$ 7. $B = \dfrac{3V}{h}$ 8. $b = S - a - c$ 9. $l = \dfrac{P - 2w}{2}$ 10. $y = 2x + 18$ 11. $y = 2x - 3$

12. $y = -\dfrac{5}{6}x + 3$ 13. $y = \dfrac{4}{15}x - 2$ 14a. $R = P + C$ 14b. \$6000 15a. $t = \dfrac{I}{Pr}$ 15b. 2 years

Section 2.5

Five-Minute Warm-Up 1. $\{-54.72\}$ 2. $\{32\}$ 3a. × 3b. − 3c. × 3d. + 3e. ÷ 3f. + 3g. −
3h. ÷ 3i. − 3j. + 3k. ÷ 3l. − 3m. × 3n. + 3o. + 3p. × 3q. − 3r. × 3s. × 3t. +

Guided Practice 1a. $-5 + 3x$ or $3x - 5$ 1b. $5(r + 25)$ 1c. $3x + 12$ 1d. $\dfrac{2p}{7}$ 1e. $\dfrac{z}{2} - 18$ or $\dfrac{1}{2}z - 18$

1f. $w - 900$ 2a. $b - a$ 2b. $a - b$ 2c. $b - a$ 2d. $\dfrac{a}{b}$ 2e. $\dfrac{b}{a}$ 3a. $x + 5 = 19$ 3b. $2y - 14 = 30$

3c. $z - 9 = \dfrac{2}{5}z + 6$ 4. Identify what you are looking for; Give names to the unknowns; Translate the problem into the language of mathematics; Solve the equation(s) found in Step 3; Check the reasonableness of your answer; Answer the question 5a. $n + 1$ 5b. $n + 2$ 6a. $x + 2$ 6b. $x + 4$ 6c. $x + 6$ 7a. $p + 2$ 7b. $p + 4$

Guided Practice (continued) 7c. $p+6$ 8a. $n+2; n+4$ 8b. $n+(n+2)+(n+4)=453$ 8c. $n=149$
8d. 149, 151, 153 are the three odd integers 9a. $x-5000$ 9b. $x+(x-5000)=20{,}000$ 9c. $x=12{,}500$
9d. $12,500 in stocks and $7500 in bonds

Do the Math 1. $2x$ 2. $x-8$ 3. $\dfrac{-14}{x}$ 4. $4x+21$ 5. Indians: r; Blue Jays: $r-3$ 6. Ralph's amount: R; Beryl's amount: $3R+0.25$ 7. number paid tickets: p; number promotion tickets: $12{,}765-p$ 8. $x+43=-72$
9. $49=2x-3$ 10. $\dfrac{x}{-6}-15=30$ 11. $2(x+5)=x+7$ 12. 25, 27, 29 13. Abraj: 120 fkoors; Burj Khalifa: 163 floors 14. Stocks: $24,000; bonds: $16,000 15. 240 minutes

Section 2.6
Five-Minute Warm-Up 1a. 0.62 1b. 0.0175 2a. 5.5% 2b. 150% 3. $1.75p$ 4. $\{13.25\}$
Guided Practice 1a. $\dfrac{3}{4}$ 1b. 1 1c. $\dfrac{21}{400}$ 2a. 0.68 2b. 0.025 2c. 1.5 3a. 8% 3b. 70% 3c. 30%
4a. $n=0.72(30)$ 4b. 21.6 5a. $150=x\bullet(120)$ 5b. 125% 6a. $0.075p$ 6b. $p+0.075p=1343.75$ or $1.075p=1343.75$ 6c. $p=1250$ 6d. The computer was $1250 before sales tax. 7a. $0.60p$
7b. $p-0.6p=224$ or $0.4p=224$ 7c. $p=560$ 7d. The price of the skis, before the discount, was $560.
Do the Math 1. 40 2. 15 3. 13.5 4. 305 5. 2000 6. 200 7. 16% 8. 75% 9. $8000 10. $28
11. $200,000 12. $23,800 13. $90 14. $1627 15. 28.32 million females

Section 2.7
Five-Minute Warm-Up 1. $\{50\}$ 2. $\left\{\dfrac{7}{2}\right\}$ 3a. $2x+12$ 3b. $\dfrac{x}{2}-25$ 3c. $2(45-x)$ 3d. $2x+15$
3e. $2(x+30)$ 4. $8-p$
Guided Practice 1. 90° 2. 180° 3a. $x-20$ 3b. $x+x-20=180$ 3c. $x=100$ 3d. The large angle is 100° and the smaller angle is 80°. 4. 180° 5a. $2x$ 5b. $2x+30$ 5c. $x+2x+2x+30=180$ 5d. $x=30$
5e. The first angle measures 30°, the second angle measures 60°, and the third angles measures 90°.
6a. $x+20$ 6b. $P=2l+2w$ 6c. $200=2(x+20)+2x$ 6d. $x=40$; The dimensions of the tablecloth are 40" by 60". 7a. The distances add to 50 miles. 7b. See Graphing Answer Section 7c. $2.5r+2.5(r+4)=50$; $r=8$ 7d. The slower boat is traveling at 8 mph and the faster boat is traveling at 12 mph.
Do the Math 1. 57.5° and 32.5° 2. 18°, 72°, 90° 3. length = 19 meters; width = 9 meters
4. 7.5 hr 4a. $72t$ 4b. $66t$ 4c. $72t-66t$ 4d. $72t-66t=45$ 5. See Graphing Answer Section; $2r+3(r-10)=580$ 6. 4 feet 7. 12 feet 8. length = 54 inches; width = 36 inches 9. 18 mph 10. $2\dfrac{1}{2}$ hours

Section 2.8
Five-Minute Warm-Up 1. < 2. < 3. > 4. = 5. > 6. < 7. See Graphing Answer Section
Guided Practice 1. square bracket; parenthesis 2. See Graphing Answer Section 3a. $(-\infty,-5)$
3b. $[10,\infty)$ 4. true 5. true 6. false 7a. $3x+10-10>1-10$ 7b. $3x>-9$ 7c. $x>-3$ 7d. $\{x\,|\,x>-3\}$
7e. $(-3,\infty)$ 7f. See Graphing Answer Section 8a. $(12,\infty)$ 8b. $(-\infty,-25]$ 9a. $6\cdot\dfrac{1}{6}(4x-5)\le 6\cdot\dfrac{1}{2}(2x+3)$
9b. $1\bullet(4x-5)\le 3\bullet(2x+3)$ 9c. $4x-5\le 6x+9$ 9d. $-2x-5\le 9$ 9e. $-2x\le 14$ 9f. $x\ge -7$
9g. $\{x\,|\,x\ge -7\}$ 9h. $[-7,\infty)$ 10a. $\ge$ 10b. $\ge$ 10c. > 10d. > 10e. $\le$ 10f. $\le$ 10g. < 10h. <
Do the Math 1–4. See Graphing Answer Section 1. $(5,\infty)$ 2. $(-\infty,6]$ 3. $[-2,\infty)$ 4. $(-\infty,-3)$ 5. $(-\infty,3]$
6. $(-2,\infty)$ 7. >; Addition Property of Inequality 8. $\ge$; Multiplication Property of Inequality
9–17. See Graphing Answer Section 9. $\{x\,|\,x<3\};(-\infty,3)$ 10. $\{x\,|\,x>3\};(3,\infty)$ 11. $\{x\,|\,x\le -4\};(-\infty,-4]$
12. $\{x\,|\,x>-2\};(-2,\infty)$ 13. $\{x\,|\,x\ge 5\};[5,\infty)$ 14. $\{x\,|\,x\ge -1\};[-1,\infty)$ 15. $\left\{x\,\Big|\,x>\dfrac{11}{2}\right\};\left(\dfrac{11}{2},\infty\right)$ 16. $\varnothing$ or $\{\ \}$
17. $\{p\,|\,p$ is any real number$\}; (-\infty,\infty)$ 18. $x\ge 250$ 19. $x<25$ 20. at most 71 miles

Chapter 3 Answers

Section 3.1

Five-Minute Warm-Up 1. See Graphing Answer Section 2a. 7 2b. −3 3a. −25 3b. 9 4. $\{5\}$ 5. $\{3\}$

Guided Practice 1. See Graphing Answer Section 2. the origin 3. left 4A $(4, -3)$ 4B $(-3, -1)$

4C $(-2, 1)$ 5a. No 5b. Yes 6a. $2y = -5$ 6b. $y = -\frac{5}{2}$ 6c. $\left(3, -\frac{5}{2}\right)$ 7a. −2; −7; $(-2, -7)$

7b. −1; −4; $(-1, -4)$ 7c. 0; −1; $(0, -1)$ 7d. 1; 2; $(1, 2)$ 7e. 2; 5; $(2, 5)$

Do the Math 1. See Graphing Answer Section; Quadrant I: R; Quadrant II: S; Quadrant III: P, T; Quadrant IV: Q, U 2. See Graphing Answer Section; Quadrant I: U; Quadrant III: T; Quadrant IV: P, V; x-axis: R, S; y-axis: Q, S 3. $P(0, 2)$: y–axis; $Q(-3, -1)$: Quadrant III; $R(-2, 4)$: Quadrant II; $S(2, 2)$: Quadrant I; $T(3, -2)$: Quadrant IV; $U(-5, 0)$: x–axis 4. A Yes B No C Yes 5. A No B No C Yes 6. $(2, 5)$ 7. $(4, 3)$

8–9. See Graphing Answer Section 10. A −9 B 7 C −6 11. A $\frac{1}{6}$ B $\frac{3}{2}$ C $-\frac{1}{2}$ 12a. $11.70 12b. $41.70

12c. 15.5 mi 12d. It costs $29.70 to take a taxi 14 miles.

Section 3.2

Five-Minute Warm-Up 1. $\{-9\}$ 2. $\left\{-\frac{10}{3}\right\}$ 3. $\{-2\}$ 4a. 22 4b. 12 5a. $y = -1$ 5b. $x = -4$

6. See Graphing Answer Section

Guided Practice 1. See Graphing Answer Section 2. standard 3. $y = 0$ 4. $x = 0$ 5a. −2 5b. 3

5c–7. See Graphing Answer Section

Do the Math 1. Not linear 2. Linear 3–4. See Graphing Answer Section 5. $(0, 2); (2, 0)$ 6. $(0, -3); (6, 0)$
7. $(0, 6)$ 8. $(7, 0)$ 9. $(0, -6); (10, 0)$ 10. $(0, -8); (2, 0)$ 11–14. See Graphing Answer Section

Section 3.3

Five-Minute Warm-Up 1. $\frac{1}{2}$ 2. 3 3. $-\frac{5}{3}$ 4. −2 5. undefined 6. 0 7. $-\frac{5}{6}$

Guided Practice 1. slope 2. $m = \frac{y_2 - y_1}{x_2 - x_1}$ 3. −5 4. undefined 5. vertical 6. 0 7. horizontal 8a. line goes up and to the right 8b. line goes down and to the right 8c. horizontal 8d. vertical

9. See Graphing Answer Section 10. average rate of change 11a. $\frac{1}{20}$ 11b. 5%

Do the Math 1. $\frac{3}{4}$ 2. $-\frac{7}{4}$ 3a – 3b. See Graphing Answer Section 3c. $m = \frac{5}{2}$; the value of y increases by 5 units when x increases by 2 units. 4a – 4b. See Graphing Answer Section 4c. $m = -\frac{1}{6}$; the value of y decreases by 1 unit when x increases by 6 units. 5. $m = 0$ 6. m is undefined 7–8. See Graphing Answer Section
9. $\frac{4}{45}$ 10. $m = 0.225$ million; the income increases by 0.225 million dollars or $225,000 for each year of college

Section 3.4

Five-Minute Warm-Up 1. $y = -\frac{7}{5}x - 7$ 2. $y = \frac{3}{8}x - \frac{3}{2}$ 3. $\left\{-\frac{4}{3}\right\}$ 4a. $-\frac{5}{3}$ 4b. $-\frac{1}{2}$

Guided Practice 1. $m; b$ 2. $-\frac{1}{2}; 6$ 3. −2; 5 4a. $\frac{2}{5}$ 4b. −1 4c. See Graphing Answer Section

5a. $y = -\frac{2}{3}x + 2$ 5b. $-\frac{2}{3}$ 5c. 2 5d. See Graphing Answer Section 6. $y = -\frac{1}{4}x + 3$ 7a. $y = 22 + 0.07x$
7b. $29

Do the Math 1. $m = -6$; (0, 2) 2. $m = -1$; (0, -12) 3. $m = \frac{3}{4}$; (0, 3) 4. $m = -\frac{5}{3}$; (0, 4) 5–8. See Graphing Answer Section 9. $y = x + 10$ 10. $y = \frac{4}{7}x - 9$ 11. $y = \frac{1}{4}x + \frac{3}{8}$ 12. $y = -2$ 13. $x = 4$ 14. $y = -3x$
15a. 2625 calories 15b. 9 years old 15c. The recommended caloric intake increases 125 calories per year.
15d. Age 3 is outside the given range for this equation applies. 15e. See Graphing Answer Section

Section 3.5

Five-Minute Warm-Up 1. $y = -4x + 10$ 2. $y = 3x + 4$ 3. $-\frac{1}{3}$ 4. -2 5. $-7x + 21$ 6. -1

Guided Practice 1. $y - y_1 = m(x - x_1)$ 2a. 9; -2; -1 2b. $y - y_1 = m(x - x_1)$
2c. $y - (-1) = 9(x - (-2))$ 2d. $y + 1 = 9(x + 2)$ 2e. $y + 1 = 9x + 18$ 2f. $y = 9x + 17$ 3a. $-\frac{5}{6}$; 0; 12
3b. $y - y_1 = m(x - x_1)$ 3c. $y - 12 = -\frac{5}{6}(x - 0)$ 3d. $y = -\frac{5}{6}x + 12$ 4a. 0; 5; -11
4b. $y - y_1 = m(x - x_1)$ 4c. $y - (-11) = 0(x - 5)$ 4d. $y + 11 = 0$; $y = -11$ 5. the slope 6a. $m = \frac{y_2 - y_1}{x_2 - x_1}$
6b. $x_1 = 1$, $y_1 = 5$, $x_2 = -1$, $y_2 = -1$ 14c. $m = \frac{-1 - 5}{-1 - 1} = \frac{-6}{-2} = 3$ 6d. $y - y_1 = m(x - x_1)$
6e. $m = 3$, $x_1 = 1$, $y_1 = 5$ 6f. $y - 5 = 3(x - 1)$ 6g. $y - 5 = 3x - 3$ 6h. 5; $y = 3x + 2$ 6i. yes 6j. no
6k. 3 6l. (0, 2) 7a. $\frac{3 + 1}{-5 + 5} = \frac{4}{0} =$ undefined 7b. vertical 7c. -5 7d. $x = -5$

Do the Math 1–4. See Graphing Answer Section 1. $y = 6x - 23$ 2. $y = -5x + 27$ 3. $y = -\frac{1}{2}x - 2$
4. $y = -1$ 5. $y = -1$ 6. $x = 4$ 7. $y = -\frac{3}{2}x + 1$ 8. $y = 3x + 6$ 9. $y = 5$ 10. $x = -3$ 11a. After 30 years, monthly income is $3600. 11b. See Graphing Answer Section 11c. $y = 45x + 2250$ 11d. $2925 11e. Monthly income increases by $45 for each additional year of service.

Section 3.6

Five-Minute Warm-Up 1. $-\frac{1}{5}$ 2. 3 3. $-\frac{4}{9}$ 4. $y = 4x - 12$ 5a. $\frac{1}{4}$ 5b. $\left(0, -\frac{1}{2}\right)$ 6. $y = -\frac{3}{2}x + \frac{11}{2}$

Guided Practice 1. Two lines are parallel if they never intersect. 2. slope; y-intercepts 3. $=$; $\neq$
4a. $y = 4x - 1$ 4b. 4; (0, -1) 4c. $y = 4x + 5$ 4d. 4; (0, 5) 4e. Since the slopes are equal and the y-intercepts are different, we know that L_1 is parallel to L_2. 5a. $y = \frac{5}{2}x + \frac{1}{2}$ 5b. $\frac{5}{2}$ 5c. $\frac{5}{2}$ 5d. $\frac{5}{2}$; -3; -4
5e. $y - y_1 = m(x - x_1)$ 5f. $y - (-4) = \frac{5}{2}(x - (-3))$ 5g. $y + 4 = \frac{5}{2}x + \frac{15}{2}$; $y = \frac{5}{2}x + \frac{7}{2}$
6. Two lines are perpendicular if they intersect at right angles $(90°)$. 7. -1 8. -1; $m_1 = -\frac{1}{m_2}$ 9a. $\frac{3}{4}$; (0, 6)
9b. $\frac{4}{3}$; (0, -2) 9c. $\frac{3}{4} \cdot \frac{4}{3} = 1$ 9d. Since the product of the slopes is not -1, L_1 is not perpendicular to L_2.
10a. $y = 2x + 7$ 10b. 2 10c. $-\frac{1}{2}$ 10d. $-\frac{1}{2}$; 3; -12 10e. $y - y_1 = m(x - x_1)$
10f. $y - (-12) = -\frac{1}{2}(x - (3))$ 10g. $y + 12 = -\frac{1}{2}x + \frac{3}{2}$; $y = -\frac{1}{2}x - \frac{21}{2}$

Do the Math 1-3. See Graphing Answer Section 4. neither 5. perpendicular 6. parallel 7. neither
8. $y = 2x - 19$ 9. $x = -4$ 10. $y = -8$ 11. $y = -\frac{2}{3}x + 11$ 12. $y = -3x + 19$ 13. $y = \frac{1}{2}x - 4$ 14. $y = -6$
15. $x = 11$

Section 3.7
Five-Minute Warm-Up 1. $\{x \mid x < -8\}$ 2. $\{x \mid x \geq -3\}$ 3. $\{x \mid x \leq -5\}$ 4. See Graphing Answer Section
Guided Practice 1. True 2. yes 3. dashed; solid 4. half-planes 5. shade 6a. $y = x + 2$ 6b. 1 6c. (0, 2)
6d. solid 6e. (0, 0) 6f. false 6g. opposite 6h. See Graphing Answer Section 7a. $4x - y = 0$; dashed 7b. no
7c. See Graphing Answer Section 8a. $4.32x + 2.50y \geq 5000$ 8b. no
Do the Math 1. A No B Yes C No 2. A No B No C Yes 3. A Yes B Yes C Yes 4. A No B Yes C No
5–12. See Graphing Answer Section 13a. $160a + 75c \leq 500$ 13b. Yes 13c. Yes

Chapter 4 Answers
Section 4.1
Five-Minute Warm-Up 1a–b. See Graphing Answer Section 2. Yes 3. No
Guided Practice 1a. No 1b. Yes 2a. $(-5, 0)$ 2b. $(0, 5)$ 2c. $y = -2x - 1$ 2d. -2 2e. $(0, -1)$
2f. $(-2, 3)$ 2g. See Graphing Answer Section 2h. $(-2, 3)$ 3. See Graphing Answer Section; $\emptyset$
4a. inconsistent; the lines are parallel 4b. consistent and dependent; the lines are coincident 4c. consistent
and independent; the lines intersect 5a. infinitely many 5b. consistent 5c. dependent
Do the Math 1a. Yes 1b. No 1c. No 2a. No 2b. No 2c. Yes 3 – 6. See Graphing Answer Section
3. $(3, -6)$ 4. no solution; $\{\ \}$ or $\emptyset$ 5. $(1, 0)$ 6. infinitely many solutions 7. no solution; inconsistent
8. infinitely many solutions; consistent; dependent 9. one solution; consistent; independent 10. infinitely
many solutions; consistent; dependent 11. See Graphing Answer Section; $\begin{cases} y = 60 + 0.14x \\ y = 30 + 0.2x \end{cases}$ The cost for
driving 500 miles is the same for both companies ($130). Choose Slow-but-Cheap to drive 400 miles.

Section 4.2
Five-Minute Warm-Up 1. $y = \frac{5}{3}x + 6$ 2. $x = -\frac{8}{3}y - 4$ 3. $\{-7\}$ 4. $\{3\}$
Guided Practice 1a. $4x + 3y = -23$ 1b. $4x + 3(-x - 7) = -23$ 1c. $4x - 3x - 21 = -23$
1d. $x - 21 = -23$ 1e. $x = -2$ 1f. $y = -x - 7$ 1g. $y = -(-2) - 7$ 1h. $y = -5$ 1i. $(-2, -5)$ 2. $(4, 10)$
3. infinitely many solutions; consistent/dependent 4a. $x + y = -11$ 4b. $x - y = -101$ 4c. $\begin{cases} x + y = -11 \\ x - y = -101 \end{cases}$
4d. $(-56, 45)$ 4e. The two numbers are -56 and 45.
Do the Math 1. $(-3, 5)$ 2. $(18, 13)$ 3. $\left(-\frac{2}{5}, -5\right)$ 4. $\left(-\frac{1}{2}, -2\right)$ 5. no solution; $\emptyset$ or $\{\ \}$; inconsistent
6. infinitely many solutions; consistent and dependent 7. no solution; $\emptyset$ or $\{\ \}$; inconsistent
8. no solution; $\emptyset$ or $\{\ \}$; inconsistent 9. $\left(2, \frac{3}{2}\right)$ 10. infinitely many solutions 11. $\left(-\frac{3}{2}, \frac{2}{3}\right)$
12. no solution; $\emptyset$ or $\{\ \}$ 13. $\left(-\frac{1}{3}, \frac{1}{6}\right)$ 14. infinitely many solutions 15. $8000 in international stock fund;
$16,000 in domestic growth fund

Section 4.3
Five-Minute Warm-Up 1. $-\frac{5}{4}$ 2. 6 3. $21x - 14y$ 4. $\emptyset$ 5. $-9y - 9$ 6a. 24 6b. 3 6c. 2

Guided Practice **1.** $(-5, 7)$ **2a.** $(1) -10x - 5y = 20$; $(2)\ 3x + 5y = 29$ **2b.** $-7x = 49$ **2c.** $x = -7$
2d. $2(-7) + y = -4$; $y = 10$ **2e.** $(-7, 10)$ **3a.** $5x - 2y = 4$ **3b.** $-3x + 10y = -20$ **3c.** $(0, -2)$ **4.** $\varnothing$
5. infinitely many solutions **6a.** $a + c = 525$ **6b.** $7.50a + 4.50c = 3337.50$ **6c.** $(325, 200)$ **6d.** 325 adult tickets and 200 children's tickets were sold.
Do the Math **1.** $(-3, 2)$ **2.** $\left(\dfrac{4}{3}, \dfrac{3}{2}\right)$ **3.** no solution; $\{\ \}$ or $\varnothing$; inconsistent
4. infinitely many solutions; dependent **5.** no solution; $\{\ \}$ or $\varnothing$ **6.** $(-14, 10)$ **7.** infinitely many solutions
8. $\left(\dfrac{95}{29}, -\dfrac{43}{29}\right)$ **9.** $\left(5, -\dfrac{1}{3}\right)$ **10.** $(-0.1, -0.2)$ **11.** $(1, 4)$ **12.** $(8, 17)$ **13.** biscuit 28 g; orange juice 42 g

Section 4.4

Five-Minute Warm-Up **1a.** 325 miles **1b.** 3 hours **2a.** \$945 **2b.** \$52.50
3a. $2x - 7 = 45$ **3b.** $5(x + 3) = \dfrac{x}{2}$ **3c.** $\dfrac{x}{3} = 20 - x$

Guided Practice **1a.** $4x + y = 68$ **1b.** $x - 2y = -1$ **1c.** $(15, 8)$ **1d.** The numbers are 15 and 8.
2. $180°$ **3a.** $x + y = 180$ **3b.** $x = y + 30$ **3c.** $(105, 75)$ **3d.** The two supplementary angles are
$105°$ and $75°$. **4a.** $a + w$ **4b.** $a - w$ **4c–d.** See Graphing Answer Section **4e.** $\begin{cases} 6a + 6w = 2400 \\ 8a - 8w = 2400 \end{cases}$

4f. $(350, 50)$ **4g.** The air speed of the plane is 350 mph, and the speed the wind is moving at is 50 mph.
Do the Math **1.** $20 + 3a = 2b + 50$ **2.** $l = 3w - 8$ **3.** $4(a - w) = 1200$ **4.** 19, 36 **5.** 18, 14
6. length 36.25 ft; width 26.25 ft **7.** $25°, 65°$ **8.** $96°, 84°$ **9.** airplane 485 mph; wind 30 mph
10. still water rate 5 mph; current speed 1 mph **11.** Gabriella's horse 2 mph; Monica's horse 6 mph

Section 4.5

Five-Minute Warm-Up **1a.** \$30 **1b.** \$2030 **2a.** 0.4 **2b.** 0.0015 **3.** $(7, -5)$

Guided Practice **1a–c.** See Graphing Answer Section **1d.** $\begin{cases} a + s = 40 \\ 7.5a + 4s = 202 \end{cases}$ **2a–c.** See Graphing
Answer Section **2d.** $\begin{cases} d + q = 42 \\ 0.1d + 0.25q = 6.75 \end{cases}$ **2e.** There are 25 dimes and 17 quarters. **3a–c.** See Graphing
Answer Section **3d.** $\begin{cases} a + p = 50 \\ 6.5a + 4p = 300 \end{cases}$ **4a–c.** See Graphing Answer Section **4d.** $\begin{cases} x + y = 90 \\ 0.25x + 0.4y = 27 \end{cases}$

Do the Math **1.** See Graphing Answer Section $\begin{cases} b + n = 69 \\ 10b + 15n = 895 \end{cases}$ **2.** See Graphing Answer Section
3. $\begin{cases} s + c = 1700 \\ 0.0275s + 0.02c = 37.75 \end{cases}$ **3.** See Graphing Answer Section $\begin{cases} p + t = 40 \\ 5p + 2t = 120 \end{cases}$ **4.** $\begin{cases} a + c = 22 \\ 15a + 7c = 274 \end{cases}$
5. $\begin{cases} A + B = 2650 \\ 0.05A + 0.065B = 155 \end{cases}$ **6.** $\begin{cases} 8.60B + 5.75W = 143.45 \\ B = 2W - 2 \end{cases}$ **7.** adult \$4.50; child \$3.25
8. \$1500 in 4.5% account; \$3500 in 9% account **9.** 14 liters

Section 4.6

Five-Minute Warm-Up **1.** $(-\infty, -2]$ **2.** $(-1, \infty)$ **3a–b.** See Graphing Answer Section
Guided Practice **1.** (c) **2.** graphing **3.** dashed **4.** solid **5.** the opposite half-plane
6. intersection **7.** See Graphing Answer Section **8a.** $5x + 8y \le 40$ **8b.** $y \ge 2x$ **8c.** $x \ge 0$; $y \ge 0$
8d. See Graphing Answer Section **8e.** No

Chapter 5 Answers
Section 5.1

Five-Minute Warm-Up **1.** -1 **2.** $-2x^2 + 4$ **3.** $-10x - 6y + 9$ **4.** $-15x + 10$ **5.** $-x^2y + 3xy^2$ **6.** 7

Guided Practice **1a.** A single term which is the product of a constant and a variable with whole number exponents. **1b.** The constant multiplied with the variable. **1c.** The degree is the same as the exponent on the variable. **1d.** The degree is the sum of the exponents on the variables. **2a.** A polynomial is the sum of two or more monomials. **2b.** The degrees of the terms are in descending order. **2c.** The degree of the polynomial is the same as the highest degree of any of the terms in the polynomial. **3a.** monomial **3b.** binomial **3c.** trinomial **3d.** polynomial **4.** Like terms have the same variables, raised to the same powers.
5. $7x^3 - x^2 - 5x + 5$ **6.** $-x^2y - 14x^2y^2 + xy^2$ **7.** $9z^3 - 7z^2 + z - 6$ **8.** $-n^2 + 2n$ **9a.** 5 **9b.** 41
10. -7

Do the Math **1.** Yes; coefficient -1; degree 7 **2.** No **3.** No **4.** Yes; $p^5 - 3p^4 + 7p + 8$; degree 5; polynomial **5.** $7x^2 - x - 3$ **6.** $-10w^2 + 6w - 2$ **7.** $\frac{29}{24}b^2 - \frac{7}{15}b$ **8.** $-2a^2 - 3ab - 8b^2$ **9.** $2x^2 + 3x - 7$
10. $-2m^4 + 2m^2 + 7$ **11.** $\frac{7}{12}x^2 - \frac{25}{24}x - 6$ **12.** $-6m^2n + 4mn - 1$ **13a.** 10 **13b.** 9 **13c.** 9
14a. 10 **14.** 2.125 **14c.** $\frac{65}{32}$ **15.** 4 **16.** $-\frac{3}{2}$

Section 5.2

Five-Minute Warm-Up **1a.** $\left(\frac{3}{4}\right)^3$ **1b.** $(-5)^4$ **2a.** -9 **2b.** $\frac{27}{8}$ **2c.** 16 **2d.** $-\frac{64}{27}$ **2e.** -32 **2f.** -36
3. -12 **4.** $20 - 10x$

Guided Practice **1.** sum **2a.** r^8 **2b.** m^3n^5 **3.** multiply **4a.** p^{18} **4b.** $(-x^{12}) = x^{12}$ **5.** positive **6.** negative
7. raised to that power **8a.** $125x^3y^6$ **8b.** m^8n^{12} **9a.** $-12x^{11}$ **9b.** q^7 **10a.** $10m^5n^7$ **10b.** $48p^4q^{10}$

Do the Math **1.** 81 **2.** a^{10} **3.** b^{13} **4.** $(-z)^6 = z^6$ **5.** 81 **6.** $(-n)^{21} = -n^{21}$ **7.** k^{24} **8.** $(-a)^{18} = a^{18}$ **9.** $16y^6$
10. $\frac{27}{64}n^6$ **11.** $-64a^3b^6$ **12.** $81a^{24}b^4c^{16}$ **13.** $-14b^9$ **14.** $\frac{1}{6}y^{11}$ **15.** $12a^7b^4$ **16.** $\frac{20}{3}a^3b^2$ **17a.** $24xy^2$ units
17b. $12x^2y^4$ sq. units

Section 5.3

Five-Minute Warm-Up **1.** x^7 **2.** $20y^8$ **3.** $64z^6$ **4.** $9a^8$ **5.** $\frac{25}{4}x^2$ **6.** $-16x^{11}y^{16}$ **7.** $44a - 33b$
8. $\frac{4x}{5} + \frac{3}{4}$

Guided Practice **1.** the Distributive Property **2a.** $-15a^3b^2 + 6a^2b^3 - 9ab^4$ **2b.** $42x^7 - 3x^6 - 8x^5$
3a. $x^2 - 6x + 8$ **3b.** $6x^2 - 7x - 5$ **4a.** multiply two binomials **4b.** First, Outer, Inner, Last **4c.** no
5a. $u^2 - 2u - 15$ **5b.** $7x^2 - 23xy + 6y^2$ **6.** $A^2 - B^2$ **7a.** $x^2 - 81$ **7b.** $4a^2 - 25b^2$ **8a.** $A^2 + 2AB + B^2$
8b. $A^2 - 2AB + B^2$ **9a.** $(A - B)^2 = A^2 - 2AB + B^2$ **9b.** x **9c.** 3 **9d.** x^2 **9e.** $6x$ **9f.** 9 **9g.** $x^2 - 6x + 9$
10a. $144 - 24x + x^2$ **10b.** $16x^2 + 40xy + 25y^2$ **11a.** $3x^3 + 5x^2 - 14x + 4$ **11b.** $8x^3 - 14x^2 - 27x - 9$

Do the Math **1.** $6m^2 - 21m$ **2.** $9b^2 - 3b$ **3.** $8w^3 + 12w^2 - 20w$ **4.** $14r^3s + 6r^2s^2$ **5.** $n^2 - 3n - 28$
6. $10n^2 - 19ny + 6y^2$ **7.** $x^2 + 10x + 21$ **8.** $12z^2 - 5z - 2$ **9.** $x^4 - 7x^2 + 10$ **10.** $18r^2 + 51rs + 35s^2$
11. $36r^2 - 1$ **12.** $64a^2 - 25b^2$ **13.** $x^2 + 8x + 16$ **14.** $36b^2 - 60b + 25$ **15.** $9x^2 + 12xy + 4y^2$
16. $y^2 - \frac{2}{3}y + \frac{1}{9}$ **17.** $6a^3 - 17a^2 - 4a + 3$ **18.** $-4k^3 - 16k^2 + 84k$ **19.** $-2m^5 + m^4 - 8m^3 - 7m - 4$
20. $2a^3 + 9a^2 + 3a - 4$ **21.** $4x^2 - 22x + 22$

Section 5.4

Five-Minute Warm-Up **1.** $49x^6y^2$ **2.** $\frac{36}{25}$ **3a.** $-\frac{1}{3}$ **3b.** $\frac{8}{7}$ **4.** $\frac{9}{5}$ **5.** $-\frac{11}{24}$

Guided Practice 1. subtract 2a. x^7 2b. $-\dfrac{9}{4}a^2b$ 3. To raise a quotient to a power, both the numerator and the denominator are raised to the indicated power. The answer is (c) it does not matter which occurs first, although generally it is easier to simplify first. Be sure variables are nonzero. 4a. $-\dfrac{v^3}{8}$ 4b. $\dfrac{49a^{10}}{16b^8}$ 5. 1
6a. 1 6b. 1 6c. 12 6d. -1 6e. 1 6f. -4 7a. $\dfrac{1}{a^n}$ 7b. a^n 7c. $\left(\dfrac{b}{a}\right)^n = \dfrac{b^n}{a^n}$ 8a. $-\dfrac{1}{32}$ 8b. $-\dfrac{1}{12}$ 8c. $\dfrac{5}{a^2}$
9a. $\left(\dfrac{4}{3}\cdot(-18)\right)(x^{-2}\cdot x^5)(y^4\cdot y^{-7})$ 9b. -24 9c. x^3 9d. y^{-3} 9e. $-\dfrac{24x^3}{y^3}$

Do the Math 1. 1000 2. x^6 3. $\dfrac{-3xy}{2}$ 4. $3rs$ 5. $\dfrac{16}{81}$ 6. $\dfrac{49}{x^4}$ 7. $-\dfrac{a^{15}}{b^{50}}$ 8. $\dfrac{16m^4n^8}{q^{12}}$ 9. -1 10. 1 11. 1
12. $-11x$ 13. $-\dfrac{7}{z^5}$ 14. $\dfrac{p^4}{16}$ 15. $4b^3$ 16. $36t$ 17. 1 18. $\dfrac{2a^2}{b^2}$ 19. $\dfrac{x^8}{25y^6}$ 20. $20a^5$ 21. This quotient has expressions with different bases so the quotient rule does not apply. To simplify, prime factor the numerator and denominator and then divide out common factors: $\dfrac{6^5}{2^2} = 2^3\cdot 3^5$.

Section 5.5

Five-Minute Warm-Up 1. $-27x^4$ 2. $\dfrac{1}{9n^3}$ 3. $-27z^3 - 18z^2 + 9z$ 4a. 3 4b. $-7x^3 - 3x^2 + 5x + 13$
5. -1 6. $-x - 2$

Guided Practice 1a. $2x^2 + 9x$ 1b. $x + 4y^2 - 1$ 2. Multiply the quotient by the divisor. Add the remainder to this product. The result will be equal to the dividend if the problem was divided correctly.
3a. 71 3b. 8047 3c. 113 3d. 24 3e. $113\dfrac{24}{71}$ 4a. x^2 4b. $x^3 + x^2$ 4c. $-3x^2 + x + 6$
4d. $x^2 - 3x + 4 + \dfrac{2}{x+1}$ 5. standard 6. False

Do the Math 1. $x - 2$ 2. $2m + 1 - \dfrac{1}{2m^2}$ 3. $\dfrac{4}{5}a - \dfrac{3}{5} + \dfrac{2}{5a}$ 4. $-\dfrac{7}{6} - \dfrac{7}{2p} + \dfrac{1}{2p^2}$ 5. $-1 + 3y^2 - \dfrac{16y^3}{5}$
6. $-7x - 4$ 7. $-\dfrac{3y^2}{x^2} - 5$ 8. $-\dfrac{2}{mn} + \dfrac{3m}{n}$ 9. $x + 8$ 10. $x^2 - 7x + 2$ 11. $x^3 - x^2 + 1$ 12. $x^2 - 4x + 3 - \dfrac{2}{x-3}$
13. $3x - 2 - \dfrac{10}{2+3x}$ 14. $16x^4 + 12x^2 + 9$

Section 5.6

Five-Minute Warm-Up 1. $19x^5$ 2. $2.56n^{-6}$ or $\dfrac{2.56}{n^6}$ 3. $0.2p^4$ 4a. 4200 4b. 0.00305 4c. 2.1 5a. 10^8
5b. 10^{-9} 5c. 10^5 5d. 10^{-8} 5e. 10^5
Guided Practice 1a. $1 \le x < 10$ 1b. any integer 1c. positive 1d. negative 2a. 7 2b. positive
2c. 4.5×10^7 3a. 5 3b. negative 3c. 3×10^{-5} 4a. right 4b. left 5a. 4 5b. right 5c. 60,200 6a. -3
6b. left 6c. 0.0091 7a. 8×10^{13} 7b. 5.4×10^{-4} 8a. 4×10^5 8b. 9×10^{36}
Do the Math 1. 8×10^9 2. 1×10^{-7} 3. 2.83×10^{-5} 4. 4.01×10^8 5. 8×10^0 6. 1.2×10^2 7. 375
8. 6,000,000 9. 0.0005 10. 0.49 11. 540,000 12. 0.005123 13. 2.4×10^{-8} 14. 1×10^4 15. 5×10^{-2}
16. 4×10^5 17. 1.1×10^{-7} 18. 2×10^{11}

Chapter 6 Answers
Section 6.1

Five-Minute Warm-Up 1a. $2 \cdot 2 \cdot 3 \cdot 3$ 1b. $3 \cdot 3 \cdot 5 \cdot 5$ 2. $-12x + 27$ 3. $2x^2 + 5x - 12$
4. $4x^4 - 36x^2$ 5a. $\frac{1}{2}, 10$ 5b. 5 6. $4x^2y$

Guided Practice 1a. $3 \cdot 3 \cdot 3$ 1b. $2 \cdot 2 \cdot 3 \cdot 3$ 1c. $3 \cdot 3 \cdot 5$ 1d. $3 \cdot 3$ 1e. 9 2a. $6x^2$ 2b. $3ab$ 3a. $2m^2n^2$
3b. $2m^2n^2 \cdot 3m^2 + 2m^2n^2 \cdot 9mn^2 - 2m^2n^2 \cdot 11n^3$ 3c. $2m^2n^2(3m^2 + 9mn^2 - 11n^3)$ 4a. $-3x^3(8x^2 + 3)$
4b. $(x+3)(4x+7)$ 5. 4 6. True 7a. $2x$ 7b. $3y$ 7c. $2x(x-2) + 3y(x-2)$ 7d. $(x-2)(2x+3y)$
8. $(x-2y)(9-4a)$ 9. factor out the GCF from every term 10. $2(2x+3)(x^2-3)$

Do the Math 1. 7 2. $13xy$ 3. xyz 4. $x+y$ 5. $3(a-b)$ 6. $b(b-6)$ 7. $4a^3b^2(2+3a^2)$ 8. $5x^2(x^2+2x-5)$
9. $-2(y^2-5y+7)$ 10. $-2n(11n^3-9n-7)$ 11. $(a-5)(a+6)$ 12. $(x+a)(x+2)$ 13. $(m-3)(n+2)$
14. $(z+4)(z^2+3)$ 15. $(x-1)(x^2-5)$ 16. $x^2 - 3x + 2$ 17. $3x + 2$

Section 6.2

Five-Minute Warm-Up 1.a. $-1, 12; -2, 6; -3, 4; 1, -12; 2, -6; 3, -4$ 1b. $1, 36; 2, 18; 3, 12; 4, 9;$
1b. $-1, -36; -2, -18; -3, -12; -4, -9$ 2a. -15 2b. 36 3a. 6 3b. -16 4a. $4, -9$ 4b. $-3, -2$
4c. $-5, 1$ 4d. $3, 6$ 5. $-1, 7, -2$ 6. $x^2 - 10x + 24$

Guided Practice 1a. $19; 11; 9; -19; -11; -9$ 1b. $2, 9$ 1c. $(y+2)(y+9)$
2a. $37; 20; 15; 13; -37; -20; -15; -13$ 2b. $(z-3)(z-12)$ 3a. See Graphing Answer Section
3b. $(x-7)(x+8)$ 4a. See Graphing Answer Section 4b. $(n+4)(n-6)$ 5. Factors of -6 are:
$-1, 6; 1, -6; -2, 3; 2, -3$. The sums are $5, -5, 1, -1$. Since there are no two factors of -6 whose sum is 3,
the polynomial is prime. 6. $3p(p-6)(p+3)$

Do the Math 1. $(n+2)(n+10)$ 2. $(y+1)(y-9)$ 3. $(y-4)(y+10)$ 4. prime 5. $(x-2y)(x-12y)$
6. $4p^2(p-2)(p+1)$ 7. $-2(x-10)(x-5)$ 8. $-3(x+1)(x+5)$ 9. $x(x+15)(x-5)$ 10. $-2z(z+4)(z-3)$
11. $-(r+6)^2$ 12. prime 13. $x(x-8)(x+2)$ 14. $(x+3y)(x+4y)$ 15. $(x+5)^2$ 16. Correct. Both answers
multiply to the given polynomial. $(1-x)(2-x) = (-1)(x-1)(-1)(x-2)$. Since $(-1)^2 = 1$,
$(1-x)(2-x) = (x-1)(x-2)$.

Section 6.3

Five-Minute Warm-Up 1. $4, 1, -1$ 2a. $2 \cdot 3 \cdot 3$ 2b. $2 \cdot 3 \cdot 5 \cdot 5$ 3. $6x^2 - x - 2$ 4. $(x-y)(4+a)$
5. $3rs(4r^2s + 1 - 2s^3)$ 6. $(2z+1)(7z-4)$

Guided Practice 1a. $3; 12$ 1b. 36 1c. $-37; -20; -15; -13; -12$ 1d. $-4, -9$
1e. $3x^2 - 4x - 9x + 12$ 1f. $x(3x-4) - 3(3x-4)$ 1g. $(x-3)(3x-4)$ 2. $(13x-4)(x+1)$
3a. $(3x + ?)(x + ?)$ 3b. $1, 2; -1,-2$ 3c. $1, 2$ 3d. $(3x+1)(x+2)$ 4. $(2a-3b)(9a-4b)$
5. $-(3x+2)(2x-7)$ 6a. $-2(8x^2+2x-3)$ 6b. $8; 2; -3; -24$; See Graphing Answer Section 6c. $6, -4$
6d. $\frac{8}{6}; \frac{4}{3}; 4; 3$ 6e. $\frac{8}{-4}; \frac{2}{-1}; 2; -1$ 6f. $(4x+3)(2x-1)$ 6g. $-2(4x+3)(2x-1)$ 7. $(3x+1)(x+7)$
8. $(5x-4)(2x+1)$ 9. $-3n(2n-3)(3n+1)$

Do the Math 1. $(3x-2)(x+6)$ 2. $(5x+1)(x+3)$ 3. $(11p-2)(p-4)$ 4. $(3x+y)(x+2y)$
5. $2(3x+2y)(x-3y)$ 6. $-3(2x-5)(x+3)$ 7. $(7n+1)(n-4)$ 8. $(5t-1)(5t+2)$ 9. $(5t+4)(4t+1)$
10. $(4p-1)(3p-5)$ 11. $2(3x+2y)(3x-y)$ 12. $-(10y+3)(y-5)$ 13. $2(x+5)(9x-1)$ 14. $3xy(3x^2+2x+1)$
15. prime 16. $(x-1)(2x-1)(5x+2)$

Section 6.4

Five-Minute Warm-Up 1. 100 2. −27 3. $64x^6$ 4. $4p^2 + 20p + 25$ 5. $16a^2 − 81b^2$
6. 1, 4, 9, 16, 25, 36, 49, 64, 81 7. 1, 8, 27, 64

Guided Practice 1a. $A^2 + 2AB + B^2$ 1b. $A^2 − 2AB + B^2$ 2a. z^2, z 2b. 16; 4 2c. $−8z$, yes, $−8z = −2(z)(4)$ 2d. $(z − 4)^2$ 3a. $(p + 11)^2$ 3b. $3n(2m − 3n)^2$ 4. $A^2 − B^2$ 5a. $(x − 7)(x + 7)$
5b. $(5m^3 − 6n^2)(5m^3 + 6n^2)$ 5c. $(a − 2)(a + 2)(a^2 + 4)(a^4 + 16)$ 5d. $3y(2x − 5y)(2x + 5y)$
6a. $A^3 + B^3$ 6b. $A^3 − B^3$ 6c. $A^3 + 3A^2B + 3AB^2 + B^3$ 7a. $(p − 4)(p^2 + 4p + 16)$
7b. $(2x^2 + 3y)(4x^4 − 6x^2y + 9y^2)$ 8a. $3x(2x + 5)(4x^2 − 10x + 25)$
8b. $2xy(5x − 2y)(25x^2 + 10xy + 4y^2)$

Do the Math 1. $(m+6)^2$ 2. $(3a−2)^2$ 3. $(4y−9)^2$ 4. $(2a+5b)^2$ 5. $(6m−5n)(6m+5n)$
6. $(a−2)(a+2)(a^2+4)$ 7. $(4r−5s)(16r^2+20rs+25s^2)$ 8. $(m^3−3n^2)(m^6+3m^3n^2+9n^4)$
9. $(5y+3z^2)(25y^2−15yz^2+9z^4)$ 10. $5(2x+3y^2)(4x^2−6xy^2+9y^4)$ 11. $3n(2n−3)^2$
12. $4(2a+b^2)(4a^2−2ab^2+b^4)$ 13. $x^2(x−15)(x+15)$ 14. prime 15. $3(x+3)^2$ 16. $(x−2)(x+8)$

Section 6.5

Five-Minute Warm-Up 1. $20x^2 + 7xy − 6y^2$ 2. $a^2 + 8ab + 16b^2$ 3. $−9a(3a^2 − a + 2)$
4. $3(2n − 1)(2m − 3)$ 5. $(p − 2q)(4p + 3)$

Guided Practice 1. factor out the GCF 2a. 4 2b. $4(x^2 + 4x − 21)$ 2c. 3 2d. $4(x + 7)(x − 3)$ 3a. 2
3b. $(8a − 9b)(8a + 9b)$ 4a. 2 4b. $2(4a^2 + 12ab + 9b^2)$ 4c. 3 4d. $4a^2; 2a$ 4e. $9b^2; 3b$
4f. $12ab$; yes, $12ab = 2(2a)(3b)$ 4g. $2(2a + 3b)^2$ 5. $(3p^2 + 1)(9p^4 − 3p^2 + 1)$ 6. $2(n^2 − 3)(n − 5)$
7. $−3x(y + 1)(y + 3)$

Do the Math 1. $(x − 16)(x + 16)$ 2. $(1 − y)(1 + y + y^2)(1 + y^3 + y^6)$ 3. $(x + 6)(x − 1)$
4. $(2x − 1)(x − 3)(x + 3)$ 5. $25(2x − y)(2x + y)$ 6. $(3st − 1)(2st + 1)$ 7. $(2x^2 + 5y)(4x^4 − 10x^2y + 25y^2)$
8. $3m(2 − m^2)(2 + m^2)(4 + m^4)$ 9. $4t^2(t^2 + 4)$ 10. $2n(2n − 3)(2n + 3)$ 11. prime
12. $3q(2p + 3q)(4p^2 − 6pq + 9q^2)$ 13. $(x + y)(x − 4)(x + 4)$ 14. $−3x(x + 3)(x + 1)$ 15. $a(2a − b)(3a − b)$
16. $(x^2 + 1)(x^2 + 2)$

Section 6.6

Five-Minute Warm-Up 1. $\left\{\dfrac{1}{2}\right\}$ 2. $\{7\}$ 3a. 15 3b. 10 4. $(x + 9)(x − 7)$ 5. $(2p + 3)(2p − 1)$
6. $(x − 9)(x + 9)$ 7. $(n − 8)^2$

Guided Practice 1. If the product of two numbers is zero, then one of the factors must be zero. That is, if $a \cdot b = 0$, then $a = 0$ or $b = 0$ or both a and b are zero. 2. $\left\{-1, \dfrac{3}{2}\right\}$ 3. quadratic equation

4. List the terms from the highest power to the lowest power. 5a. $(x + 8)(x − 3) = 0$ 5b. $x + 8 = 0$
5c. $x − 3 = 0$ 5d. $x = −8$ 5e. $x = 3$ 5f. $\{−8, 3\}$ 6a. $3t^2 + 11t − 4 = 0$ 6b. $(3t − 1)(t + 4) = 0$
6c. $\left\{\dfrac{1}{3}, −4\right\}$ 7a. multiply the binomials 7b. standard form 7c. $(x − 10)(x + 5) = 0$ 7d. $\{10, −5\}$
8a. $4x^2 − 24x + 36 = 0$ 8b. $4(x − 3)^2 = 0$ 8c. $\{3\}$ 9a. no 9b. 4 9c. factor by grouping
9d. $(p + 2)(p − 3)(p + 3) = 0$ 9e. $p + 2 = 0$ 9f. $p − 3 = 0$ 9g. $p + 3 = 0$ 9h. $p = −2$ 9i. $p = 3$
9j. $p = −3$ 9k. $\{2, 3, −3\}$

Do the Math 1. $\{-9, 0\}$ 2. $\left\{-4, \frac{3}{4}\right\}$ 3. linear 4. quadratic 5. $\{-3, 8\}$ 6. $\left\{0, \frac{2}{7}\right\}$ 7. $\left\{-\frac{7}{3}, 2\right\}$ 8. $\{-6\}$
9. $\{-2, 8\}$ 10. $\{-3, 10\}$ 11. $\left\{-\frac{3}{2}, \frac{1}{2}\right\}$ 12. $\{-6, 4\}$ 13. $\left\{-\frac{7}{3}, 0, 2\right\}$ 14. $\{-3, -2, 3\}$ 15. 1 sec, 3 sec

Section 6.7

Five-Minute Warm-Up 1. $\{-9, 5\}$ 2. $\{-6\}$ 3. $\left\{-\frac{5}{2}, -\frac{1}{3}\right\}$ 4. 15 in. 5. $x^2 - 10x + 25$
6. $-9h(2h - 1)(h - 1)$

Guided Practice 1a. 0 sec or 3 sec 1b. 4 sec 2a. $n + 2$ 2b. $A = l \cdot w$ 2c. n; $n + 2$ 2d. $255 = n(n + 2)$
2e. $n = 15$ or $n = -17$ 2f. The rectangle has dimensions 15 cm $\times$ 17 cm 3. In a right triangle, the square of the length of the hypotenuse is equal to the sum of the squares of the lengths of the legs. 4. $z^2 = x^2 + y^2$
5a. 30 cm 5b. 7 ft 6a. $x = 0$ or $x = 9$ 6b. leg: 9 m; leg: 12 m; hypotenuse: 15 m

Do the Math 1. 10, 25 2. 14, 26 3. base = 6; height = 14 4. base = 12; height = 14 5. base = 11; height = 7
6. $B = 17$; $h = 5$; $b = 11$ 7. 9, 12, 15 8. 15, 8, 17 9. 3 sec 10. width = 3 ft; length = 7 ft
11. base = 8 m; height = 6 m 12. length = 13 yd; width = 8 yd 13a. height = 26 in.; width = 36 in. 13b. No

Chapter 7 Answers

Section 7.1

Five-Minute Warm-Up 1. -2 2. $(4x - 3)(x + 1)$ 3. $\left\{-\frac{5}{2}, 3\right\}$ 4. $\frac{10}{21}$ 5. $\frac{2x}{3y^2}$

Guided Practice 1a. $\frac{3}{10}$ 1b. 4 2a. $-\frac{2}{9}$ 2b. $\frac{7}{4}$ 3. the denominator is equal to zero. 4a. $x = -8$
4b. $x = 3$ or $x = 6$ 5a. factor 5b. divide out 6a. $\frac{6(x-3)}{(x-2)(x-3)}$ 6b. $\frac{6\cancel{(x-3)}}{(x-2)\cancel{(x-3)}}$ 6c. $\frac{6}{x-2}$
7. $\frac{x+1}{2(x+2)}$ 8. $-\frac{x-5}{2(x-3)}$

Do the Math 1a. $\frac{2}{3}$ 1b. undefined 1c. 0 2a. $\frac{7}{4}$ 2b. 1 2c. 3 3a. 15 3b. $-\frac{3}{5}$ 3c. -3 4a. -5 4b. -9 4c. 0
5. -8 6. $\frac{3}{4}$ 7. $-2, 1$ 8. $-4, -1, 0$ 9. $2p + 1$ 10. -1 11. $\frac{x-3}{x+2}$ 12. $\frac{x+2}{x-3}$ 13. $-\frac{1}{b+a}$ 14. $\frac{z^2+2}{z+2}$
15. $\frac{x+3}{2x-3}$ 16. $\frac{-(x+1)}{4x+3}$ 17. The property $\frac{p \cdot r}{q \cdot r} = \frac{p}{q}$ for $q \neq 0$, $r \neq 0$ enables us to divide out like factors.
But there is no equivalent property of algebra for terms: $\frac{p+r}{q+r} \neq \frac{p}{q}$.

Section 7.2

Five-Minute Warm-Up 1. $\frac{27}{14}$ 2. $-\frac{2}{5}$ 3. $\frac{1}{7}$ 4. $-5x(x - 5)$ 5. $\frac{x-3}{x-6}$

Guided Practice 1a. $\frac{x+3}{8} \cdot \frac{4(x-3)}{(x+3)(x-3)}$ 1b. $\frac{4(x+3)(x-3)}{8(x+3)(x-3)}$ 1c. $\frac{1 \cdot \cancel{4}\cancel{(x+3)}\cancel{(x-3)}}{2 \cdot \cancel{4}\cancel{(x+3)}\cancel{(x-3)}}$ 1d. $\frac{1}{2}$
2a. $\frac{x(x-4)}{(x-2)(x+2)} \cdot \frac{(x-3)(x+2)}{(x-4)}$ 2b. $\frac{x(x-4)(x-3)(x+2)}{(x-2)(x+2)(x-4)}$ 2c. $\frac{x\cancel{(x-4)}(x-3)\cancel{(x+2)}}{(x-2)\cancel{(x+2)}\cancel{(x-4)}}$
2d. $\frac{x(x-3)}{x-2}$ 3. $-\frac{2(x-y)}{x(x+y)}$ 4a. $\frac{3x+15}{36} \cdot \frac{24}{5x+25}$ 4b. $\frac{3(x+5)}{2 \cdot 2 \cdot 3 \cdot 3} \cdot \frac{2 \cdot 2 \cdot 2 \cdot 3}{5(x+5)}$

Guided Practice 4c. $\dfrac{3\cdot(x+5)\cdot 2\cdot 2\cdot 2\cdot 3}{3\cdot 2\cdot 2\cdot 3\cdot 5\cdot(x+5)}$ 4d. $\dfrac{\cancel{3}\cdot(x+5)\cdot\cancel{2}\cdot\cancel{2}\cdot 2\cdot\cancel{3}}{\cancel{3}\cdot\cancel{2}\cdot\cancel{2}\cdot\cancel{3}\cdot 5\cdot\cancel{(x+5)}}$ 4e. $\dfrac{2}{5}$ 5. $\dfrac{1}{2x-3}$ 6. $-\dfrac{3}{2x}$

Do the Math 1. $\dfrac{3x}{x+4}$ 2. $\dfrac{2(n+2)}{3(n-2)}$ 3. $z+2$ 4. 1 5. $\dfrac{2}{p}$ 6. $\dfrac{z+5}{2(z-5)}$ 7. $\dfrac{9y(y-1)}{2}$ 8. $\dfrac{3}{2}$ 9. $\dfrac{1}{(2x-3)(x+4)}$
10. $\dfrac{2(x+5)^2}{45x}$ 11. $r-3s$ 12. $\dfrac{x^2+1}{(x+1)^2}$ 13. $\dfrac{2x^3(x-3)}{(x+3)}$ 14. $-\dfrac{(p-7)}{(p+2)(p-2)}$ 15. $\dfrac{b-9}{b-3}$ 16. $\dfrac{t(t+3)}{(t+4)(t-3)}$

Section 7.3

Five-Minute Warm-Up 1. $\dfrac{3}{5}$ 2a. $\dfrac{13}{3}$ 2b. $\dfrac{5}{2}$ 3a. 1 3b. $\dfrac{5}{8}$ 4. $-\dfrac{1}{2}$ 5. $-2x+3$

Guided Practice 1a. $\dfrac{2x+5x}{x^2-3x}=\dfrac{7x}{x^2-3x}$ 1b. $\dfrac{7x}{x(x-3)}$ 1c. $\dfrac{7\cancel{x}}{\cancel{x}(x-3)}$ 1d. $\dfrac{7}{x-3}$ 2. $\dfrac{5}{x+5}$
3a. $\dfrac{x^2-1}{x^2-5x-6}$ 3b. $\dfrac{(x-1)(x+1)}{(x-6)(x+1)}$ 3c. $\dfrac{(x-1)\cancel{(x+1)}}{(x-6)\cancel{(x+1)}}$ 3d. $\dfrac{x-1}{x-6}$ 4. $x+2$ 5. $\dfrac{(n+5)(n-1)}{n-5}$ 6. $\dfrac{x+1}{x-3}$

Do the Math 1. $2n$ 2. $\dfrac{5p}{p-1}$ 3. $2(x+2)$ 4. $\dfrac{1}{x-3}$ 5. a 6. $\dfrac{2(x+2)}{x}$ 7. $\dfrac{z+6}{z-1}$ 8. $\dfrac{n+3}{n+1}$ 9. $\dfrac{4(1-2x)}{x-1}$
10. $\dfrac{x^2+x+6}{x-1}$ 11. $\dfrac{x+3}{x+1}$ 12. $\dfrac{m+n}{n-m}$ 13. 0 14. $\dfrac{x+3}{x-3}$ 15. $\dfrac{4s-t}{t-s}$ or $\dfrac{-4s+t}{s-t}$ 16. $\dfrac{n+3}{2n-3}$

Section 7.4

Five-Minute Warm-Up 1. $\dfrac{21}{45}$ 2. 180 3. $\dfrac{9}{20}=\dfrac{81}{180}$; $\dfrac{2}{45}=\dfrac{8}{180}$ 4. $3y^2$ 5. $(x-2)^2$ 6. $-3x(x+3)$

Guided Practice 1a. $2^3\cdot 3$ 1b. $2^2\cdot 5$ 1c. $2^3\cdot 3\cdot 5$ 1d. LCD $=120$ 2a. $2\cdot 3\cdot x^2$ 2b. $3^2\cdot x$ 2c. $2\cdot 3^2\cdot x^2$
2d. LCD $=18x^2$ 3. $40a^3b^2$ 4. $14a^2(a-2)$ 5. $(x-3)^2(x+3)$ 6a. $x(3x-1)$ 6b. $3x^2(3x-1)$ 6c. $3x$
6d. $\dfrac{3x}{3x}$ 6e. $\dfrac{x-2}{x(3x-1)}\cdot\left[\dfrac{3x}{3x}\right]=\dfrac{3x^2-6x}{3x^2(3x-1)}$ 7. LCD $=36x^2y^2$; $\dfrac{4}{9xy^2}=\dfrac{16x}{36x^2y^2}$; $\dfrac{5}{12x^2}=\dfrac{15y^2}{36x^2y^2}$

Do the Math 1. $49y^2$ 2. $24(x-3)$ 3. $4(x-3)^2(x+3)$ 4. $(x-1)^2(x+1)$ 5. $-3(c-7)(c+7)$
6. $-(z-4)(z+5)$ 7. $\dfrac{7a^2b+ab}{a^2b^2c}$ 8. $\dfrac{3x+9}{(x+2)(x+3)}$ 9. $\dfrac{14a}{6(a+1)(a+2)}$ 10. $\dfrac{7t^2+7}{t^2+1}$ 11. $\dfrac{28}{35n^2};\dfrac{15n}{35n^2}$
12. $\dfrac{8p^3-4p}{24p^2};\dfrac{9p^3+6}{24p^2}$ 13. $\dfrac{(x+2)^2}{x(x+2)};\dfrac{x^2}{x(x+2)}$ 14. $\dfrac{3ab+6b^2}{7a(a+2b)};\dfrac{2a^2}{7a(a+2b)}$ 15. $\dfrac{5}{m-6};\dfrac{-2m}{m-6}$
16. $\dfrac{4a}{(a+1)(a-1)};\dfrac{7a-7}{(a+1)(a-1)}$ 17. $\dfrac{4n^2+8n}{(n+2)^2(n-3)};\dfrac{2n-6}{(n+2)^2(n-3)}$ 18. $\dfrac{3}{(n-3)(2n-1)};\dfrac{4n^2-2n}{(n-3)(2n-1)}$

Section 7.5

Five-Minute Warm-Up 1a. LCD $=6$; $\dfrac{3}{2}=\dfrac{9}{6}$; $\dfrac{5}{3}=\dfrac{10}{6}$ 1b. LCD $=84$; $\dfrac{7}{12}=\dfrac{49}{84}$; $\dfrac{5}{28}=\dfrac{15}{84}$
2. $(3z-2)(2z-1)$ 3. $2x-1$ 4. $-\dfrac{x+4}{2x}$ 5. $\dfrac{1}{2x}$

Guided Practice 1a. $3\cdot 5$ 1b. 5^2 1c. $3\cdot 5^2=75$ 1d. $\dfrac{7}{15}\cdot\left[\dfrac{5}{5}\right]=\dfrac{35}{75}$ 1e. $\dfrac{10}{25}\cdot\left[\dfrac{3}{3}\right]=\dfrac{30}{75}$ 1f. $\dfrac{35+30}{75}=\dfrac{65}{75}$
1g. $\dfrac{13}{15}$ 2a. $2^2\cdot 3\cdot x$ 2b. $2\cdot 3^2\cdot x^2$ 2c. $2^2\cdot 3^2\cdot x^2=36x^2$ 2d. $\dfrac{5}{12x}\cdot\left[\dfrac{3x}{3x}\right]=\dfrac{15x}{36x^2}$ 2e. $\dfrac{7x}{18x^2}\cdot\left[\dfrac{2}{2}\right]=\dfrac{14x}{36x^2}$
2f. $\dfrac{15x+14x}{36x^2}=\dfrac{29x}{36x^2}$ 2g. $\dfrac{29}{36x}$ 3. $\dfrac{6x+8}{(x+2)(x+1)}$ 4a. $x(x-2)$ 4b. $\dfrac{2}{x}\cdot\left[\dfrac{x-2}{x-2}\right]=\dfrac{2x-4}{x(x-2)}$

4c. $\dfrac{4}{x-2} \cdot \left[\dfrac{x}{x}\right] = \dfrac{4x}{x(x-2)}$ 4d. $\dfrac{2x-4-4x}{x(x-2)} = \dfrac{-2x-4}{x(x-2)}$ 4e. $\dfrac{-2(x+2)}{x(x-2)}$ 5. $\dfrac{3}{y-3}$ 6. $\dfrac{2x-10}{x-4}$

Do the Math 1. $\dfrac{4}{3}$ 2. $\dfrac{25+12x}{10x}$ 3. $\dfrac{x^2+3}{(x-1)(x+1)}$ 4. $-\dfrac{1}{n-4}$ 5. $\dfrac{x+6}{x+2}$ 6. $\dfrac{4}{x-4}$ 7. $\dfrac{22}{105}$ 8. $\dfrac{-8x}{(x+2)(x-2)}$

9. $\dfrac{(x+2)^2}{(x+4)^2}$ 10. $\dfrac{2p^2+4p+1}{p(p-2)}$ 11. $\dfrac{5n-12}{4n^2}$ 12. $\dfrac{3x+4}{x+3}$ 13. $\dfrac{4}{3n(n-3)}$ 14. $\dfrac{a^2-5a+3}{2a^2(a-3)}$ 15. $\dfrac{x^2-2x-11}{(x-3)(x-2)}$

16. $\dfrac{15}{w(w-3)}$

Section 7.6

Five-Minute Warm-Up 1. $(2x+3)(x-4)$ 2. $3(x+5)(x-5)$ 3. $\dfrac{4}{3(x+2)}$ 4. x^2-2x+1

5. $3x^2+2x$

Guided Practice 1. When sums and/or differences of rational expressions occur in the numerator or denominator of a quotient, the quotient is called a complex rational expression. Rational expressions with more than one fraction bar are complex and need to be simplified. **2a.** $5x$ **2b.** $\dfrac{25-x^2}{5x}$ **2c.** $5x^2$ **2d.** $\dfrac{x^2-25}{5x^2}$

2e. $\dfrac{\dfrac{25-x^2}{5x}}{\dfrac{x^2-25}{5x^2}}$ **2f.** $\dfrac{25-x^2}{5x} \cdot \dfrac{5x^2}{x^2-25}$ **2g.** $\dfrac{^{(-1)}(25-x^2) \cdot \cancel{5} \cdot \cancel{x} \cdot x}{\cancel{5} \cdot \cancel{x} \cdot (25-x^2) \cdot 1} = \dfrac{-x}{1} = -x$ **3a.** Multiplicative Inverse

(a number multiplied by its reciprocal is one) or alternately, any quotient with the same numerator and denominator is a representation of one whole. **3b.** Multiplicative Identity   **3c.** Distributive Property   **4a.** x^2

4b. $\left(\dfrac{1+\dfrac{1}{x}}{1-\dfrac{1}{x^2}}\right)\left[\dfrac{x^2}{x^2}\right]$ **4c.** $\dfrac{1 \cdot x^2 + \dfrac{1}{x} \cdot x^2}{1 \cdot x^2 - \dfrac{1}{x^2} \cdot x^2} = \dfrac{x^2+x}{x^2-1}$ **4d.** $\dfrac{x(\cancel{x+1})}{(x-1)(\cancel{x+1})}$ **4e.** $\dfrac{x}{x-1}$

Do the Math 1. $-t(t-2)$ 2. $\dfrac{1}{x}$ 3. $\dfrac{3n+2}{(n-2)(n+1)}$ 4. $\dfrac{2ab}{(a+b)^2}$ 5. $\dfrac{5c+2d}{2(5c-4d)}$ 6. $\dfrac{a+8}{4-a}$ 7. $\dfrac{x-1}{x+1}$

8. $\dfrac{2xy+1}{3x+1}$ 9. $\dfrac{x-2}{x-3}$ 10. $\dfrac{6}{y+5}$ 11. $\dfrac{2x+1}{x-3}$ 12. $\dfrac{5m+n}{5m}$ 13. $\dfrac{y}{x}$ 14. $\dfrac{4(b-1)}{b}$ 15. $\dfrac{13z-11}{36}$

Section 7.7

Five-Minute Warm-Up 1. $\left\{\dfrac{8}{3}\right\}$ 2. $(x-7)(x+4)$ 3. $\left\{-\dfrac{1}{2}, \dfrac{1}{2}\right\}$ 4. $x=-8$ or $x=2$ 5. $P=\dfrac{A}{1+rt}$

Guided Practice 1. the denominator being equal to zero.   **2a.** $x=7, x=-7$ **2b.** $(x-7)(x+7)$

2c. $(x-7)(x+7)\dfrac{x+5}{x-7} = \dfrac{x-3}{x+7}(x-7)(x+7)$ **2d.** $(x+7)(x+5) = (x-3)(x-7)$

2e. $x^2+12x+35 = x^2-10x+21$ **2f.** $x=-\dfrac{7}{11}$ **2g.** $\left\{-\dfrac{7}{11}\right\}$ **3a.** $x=0$ **3b.** $10x$ **3c.** $90+8=7x$

3d. $\{14\}$ **4a.** $x=1, x=-1$ **4b.** $(x-1)(x+1)$ **4c.** $x=-1$ **4d.** $\varnothing$ **5a.** $x=0, y=0, z=0$ **5b.** xyz

5c. $(xyz)\left(\dfrac{6}{z}\right) = \dfrac{2}{x}(xyz) + \dfrac{1}{y}(xyz)$ **5d.** $6xy = 2yz+xz$ **5e.** $6xy-2yz = xz$ **5f.** $y(6x-2z) = xz$

5g. $y = \dfrac{xz}{6x-2z}$ **5h.** $y = \dfrac{xz}{6x-2z}$

Do the Math 1. $\left\{\dfrac{12}{13}\right\}$ 2. $\{36\}$ 3. $\{4\}$ 4. $\{\ \}$ or $\varnothing$ 5. $\{0\}$ 6. $\{7\}$ 7. $\left\{-\dfrac{3}{2},\dfrac{2}{3}\right\}$ 8. $\{\ \}$ or $\varnothing$
9. $\{-3, 4\}$ 10. $\{-5\}$ 11. $b = \dfrac{a}{c} - 2$ or $b = \dfrac{a-2c}{c}$ 12. $j = \dfrac{4ik}{3k-8i}$ 13. $a = \dfrac{bX}{X-b}$ 14. $k = 6$
15. after 1 hour and after 3 hours 16. 40 or 125 bicycles

Section 7.8

Five-Minute Warm-Up 1. $\{12\}$ 2. $\left\{\dfrac{2}{3}, 4\right\}$ 3. $\{3, -1\}$

Guided Practice 1. $\dfrac{3}{x}$ Answers may vary 2. $\dfrac{3}{x} = \dfrac{x+2}{9}$ Answers may vary 3. $\left\{\dfrac{13}{3}\right\}$ 4. about 26 million flight hours in 2001. 5. The figures are the same shape, but they are a different size. The corresponding angles are equal and the corresponding sides are proportional. 6a. $\dfrac{6}{3.2} = \dfrac{x}{8}$ 6b. 15 ft
7a. $\dfrac{1}{15}$ 7b. $\dfrac{1}{t}$ 7c. $\dfrac{1}{t+3}$ 8a. $\dfrac{1}{3}$ 8b. $\dfrac{1}{5}$ 8c. $\dfrac{1}{t}$ 8d. $\dfrac{1}{3} + \dfrac{1}{5} = \dfrac{1}{t}$ 8e. It takes 1.875 hours to clean the building when Josh and Ken work together. 9a. $180 + w$ 9b. $180 - w$ 9c. $\dfrac{1000}{180+w}$ 9d. $\dfrac{600}{180-w}$
9e. $\dfrac{1000}{180+w} = \dfrac{600}{180-w}$ 9f. The speed of the wind is 45 mph.

Do the Math 1. $\left\{\dfrac{20}{9}\right\}$ 2. $\{2\}$ 3. $\{-9\}$ 4. $\{-4, -3\}$ 5. $\{-2, 6\}$ 6. $\{6\}$ 7. $\dfrac{n}{2}$ 8. $r+3$ 9. 5 lb
10. 1600 pesos 11. 20 m 12. $\dfrac{28}{11} \approx 2.5$ hr 13. The speed of the boat in still water is 7 mph. 14. 9 mph

Section 7.9

Five-Minute Warm-Up 1a. $\left\{\dfrac{15}{2}\right\}$ 1b. $\{32\}$ 1c. $\left\{\dfrac{2}{3}\right\}$ 1d. $\{6\}$ 2. 15.50 3. $\dfrac{1}{2}$

Guided Practice 1. $y = kx$ 2. constant of proportionality 3a. $y = -\dfrac{2}{3}x$ 3b. $y = -\dfrac{25}{3}$ 3c. $C = 46.99$
4a. $p = 0.008b$ 4b. $\$1200$ 5. $y = \dfrac{k}{x}$ 6. $y = \dfrac{8}{3}$ 7a. $I = \dfrac{k}{R}$ 7b. $k = 9000$ 7c. $I = \dfrac{9000}{R}$; $I = 75$ amps

Do the Math 1. $y = \dfrac{1}{3}x$ 2. $y = -\dfrac{5}{12}x$ 3. (a) $d = 40t$ (b) $d = 200$ 4. (a) $m = \dfrac{4}{3}n$ (b) $m = -\dfrac{32}{15}$
5. $y = \dfrac{70}{x}$ 6. $y = -\dfrac{1}{6x}$ 7. (a) $y = \dfrac{12}{x}$ (b) $y = \dfrac{2}{3}$ 8. (a) $f = \dfrac{5}{3d}$ (b) $d = \dfrac{4}{9}$ 9. $r = 64$ 10. $x = 84$
11. 27 representatives 12. $\$143.50$ 13. 13.5 meters 14. 28 min

Chapter 8 Answers
Section 8.1

Five-Minute Warm-Up 1. See Graphing Answer Section 2. (b) only 3a. -14 3b. 6 3c. 4
4. $y = \dfrac{4}{3}x + 4$ 5a. 12 5b. 0 5c. 125

Guided Practice 1. See Graphing Answer Section 2. origin 3. left 4. 2 5. See Graphing Answer Section 5a. $-2; -7; (-2, -7)$ 5b. $-1; -4; (-1, -4)$ 5c. $0; -1; (0, -1)$ 5d. $1; 2; (1, 2)$ 5e. $2; 5; (2, 5)$
6. $(a, 0)$ and $(0, b)$ 7. $(0, c)$ 8. If one unit is manufactured and sold, there will be a gain (profit) of $\$50$.
9. 2 sec after the ball leaves the hand of the thrower, it will be 8 ft above the ground.

Do the Math 1A. $(4, -3)$; IV 1B. $(-3, -1)$; III 1C. $(-2, 1)$; II 2. See Graphing Answer Section
2D. y-axis 2E. I 2F. x-axis 3. (b) only 4. (a) and (d) 5. $(2, 0)$ and $(0, -4)$
6 – 9. See Graphing Answer Section 10. $a = \dfrac{7}{3}$ 11. $b = -13$

Section 8.2

Five-Minute Warm-Up 1–4. See Graphing Answer Section
Guided Practice 1. A relation is a mapping that pairs elements of one set with elements of a second set.
2. (AR, 75); (DE, 3); (FL, 67); (IL, 102); (NM, 33); (NY, 58); (PA, 67); (TX, 254); (VT, 14);
Domain: {Arkansas, Delaware, Florida, Illinois, New Mexico, New York, Pennsylvania, Texas, Vermont};
Range: {3, 14, 33, 58, 67, 75, 102, 254} 3. inputs; x 4. outputs; y 5a. $\{x \mid x \text{ is any real number}\}$
5b. $\{y \mid y \geq -3\}$ 6a. $\{x \mid -3 \leq x \leq 3\}$ 6b. $\{y \mid -4 \leq y \leq 4\}$ 7. See Graphing Answer Section 7a. 1; $(-2, 1)$
7b. -2; $(-1, -2)$ 7c. -3; $(0, -3)$ 7d. -2; $(1, -2)$ 7e. 1; $(2, 1)$ 7f. $\{x \mid x \text{ is any real number}\}$ 7g. $\{y \mid y \geq -3\}$
Do the Math 1. {(Sandy, 426–555), (Kathleen, 278–896), (Shawn, 377–204), (Roberto, 368–110)};
Domain: {Sandy, Kathleen, Shawn, Roberto}; Range: {426–555, 278–896, 377–204, 368–110}
2. {(Christian, 8), (Maria, 11), (Irma, 5), (Dwayne, 11), (Eleazar, 20)}; Domain: {Christian, Maria, Irma, Dwayne, Eleazar}; Range: {5, 8, 11, 20}
3. See Graphing Answer Section; Domain: {–2, –4, 1}; Range: {3, 2, 1}
4. See Graphing Answer Section; Domain: {3, –1, 1}; Range: {0, –6, 6}
5. Domain: {–2, 1, 3}; Range: {–2, 0, 3, 4}
6. Domain: $\{x \mid x \text{ is any real number}\}$; Range: $\{y \mid y \text{ is any real number}\}$
7. Domain: $\{x \mid x \text{ is any real number}\}$; Range: $\{y \mid y \geq 0\}$
8. Domain: $\{x \mid -3 \leq x \leq 1\}$; Range: $\{y \mid -1 \leq y \leq 3\}$
9. See Graphing Answer Section; Domain: $\{x \mid x \text{ is any real number}\}$; Range: $\{y \mid y \text{ is any real number}\}$
10. See Graphing Answer Section; Domain: $\{x \mid x \text{ is any real number}\}$; Range: $\{y \mid y \text{ is any real number}\}$
11. See Graphing Answer Section; Domain: $\{x \mid x \text{ is any real number}\}$; Range: $\{y \mid y \leq 0\}$
12. See Graphing Answer Section; Domain: $\{x \mid x \text{ is any real number}\}$; Range: $\{y \mid y \geq -3\}$
13. See Graphing Answer Section; Domain: $\{x \mid x \text{ is any real number}\}$; Range: $\{y \mid y \leq \dfrac{33}{4}\}$
14. See Graphing Answer Section; Domain: $\{x \mid x \text{ is any real number}\}$; Range: $\{y \mid y \geq -\dfrac{21}{4}\}$

Section 8.3

Five-Minute Warm-Up 1a. 53 1b. 5 2a. -4 2b. $\dfrac{11}{3}$ 3. $\dfrac{34}{3}$ 4. $2\sqrt{2}$ 5a. $\{-2, -1, 2, 3\}$ 5b. $\{4, 2, 6\}$
Guided Practice 1. A function is a relation in which each element of the domain (inputs) corresponds to exactly one element in the range (outputs). 2a. yes; domain: $\{-1, 4, 5\}$; range: $\{-1, 4, 5\}$ 2b. no
3a. $2x$; $-2x$ 3b. no 4a. yes 4b. no 4c. no 4d. yes 5. A graph represents a function if and only if every possible vertical line intersects the graph in at most one point. 6. (c) and (d) only 7a. -44 7b. 0 8a. 14
8b. -1 9a. 4 9b. 4 10a. 8 10b. 29
Do the Math 1. function; Domain: {English, Math, Psychology, History}; Range: {125, 85, 120}
2. not a function 3. function; Domain: {3, 2, 0, –1}; Range: {1} 4. not a function 5. not a function
6. function 7. function 8. function 9. not a function 10. function 11a. 4 11b. 10 11c. 0 12a. 1 12b. -2
12c. 3 13a. -1 13b. -1 13c. -1 14a. 4 14b. 13 14c. 8 15a. -3 15b. -18 15c. -13

Section 8.4

Five-Minute Warm-Up 1. $\left\{\dfrac{1}{2}\right\}$ 2. $\{6\}$ 3 – 4. See Graphing Answer Section 5a. domain: $[-4, 4]$
5b. range: $[-1, 2]$ 6a. domain: $[-3, 2]$ 6b. range: $\left[-\dfrac{7}{6}, \dfrac{10}{3}\right]$

Guided Practice 1. $8,(-2,8); 6,(-1,6); 4,(0,4); 2,(1,2); 0,(2,0); 2,(3,2); 4,(4,4)$; See Graphing Answer Section 2a. $(-\infty, \infty)$ 2b. $[-2, \infty)$ 2c. x-intercepts: $(-2,0), (2,0)$ 2d. y-intercept: $(0,-2)$ 3a. yes 3b. $-7; (6,-7)$ 3c. $2; (2, 3)$ 4. x; zero 5. $(2,0), (-2,0)$ 6a. linear 6b. identity 6c. constant 6d. square 6e. cube 6f. absolute value

Do the Math 1. $(-\infty, \infty)$ 2. $\left(-\infty, -\frac{1}{2}\right) \cup \left(-\frac{1}{2}, \infty\right)$ 3. $(-\infty, \infty)$ 4. $\left(-\infty, -\frac{5}{6}\right) \cup \left(-\frac{5}{6}, \infty\right)$ 5. $(-\infty, \infty)$ 6 – 7. See Graphing Answer Section 8a. $(-\infty, \infty)$ 8b. $(-\infty, \infty)$ 8c. x-intercept: $(3,0)$; y-intercept: $(0,-1)$ 8d. $x = 3$ 9a. $(-\infty, \infty)$ 9b. $(-\infty, \infty)$ 9c. x-intercepts: $(-2,0), (1,0), (4,0)$; y-intercept: $(0,2)$ 9d. $x = -2, x = 1, x = 4$ 10a. 8 10b. 0, 7 10c. $(-4,10)$ 10d. $(0,5)$ 10e. yes; $x = -4$ 11a. no 11b. $17; (4,17)$ 11c. $-3; (-3,-4)$ 11d. no 12a. $[0, \infty)$ 12b. 381.7 cm^3 13a. -5 13b. 1 13c. 1

Section 8.5

Five-Minute Warm-Up 1– 2. See Graphing Answer Section 3. $y = -3x - 1$ 4. $\{-0.9\}$ 5. $\left\{y \mid y \geq -\frac{1}{24}\right\}$

Guided Practice 1. linear; line 2. slope; y-intercept 3a. $(0, -2)$ 3b. $\frac{3}{2}$ 3c. See Graphing Answer Section 4. $mx + b = 0$ 5a. $(0, \infty)$ 5b. 66 ft 5c. 15 ft 5d. when the width exceeds 3 ft 6. fixed; a 7a. $4500 7b. negative 7c. $V(x) = 22{,}500 - 4500x$ 7d. $[0, 5]$ 7e. $4500 7f. after 2.75 years 7g. time, x 7h. the value, V 8. 2 9a. slope 9b. $y - y_1 = m(x - x_1)$

Do the Math 1 – 4. See Graphing Answer Section 5. $x = 8$ 6. $x = 4$ 7a. $\{-4\}$ 7b. $-\frac{1}{3}$ 7c. $-\frac{1}{3}$ 7d. $\{x \mid x > -4\}$ 8. See Graphing Answer Section; $\left(-4, -\frac{1}{3}\right)$ 9a. $g(x) = 3x + 2$ 9b. -7 10a. age, a 10b. Birth rate, B 10c. $[15, 44]$ 10d. 23.5 per 1000 women 10e. 37 years old 11. No

Section 8.6

Five-Minute Warm-Up 1a. $\{x \mid -4 \leq x < -1\}$ 1b. $[-4, -1)$ 1c – 2. See Graphing Answer Section 3. $(-3, 5]$ 4. $\{-3\}$ 5. $\{x \mid x \leq -6\}$ 6. $\{x \mid x > -2\}$

Guided Practice 1. $A \cap B$; and 2. $A \cup B$; or 3a – 3b. See Graphing Answer Section 4a. $5x < 10$ 4b. $x < 2$ 4c. $x \geq -4$ 4d – 4f. See Graphing Answer Section 5. $a < x < b$ 6a. $-8 + 3 \leq 5x - 3 + 3 \leq 7 + 3$ 6b. $-5 \leq 5x \leq 10$ 6c. $-1 \leq x \leq 2$ 6d. See Graphing Answer Section 7a. $-\frac{3}{2}x > 3$ 7b. $x < -2$ 7c. $7x > 14$ 7d. $x > 2$ 7e – 7g. See Graphing Answer Section 7h. $(-\infty, -2) \cup (2, \infty)$

Do the Math 1. $\{2, 3, 4, 5, 6, 7, 8, 9\}$ 2. $\{4,6\}$ 3. $\emptyset$ 4. $[-2, 2]$ 5. $(-\infty, \infty)$ 6. $(0, 5]$ 7. $(-\infty, 0) \cup [6, \infty)$ 8. $\{1\}$ 9. $(-\infty, 2) \cup [5, \infty)$ 10. $\left(-2, \frac{4}{7}\right]$ 11. $(-\infty, -2] \cup (3, \infty)$ 12. $(-\infty, -10) \cup (1, \infty)$ 13. $(2, 4]$ 14. $\left(-\frac{5}{3}, \frac{11}{4}\right]$ 15. $(-4, 1)$ 16. $\left(-\frac{9}{4}, 3\right)$ 17. $\left[-\frac{7}{3}, 3\right)$ 18. $a = -15; b = -1$ 19. $60 < x < 90$ 20. Between 500 kwh and 910 kwh

Section 8.7

Five-Minute Warm-Up 1a. 12 1b. 0 1c. $\frac{3}{4}$ 1d. 5.2 2. $|45|$ 3. $|-12|$ 4a. $\{4\}$ 4b. $\{7\}$

5a. $\{x|x > -6\}$ **5b.** $\{x|x \le 3\}$

Guided Practice **1.** $u = a$; $u = -a$ **2.** isolate the absolute value expression **3a.** $|3x - 1| = 2$
3b. $3x - 1 = 2$ or $3x - 1 = -2$ **3c.** $1; -\dfrac{1}{3}$ **3d.** $\left\{1, -\dfrac{1}{3}\right\}$ **4.** $u = v$; $u = -v$ **5.** $-a < u < a$; $-a \le u \le a$
6a. $-9 \le 4x - 3 \le 9$ **6b.** $-6 \le 4x \le 12$ **6c.** $-\dfrac{3}{2} \le x \le 3$ **6d.** See Graphing Answer Section **6e.** $\left[-\dfrac{3}{2}, 3\right]$
7. $u > a$ or $u < -a$; $u \ge a$ or $u \le -a$ **8a.** $|8x + 3| > 3$ **8b.** $8x + 3 > 3$ or $8x + 3 < -3$
8c. $x > 0$ or $x < -\dfrac{3}{4}$ **8d.** See Graphing Answer Section **8e.** $\left(-\infty, -\dfrac{3}{4}\right) \cup (0, \infty)$

Do the Math **1.** $\left\{-\dfrac{7}{2}, \dfrac{13}{2}\right\}$ **2.** $\{0, 8\}$ **3.** $\{y|-10 < y < 2\}$ **4.** $\left\{x|-1 \le x \le \dfrac{7}{3}\right\}$ **5.** $\{x|x \le -11 \text{ or } x \ge 3\}$
6. $\{z|z \text{ is any real number}\}$ **7.** $\varnothing$ **8.** $\left(-\dfrac{4}{5}, 2\right)$ **9.** $\varnothing$ **10.** $(-\infty, \infty)$ **11.** $\left\{-1, -\dfrac{1}{2}\right\}$ **12.** $\left(-\infty, \dfrac{1}{2}\right) \cup (1, \infty)$
13. $\varnothing$ **14.** $(-\infty, \infty)$ **15.** $|x - (-4)| < 2$ **16.** $|2x - 7| > 3$ **17.** <234.64 days or >297.36 days **18.** The absolute value of every real is number is positive or 0 and therefore greater than any negative number.

Section 8.8

Five-Minute Warm-Up **1a.** $\{-5\}$ **1b.** $\left\{-\dfrac{4}{9}\right\}$ **2.** $\{-48\}$ **3 – 4.** See Graphing Answer Section

Guided Practice **1.** $y = kx$ **2.** constant of proportionality **3.** linear; zero **4.** $y = -\dfrac{25}{3}$ **5.** $y = \dfrac{k}{x}$
6. $y = \dfrac{8}{3}$ **7.** $r = 42$ **8a.** $V = \dfrac{k \cdot T}{P}$ **8b.** $T = 300$; $P = 15$; $V = 100$ **8c.** $k = 5$ **8d.** $V = \dfrac{5 \cdot T}{P}$ **8e.** 22.5 Atm

Do the Math **1a.** $k = 5$ **1b.** $y = 5x$ **1c.** $y = 25$ **2a.** $k = \dfrac{1}{5}$ **2b.** $y = \dfrac{1}{5}x$ **2c.** $y = 7$ **3.** $B = 80$ **4.** $r = 64$
5. $\$1225.62$ **6.** $\$80.50$ **7.** 96 feet per second **8a.** $k = 80$ **8b.** $y = \dfrac{80}{x}$ **8c.** $y = \dfrac{16}{7}$ **9a.** $k = \dfrac{1}{3}$
9b. $y = \dfrac{1}{3}xz$ **9c.** $y = 40$ **10a.** $k = \dfrac{21}{10}$ **10b.** $Q = \dfrac{21x}{10y}$ **10c.** $Q = \dfrac{28}{5}$ **11.** 24 amps **12.** ≈ 0.033 foot-candle

Chapter 9 Answers
Section 9.1

Five-Minute Warm-Up **1.** $\left\{5, \dfrac{21}{7}, \dfrac{0}{11}, -10\right\}$ **2.** $\left\{\dfrac{-2}{3}, 5, -2.3, \dfrac{21}{7}, 1.\overline{6}, \dfrac{0}{11}, -10\right\}$ **3.** $\{2.1010010001...\}$
4. $\left\{\dfrac{-2}{3}, 5, -2.3, \dfrac{21}{7}, 1.\overline{6}, \dfrac{0}{11}, -10, 2.1010010001...\right\}$ **5.** 25 **6.** 121 **7.** 0.64 **8.** 1.44 **9.** $\dfrac{16}{49}$ **10.** $\dfrac{169}{36}$

Guided Practice **1a.** positive; negative **1b.** radical **1c.** principal **1d.** radicand **2a.** 8 **2b.** 20 **2c.** $\dfrac{2}{5}$
2d. 17 **3a.** -21 **3b.** 9 **3c.** 13 **3d.** 5 **4a.** rational **4b.** irrational **4c.** real **5a.** irrational; 6.93
5b. not a real number **5c.** rational; 16 **5d.** rational; -15 **6a.** 23 **6b.** $\dfrac{1}{4}$ **6c.** $|3y - 2|$ **6d.** $|5x - 3|$

Do the Math **1.** 3 **2.** $\dfrac{2}{9}$ **3.** 0.4 **4.** 3.7 **5.** not a real number **6.** rational; $\dfrac{7}{10}$ **7.** irrational; 4.90
8. irrational; 3.46 **9.** 13 **10.** $|w|$ **11.** $|x - 8|$ **12.** $|3z - 4|$ **13.** 1 **14.** 1 **15.** not a real number

Section 9.2

Five-Minute Warm-Up 1. 0.5 2. $\dfrac{4}{9}$ 3. $|4x+3|$ 4. $|x-y|$ 5. $\dfrac{1}{49}$ 6. x^2 7. $\dfrac{9x^2}{y^4}$ 8. $8b^{12}$

Guided Practice 1. index; radicand; cube root 2. ≥ 0; any real number 3. principal root 4a. -3
4b. not a real number 4c. $\dfrac{2}{3}$ 5. ≈ 2.83 6. $|2x-1|$ 7a. $\sqrt{144}=12$ 7b. $\sqrt{-64}$ not a real number
7c. $2\sqrt[3]{x}$ 8. $\sqrt[n]{a^m}; \left(\sqrt[n]{a}\right)^m$ 9a. 216 9b. -1024 9c. 16 10. ≈ 8.55 11a. $(2x)^{\frac{3}{4}}$ 11b. $(2x^2y)^{\frac{2}{3}}$
12a. $\dfrac{1}{64^{\frac{1}{2}}}=\dfrac{1}{8}$ 12b. $2\cdot 16^{\frac{1}{2}} = 2\sqrt{16}=8$ 12c. $\dfrac{1}{(4x)^{\frac{5}{2}}}=\dfrac{1}{32x^{\frac{5}{2}}}$

Do the Math 1. 6 2. -4 3. -4 4. $\dfrac{2}{5}$ 5. 6 6. n 7. $|2x-3|$ 8. 4 9. -5 10. -3
11. not a real number 12. $-100{,}000$ 13. 8 14. $\dfrac{1}{11}$ 15. 343 16. $\dfrac{1}{81}$ 17. $x^{\frac{3}{4}}$ 18. $(3x)^{\frac{2}{5}}$ 19. $(3pq)^{\frac{7}{4}}$
20. 4.40 21. 3.16 22. 1000 23. not a real number 24. $\dfrac{1}{5}$ 25. 2 26. $-\dfrac{1}{5}$ 27. $(-9)^{\frac{1}{2}}=\sqrt{-9}$ but there is no real number whose square is -9. However, $-9^{\frac{1}{2}}=-1\cdot 9^{\frac{1}{2}}=-\sqrt{9}=-3$.

Section 9.3

Five-Minute Warm-Up 1. $54z^7$ 2. $\dfrac{3u^2}{2}$ or $\dfrac{3}{2}u^2$ 3. $\dfrac{1}{625}$ 4. $\dfrac{27}{64}$ 5. $\dfrac{3x^6}{y}$ 6. $\dfrac{1}{25p^6}$ 7. $\dfrac{64b^{10}}{a^2}$
8. $\dfrac{1}{27xy^7}$ 9. $\dfrac{9}{7}$

Guided Practice 1. All exponents are positive; each base only occurs once; there are no parentheses in the expression; there are no powers written to powers. 2a. 36 2b. $-2a^{\frac{1}{2}}$ 3. $\dfrac{x^{\frac{11}{8}}}{y^{\frac{15}{8}}}$ 4a. 9 4b. $2x^2|y|\sqrt[4]{8}$
4c. $\sqrt[4]{x}$ 4d. $\sqrt[6]{p}$ 5. $2x^{\frac{1}{3}}\left(3x^{\frac{1}{3}}+5x-2\right)$

Do the Math 1. 9 2. 10 3. $\dfrac{1}{y^{\frac{7}{10}}}$ 4. $\dfrac{2}{9}$ 5. $a^{\frac{1}{2}}b^{\frac{3}{5}}$ 6. $\dfrac{b^2}{a^{\frac{2}{3}}}$ 7. $\dfrac{5p^{\frac{1}{5}}}{q^{\frac{1}{2}}}$ 8. $\dfrac{8m^{\frac{5}{4}}}{n^{\frac{1}{6}}}$ 9. $\dfrac{x^{\frac{1}{3}}}{y^{\frac{1}{9}}}$
10. $x^2+4x^{\frac{1}{3}}$ 11. $\dfrac{6}{a^{\frac{1}{2}}}-3a^{\frac{1}{2}}$ 12. $8p^2-32$ 13. x^2 14. 25 15. $5x^2y^3$ 16. $\sqrt[12]{p^{13}}$ 17. $\sqrt[6]{5^7}$
18. $3(x-5)^{\frac{1}{2}}(5x-9)$ 19. $\dfrac{2(6x+1)}{x^{\frac{2}{3}}}$ 20. $\dfrac{6x+13}{(x+3)^{\frac{1}{2}}}$ 21. $3(x^2-1)^{\frac{1}{3}}(3x-1)(x+3)$ 22. 3 23. 125 24. 32 25. 0

Section 9.4

Five-Minute Warm-Up 1. 1, 4, 9, 16, 25, 36, 49, 64, 81, 100, 121, 144 2. 1, 8, 27, 64, 125 3. 1, 16, 81
4. not a real number 5. 10 6. $|x|$ 7. $|4x+1|$ 8. $|2p-1|$ 9. 13

Guided Practice 1a. $\sqrt{35}$ 1b. $\sqrt{x^2-25}$ 1c. $\sqrt[3]{14x^2}$ 2. the radicand does not contain factors that are perfect powers of the index. For square roots, there can be no factor with exponent two or higher in the radicand. For cube roots, there can be no factor with exponent three or higher in the radicand. 3a. 25

3b. $\sqrt{25} \cdot \sqrt{3}$ **3c.** $5\sqrt{3}$ **4a.** $16 \cdot 3$ **4b.** $\sqrt{16 \cdot 3} = \sqrt{16} \cdot \sqrt{3}$ **4c.** $\dfrac{-6+4\cdot\sqrt{3}}{2}$ **4d.** $\dfrac{2(-3+2\sqrt{3})}{2}$ **4e.** $-3+2\sqrt{3}$

5. Rewrite the variable expression as the product of two variable expressions, where one of the factors has an exponent that is a multiple of the index. **6.** $\dfrac{2x\sqrt{3}}{11}$ **7.** $\dfrac{6}{a^2}$ **8a.** Rewrite $\sqrt{6}\cdot\sqrt[4]{48}$

Guided Practice

as $6^{\frac{1}{2}} \cdot 48^{\frac{1}{4}}$. **8b.** 4 **8c.** $(6^2)^{\frac{1}{4}} \cdot (48)^{\frac{1}{4}}$ **8d.** $(6^2 \cdot 48)^{\frac{1}{4}}$ **8e.** $1728^{\frac{1}{4}}$ **8f.** $\sqrt[4]{2^6 \cdot 3^3} = 2\sqrt[4]{108}$

Do the Math **1.** $\sqrt[4]{42a^2b^2}$ **2.** $\sqrt{p^2-25}$; $|p|\geq 5$ **3.** $\sqrt[3]{-3x}$ **4.** $3\sqrt[4]{2}$ **5.** $2|a|\sqrt{5}$ **6.** $-4p$ **7.** $12\sqrt{3b}$
8. $s^4\sqrt{s}$ **9.** $x^2\sqrt[5]{x^2}$ **10.** $-3q^4\sqrt[3]{2}$ **11.** $5|x^3|\sqrt{3y}$ **12.** -1 **13.** $\dfrac{-3+2\sqrt{3}}{4}$ **14.** 6 **15.** $6x\sqrt{5}$ **16.** $3a\sqrt[3]{2}$
17. $6m\sqrt[3]{2mn^2}$ **18.** $\dfrac{\sqrt{5}}{6}$ **19.** $\dfrac{x\sqrt[4]{5}}{2}$ **20.** $-\dfrac{3x^3}{4y^4}$ **21.** 2 **22.** $3y^2\sqrt{2}$ **23.** $-4x^3$ **24.** $\sqrt[6]{392}$ **25.** $\sqrt[8]{45}$

Section 9.5

Five-Minute Warm-Up **1.** $3z^3 - 5z^2 - 4z$ **2.** $-3a^2b^2 - a^2 + 6ab - b^2$ **3.** $-8x^4 - 12x^3y + 20x^2y^2$
4. $\dfrac{2}{9}x^5 - \dfrac{4}{7}x^4 + x^2$ **5.** $18c^2 - 23c - 6$ **6.** $a^2b^2 - 4$ **7.** $49n^2 - 42n + 9$

Guided Practice **1.** Like radicals have the same index and the same radicand. **2a.** $9\sqrt{x}$ **2b.** $8\sqrt[3]{3p}$
3a. $-10\sqrt{2}$ **3b.** $-n^4\sqrt{6} + 4n^4\sqrt{6n}$ **4.** $18 + 4\sqrt{6} - 27\sqrt{2} - 12\sqrt{3}$ **5a.** $(A+B)(A-B) = A^2 + B^2$
5b. $49 - 3 = 46$ **6.** 72 square units

Do the Math **1.** $14\sqrt{3}$ **2.** $7\sqrt[4]{z}$ **3.** $11\sqrt[3]{5} - 11\sqrt{5}$ **4.** $8\sqrt{3}$ **5.** $2\sqrt[4]{3}$ **6.** $3\sqrt{3z}$ **7.** $13z\sqrt{7z}$ **8.** $2y\sqrt{3}$
9. $3(2x-1)\sqrt[3]{5}$ **10.** $5\sqrt{5} + 3\sqrt{15}$ **11.** $\sqrt[3]{12} + 2\sqrt[3]{9}$ **12.** $15 + 3\sqrt{5} + 5\sqrt{6} + \sqrt{30}$ **13.** $-141 - 22\sqrt{10}$
14. $-2\sqrt{3}$ **15.** $7 - 4\sqrt{3}$ **16.** 2 **17.** 18 **18.** $\sqrt[3]{y^2} - 3\sqrt[3]{y} - 18$ **19.** $-2\sqrt{3}$ **20.** $z + 2\sqrt{5z} + 5$ **21.** $8 - 4\sqrt{15}$
22. $19 + 2x + 8\sqrt{2x+3}$ **23.** $3a$ **24.** $9 - 4\sqrt{14}$ **25.** $\dfrac{2\sqrt{5}}{25}$

Section 9.6

Five-Minute Warm-Up **1.** 3 **2.** $11a$ **3.** $\dfrac{3\sqrt{2}}{2}$ **4.** 1 **5.** $24 - \sqrt{15}$

Guided Practice **1.** The process of rationalizing the denominator of a quotient requires rewriting the quotient, using properties of rational expressions, so that the denominator of the equivalent expression does not contain any radicals. **2.** a perfect square **3a.** $\dfrac{\sqrt{3}}{\sqrt{3}}$ **3b.** $\dfrac{\sqrt{5}}{\sqrt{5}}$ **3c.** $\dfrac{\sqrt{2a}}{\sqrt{2a}}$ **4a.** $\dfrac{\sqrt[3]{25}}{\sqrt[3]{25}}$ **4b.** $\dfrac{\sqrt[3]{18}}{\sqrt[3]{18}}$ **4c.** $\dfrac{\sqrt[4]{3x^2y^3}}{\sqrt[4]{3x^2y^3}}$

5. conjugate **6a.** $3 - \sqrt{5}$; 4 **6b.** $2\sqrt{3} + 5\sqrt{2}$; -38 **7.** $\dfrac{\sqrt{2}+3}{\sqrt{2}+3}$ **8.** $-4(\sqrt{2}+\sqrt{5})$

Do the Math **1.** $\dfrac{2\sqrt{3}}{3}$ **2.** $-\dfrac{\sqrt{3}}{2}$ **3.** $\dfrac{\sqrt{5}}{2}$ **4.** $\dfrac{\sqrt{33}}{11}$ **5.** $\dfrac{\sqrt{5z}}{z}$ **6.** $\dfrac{4\sqrt{2a}}{a^3}$ **7.** $-\dfrac{\sqrt[3]{4p^2}}{p}$ **8.** $-\dfrac{\sqrt[3]{15}}{6}$
9. $\dfrac{4\sqrt[3]{6z}}{3z}$ **10.** $\dfrac{2\sqrt[4]{9b^2}}{b} = \dfrac{2\sqrt{3b}}{b}$ **11.** $2(\sqrt{7}+2)$ **12.** $10(\sqrt{10}-3)$ **13.** $\dfrac{\sqrt{5}+\sqrt{2}}{3}$ **14.** $2+\sqrt{3}$ **15.** $\dfrac{4\sqrt{5}}{5}$
16. $\dfrac{11\sqrt{5}}{10}$ **17.** $\dfrac{\sqrt{10}-5\sqrt{5}}{5}$ **18.** $\sqrt{3}$ **19.** $\dfrac{1}{3}$ **20.** $\dfrac{3\sqrt{5}}{5}$ **21.** $\dfrac{10\sqrt{2}-27}{23}$ **22.** $\dfrac{4-\sqrt{6}}{2}$ **23.** 5
24. $\dfrac{1}{2(2-\sqrt{3})}$ **25.** $\dfrac{a-b}{\sqrt{2a}+\sqrt{2b}}$

Section 9.7

Five-Minute Warm-Up 1. 12 2. n 3. $\{x|x \leq -3\}$ 4. -40 5. See Graphing Answer Section

Guided Practice 1a. $4\sqrt{5}$ 1b. $\sqrt[3]{-9}$ 1c. $\sqrt{2}$ 2. ≥ 0 3. any real number 4a. $\left\{x\middle|x \geq \dfrac{3}{2}\right\}$ or $\left[\dfrac{3}{2}, \infty\right)$
4b. $\{x|x \text{ is any real number}\}$ or $(-\infty, \infty)$ 4c. $\{t|t \leq 2\}$ or $(-\infty, 2]$ 5a. $[-3, \infty)$ 5b. See Graphing Answer Section 5c. $[0, \infty)$ 6a. $(-\infty, \infty)$ 6b. See Graphing Answer Section 6c. $(-\infty, \infty)$

Do the Math 1a. 4 1b. $2\sqrt{3}$ 1c. 2 2a. 1 2b. -3 2c. $2\sqrt[3]{2}$ 3a. $\dfrac{1}{3}$ 3b. $\dfrac{\sqrt{3}}{3}$ 3c. $\dfrac{\sqrt{2}}{2}$ 4. $[-4, \infty)$
5. $\left(-\infty, \dfrac{5}{2}\right]$ 6. $(-\infty, \infty)$ 7. $\left[\dfrac{2}{3}, \infty\right)$ 8. $(3, \infty)$ 9a. $[1, \infty)$ 9b. $[0, \infty)$ 10a. $(-\infty, 4]$ 10b. $[0, \infty)$
11a. $[0, \infty)$ 11b. $[1, \infty)$ 12a. $(-\infty, \infty)$ 12b. $(-\infty, \infty)$ 9 – 12(c). See Graphing Answer Section

Section 9.8

Five-Minute Warm-Up 1. $\{-3\}$ 2. $\{1, 2\}$ 3. $2x$ 4. $2x + 10\sqrt{2x} + 25$ 5. $3x + 8$ 6. $4\sqrt{3}$

Guided Practice 1a. $\sqrt{4x+1} = 5$ 1b. $\left(\sqrt{4x+1}\right)^2 = 5^2$ 1c. $4x + 1 = 25$ 1d. $x = 6$ 1e. yes 1f. $\{6\}$
2. Extraneous solutions are apparent solutions; they are algebraically correct but do not satisfy the original equation. 3. Isolate the radical on one side of the equation. If there are two radicals, one radical should be on the left side and the other radical on the right side of the equation. 4a. $x = 1$ 4b. no 4c. $\varnothing$ 4d. The square root function is equal to a negative number. 5a. $x = 4$ or $x = -1$ 5b. $x = 4$, yes ; $x = -1$, no
5c. $\{4\}$ 6a. $(3x+1)^{\frac{3}{2}} = 8$ 6b. $\dfrac{2}{3}$ 6c. $x = 1$ 6d. $\{1\}$ 7a. $\sqrt{2x^2 - 5x - 20} = \sqrt{x^2 - 3x + 15}$
7b. $2x^2 - 5x - 20 = x^2 - 3x + 15$ 7c. $x = 7$, $x = -5$ 7d. yes 7e. $\{7, -5\}$

Do the Math 1. $\{14\}$ 2. $\varnothing$ 3. $\{3\}$ 4. $\{49\}$ 5. $\{13\}$ 6. $\{9\}$ 7. $\{4\}$ 8. $\{-2\}$ 9. $\{6\}$ 10. $\{-6, 3\}$
11. $\{1, 5\}$ 12. $\{12\}$ 13. $\{-6\}$ 14. $\{3\}$ 15. $\{5\}$ 16. $\{9\}$ 17. $\{12\}$ 18. $\{9\}$ 19. $a = \dfrac{v^2}{r}$ 20. $S = 4\pi r^2$
21. $U = \dfrac{CV^2}{2}$ 22. $\$1102.50$

Section 9.9

Five-Minute Warm-Up 1a. $\{8, |-5|\}$ 1b. $\left\{8, |-5|, \dfrac{0}{-12}\right\}$ 1c. $\left\{8, |-5|, \dfrac{0}{-12}, \dfrac{-6}{3}\right\}$
1d. $\left\{8, |-5|, \dfrac{0}{-12}, \dfrac{-6}{3}, 0.\overline{3}, -\dfrac{2}{5}\right\}$ 1e. $\{\pi\}$ 1f. $\left\{8, |-5|, \dfrac{0}{-12}, \dfrac{-6}{3}, 0.\overline{3}, -\dfrac{2}{5}, \pi\right\}$ 2. $3x^2 - 2$
3. $-12p^6 + 36p^5$ 4. $-6x^2 + 13x + 28$ 5. $x^4 - 16$

Guided Practice 1. $-1; \sqrt{-1}$ 2. $a + bi$; real numbers; standard; 6; -2 3. $9i$ 4. $6 + 2i$ 5. standard
6. Add the real parts, then add the imaginary parts together. This is similar to combining like terms for polynomials. 7. $12 + 9i$ 8a. It cannot be used because $\sqrt{-4}$ and $\sqrt{-9}$ are not real numbers. 8b. $2i \bullet 3i$
8c. -6 9a. $4 - 4i$ 9b. $\dfrac{3 + 6i}{4 + 4i} \cdot \dfrac{4 - 4i}{4 - 4i}$ 9c. $\dfrac{12 - 12i + 24i - 24i^2}{16 + 16}$ 9d. $\dfrac{36 + 12i}{32}$ 9e. $\dfrac{36}{32} + \dfrac{12i}{32}$
9f. $\dfrac{9}{8} + \dfrac{3}{8}i$ 10a. -1 10b. $-i$ 10c. 1 10d. i 10e. -1 10f. $-i$ 10g. 1

Do the Math 1. $-10i$ 2. $9\sqrt{2}i$ 3. $10 + 4\sqrt{2}i$ 4. $2 - i$ 5. $3 - \sqrt{2}i$ 6. $-3 + 14i$ 7. $-4 + i$ 8. $-3 + i$
9. $-4 - \sqrt{5}i$ 10. $18 - 6i$ 11. $5 + 5i$ 12. $2 - 26i$ 13. $\dfrac{5}{3} + \dfrac{5}{3}i$ 14. $-21 + 20i$ 15. $-45 - 28i$ 16. -12

17. $54 + 23i$ **18.** $-\dfrac{1}{2} - i$ **19.** $\dfrac{8}{17} - \dfrac{2}{17}i$ **20.** $\dfrac{10}{17} - \dfrac{6}{17}i$ **21.** i **22.** 1 **23.** -1 **24.** $-i$

Chapter 10 Answers
Section 10.1

Five-Minute Warm-Up 1. $9x^2 - 6x + 1$ **2.** $(x-2)^2$ **3.** $\left\{\dfrac{1}{3}\right\}$ **4.** $\left\{\dfrac{7}{5}, -\dfrac{7}{5}\right\}$ **5.** $\dfrac{9}{5}$ **6.** $|8x - 3|$

7. $2\sqrt{3}i$ **8.** $2 + \dfrac{1}{2}i$ **9.** $-15 - 7i$

Guided Practice 1. $x = \sqrt{p}$ or $x = -\sqrt{p}$ $(x = \pm\sqrt{p})$ **2.** $\{-1 + 2\sqrt{3}, -1 - 2\sqrt{3}\}$ **3.** False **4.** False

5a. $n^2 = 144$ **5b.** $\sqrt{n^2} = \pm\sqrt{144}$ **5c.** $n = \pm 12$ **5d.** $\{12, -12\}$ **6a.** integer, rational, irrational

6b. complex; $a + bi, b \neq 0$ **7.** $\left(\dfrac{1}{2}b\right)^2$ **8a.** $49; (p-7)^2$ **8b.** $\dfrac{81}{4}; \left(n + \dfrac{9}{2}\right)^2$ **8c.** $\dfrac{4}{9}; \left(z + \dfrac{2}{3}\right)^2$

9a. $p^2 - 6p = 18$ **9b.** 9 **9c.** $p^2 - 6p + 9 = 27$ **9d.** $(p-3)^2 = 27$ **9e.** $\sqrt{(p-3)^2} = \pm\sqrt{27}$

9f. $p - 3 = \pm 3\sqrt{3}$ **9g.** $p = 3 \pm 3\sqrt{3}$ **9h.** $\{3 + 3\sqrt{3}, 3 - 3\sqrt{3}\}$ **10.** divide both sides of the equation by 3.

11. In a right triangle, the square of the length of the hypotenuse is equal to the sum of the squares of the lengths of the legs; $z^2 = x^2 + y^2$

Do the Math 1. $\{-4\sqrt{3}, 4\sqrt{3}\}$ **2.** $\{-2\sqrt{5}, 2\sqrt{5}\}$ **3.** $\{-1, 5\}$ **4.** $\left\{-\dfrac{7}{2}, \dfrac{1}{2}\right\}$ **5.** $\left\{-\dfrac{3}{2} \pm \dfrac{\sqrt{3}}{2}\right\}$

6. $\{-1, 7\}$ **7.** $4; (p-2)^2$ **8.** $\dfrac{1}{36}; \left(z - \dfrac{1}{6}\right)^2$ **9.** $\dfrac{25}{16}; \left(m + \dfrac{5}{4}\right)^2$ **10.** $\{-6, 3\}$ **11.** $\left\{-\dfrac{7}{2} \pm \dfrac{\sqrt{21}}{2}\right\}$

12. $\left\{\dfrac{5}{2} \pm \dfrac{\sqrt{37}}{2}\right\}$ **13.** $\{5 \pm \sqrt{30}\}$ **14.** $\left\{-\dfrac{2}{3}, 2\right\}$ **15.** $\left\{-\dfrac{3}{2} \pm \dfrac{1}{2}i\right\}$ **16.** 25 **17.** 3 **18.** $4\sqrt{6} \approx 9.80$

19. $-1 \pm 4\sqrt{2}$ **20.** ≈ 82.284 feet **21.** right triangle; hypotenuse is 52

Section 10.2

Five-Minute Warm-Up 1. $5\sqrt{5}$ **2.** $\dfrac{5}{2} + \sqrt{2}$ **3.** $2\sqrt{7}i$ **4.** $1 - \dfrac{2}{3}i$ **5.** $3x^2 - 9x + 1$ **6.** $2\sqrt{10}$

Guided Practice 1. $\dfrac{-b \pm \sqrt{b^2 - 4ac}}{2a}$ **2.** standard form **3a.** $2x^2 - x - 4 = 0; a = 2, b = -1, c = -4$

3b. $3x^2 + 6 = 0; a = 3, b = 0, c = 6$ **3c.** $3x^2 - 6x = 0; a = 3, b = -6, c = 0$ **4a.** 8; -2; -3

4b. $n = \dfrac{-b \pm \sqrt{b^2 - 4ac}}{2a}$ **4c.** $n = \dfrac{-(-2) \pm \sqrt{(-2)^2 - 4(8)(-3)}}{2(8)}$ **4d.** 100 **4e.** $\dfrac{2 \pm 10}{16} = \dfrac{1 \pm 5}{8}$ **4f.** $\dfrac{1+5}{8}, \dfrac{1-5}{8}$

4g. $\left\{\dfrac{3}{4}, -\dfrac{1}{2}\right\}$ **5.** $b^2 - 4ac$ **6a.** 169; two rational **6b.** -20; two complex, not real **6c.** 0; one repeated real

7a. 200 **7b.** $200 = -16t^2 + 150t + 2$ **7c.** $\{1.6, 7.8\}$ **7d.** The rocket will be at a height of 200 feet at two different times. While it is going up, 1.6 seconds after launch, and again when it comes down, at 7.8 seconds. **7e.** no **7f.** At 9.4 seconds the rocket will hit the ground ($h = 0$).

Do the Math 1. $\{-4, 8\}$ **2.** $\left\{-\dfrac{1}{2}, \dfrac{2}{5}\right\}$ **3.** $\left\{1 \pm \dfrac{\sqrt{2}}{2}\right\}$ **4.** $\left\{\dfrac{3}{2} \pm \dfrac{\sqrt{5}}{2}\right\}$ **5.** $\left\{1 \pm \dfrac{\sqrt{10}}{2}i\right\}$ **6.** $\left\{-\dfrac{3}{5} \pm \dfrac{\sqrt{14}}{5}\right\}$

7. 24, 2 irrational **8.** 0, one repeated real **9.** –95, 2 complex, not real **10.** $\left\{\dfrac{7}{2} \pm \dfrac{\sqrt{21}}{2}\right\}$ **11.** $\left\{-2, \dfrac{1}{3}\right\}$

12. $\left\{\dfrac{2}{5} \pm \dfrac{\sqrt{29}}{5}\right\}$ **13.** $\left\{\dfrac{5}{2}\right\}$ **14.** $\{1 \pm \sqrt{6}\}$ **15.** $\left\{\dfrac{1}{2} \pm \dfrac{\sqrt{19}}{2}i\right\}$ **16a.** $x = -4$ or $x = 2$ **16b.** $(-2,-8); (0,-8)$

17. width ≈ 5.307 in.; length ≈ 11.307 in. **18.** base ≈ 9.426 in.; height ≈ 7.426 in. **19.** ≈ 49.0 mph

20. Let $x_1 = \dfrac{-b + \sqrt{b^2 - 4ac}}{2a}; x_2 = \dfrac{-b - \sqrt{b^2 - 4ac}}{2a}$. $x_1 + x_2 = \dfrac{-2b}{2a} = -\dfrac{b}{a}$; $x_1 \cdot x_2 = \dfrac{b^2 - (b^2 - 4ac)}{2a \cdot 2a} = \dfrac{4ac}{4a^2} = \dfrac{c}{a}$

Section 10.3

Five-Minute Warm-Up **1.** $(a-3)(a+3)(a^2+2)$ **2.** $(2x-5)(3x-5)$ **3.** $\dfrac{25}{x^2}$ **4.** $\dfrac{4}{9}x^6$ **5.** $\left\{\dfrac{3}{2}, -2\right\}$

Guided Practice **1a.** $\sqrt{y}$; $2u^2 - 11u + 15 = 0$ **1b.** $1v^2$; $u^2 + 10u + 1 = 0$ **1c.** $(x-1)$; $u^2 - 5u + 6 = 0$

1d. $x^{\frac{1}{3}}$; $u^2 - u - 3 = 0$ **1e.** x^{-1} or $\dfrac{1}{x}$; $4u^2 + u - 3 = 0$ **2a.** x^2 **2b.** $u^2 - 6u - 16 = 0$ **2c.** $(u-8)(u+2) = 0$

2d. $u = 8$ or $u = -2$ **2e.** $x^2 = 8$ or $x^2 = -2$ **2f.** $\sqrt{x^2} = \pm\sqrt{8}$ or $\sqrt{x^2} = \pm\sqrt{-2}$ **2g.** $x = \pm 2\sqrt{2}$ or $x = \pm\sqrt{2}i$

2h. $\{2\sqrt{2}, -2\sqrt{2}, \sqrt{2}i, -\sqrt{2}i\}$ **3a.** $(x^2 + 3)$ **3b.** $\{1, -1, i, -i\}$ **4a.** $\sqrt{x}$ **4b.** $\{4\}$ **5a.** $n^{\frac{1}{3}}$ **5b.** $\{-1, 27\}$

6a. For $\sqrt{x}$, $x \geq 0$. **6b.** For $\dfrac{1}{x}$, $x \neq 0$.

Do the Math **1.** $\{-3, -1, 1, 3\}$ **2.** $\left\{-1, -\dfrac{1}{2}, \dfrac{1}{2}, 1\right\}$ **3.** $\{3, -4\}$ **4.** $\{36\}$ **5.** ∅ **6.** $\left\{-\dfrac{1}{5}, \dfrac{1}{3}\right\}$ **7.** $\left\{-\dfrac{2}{5}, 5\right\}$

8. $\{-1, 27\}$ **9.** $\left\{\dfrac{1}{3}, \dfrac{1}{4}\right\}$ **10.** $\left\{-1, 2, -1 \pm \sqrt{3}i, \dfrac{1}{2} \pm \dfrac{\sqrt{3}}{2}i\right\}$ **11.** $\{2, -3\}$ **12.** $\{-2i, -1, 1, 2i\}$ **13.** $\{81\}$

14. $\left\{\dfrac{7}{8}, 2\right\}$ **15.** $0, \sqrt{5}i, -\sqrt{5}i$ **16.** $-\sqrt{7}i, \sqrt{7}i, -\sqrt{2}, \sqrt{2}$ **17.** $0, -\sqrt{3}, \sqrt{3}$ **18.** $-\sqrt{2}i, \sqrt{2}i, -\sqrt{5}, \sqrt{5}$

19. $-\sqrt{6}, \sqrt{6}, -\sqrt{7}, \sqrt{7}$ **20.** $\dfrac{49}{4}$

Section 10.4

Five-Minute Warm-Up **1–2.** See Graphing Answer Section **3.** –29 **4.** 33 **5.** $\{x | x \text{ is any real number}\}$

Guided Practice **1.** (b) **2.** (a) **3a.** vertically **3b.** 2 **3c.** down **4a.** horizontally **4b.** 1 **4c.** right **5a.** up

5b. down **6a.** >1 **6b.** $0 < |a| < 1$ **7a.** $f(x) = (3x^2 - 12x) + 7$ **7b.** $f(x) = 3(x^2 - 4x + \underline{}) + 7 + \underline{}$

7c. 4 **7d.** Add the additive inverse of $(3)(4)$ or -12 **7e.** $f(x) = 3(x^2 - 4x + 4) + 7 + (-12)$

7f. $f(x) = 3(x-2)^2 - 5$ **7g.** $(2, -5)$ **7h.** up **7i.** $x = 2$

Do the Math **1–6.** See Graphing Answer Section **7.** $h(x) = \left(x - \dfrac{7}{2}\right)^2 - \dfrac{9}{4}$; $V\left(\dfrac{7}{2}, -\dfrac{9}{4}\right)$; $x = \dfrac{7}{2}$

8. $f(x) = 3(x+3)^2 - 2$; $V(-3, -2)$; $x = -3$ **9.** $g(x) = -(x+4)^2 + 2$; $V(-4, 2)$; $x = -4$

10. $h(x) = -4\left(x - \dfrac{1}{2}\right)^2 + 1$; $V\left(\dfrac{1}{2}, 1\right)$; $x = \dfrac{1}{2}$ **11.** $f(x) = \dfrac{1}{2}(x+5)^2$ **12.** $f(x) = x^2 - 5$

Section 10.5

Five-Minute Warm-Up **1.** x-intercept: $(-4, 0)$; y-intercept: $(0, 3)$ **2.** $\left\{8, -\dfrac{3}{2}\right\}$ **3.** $(6, 0)$ and $(-1, 0)$ **4.** –12

Guided Practice 1. $-\dfrac{b}{2a}$ 2. setting $f(x)=0$ and solving for x. 3a. 1; 2; −8 3b. up 3c. $x=-1$ 3d. $y=-9$ 3e. $(-1,-9)$ 3f. $x=-1$ 3g. −8 3h. 36 3i. 2 3j. $(-4,0);(2,0)$ 3k. See Graphing Answer Section 4. y-coordinate 5. maximum 6. minimum 7a. $x \cdot y$ 7b. $2x+y=50$ 7c. $y=50-2x$ 7d. $A=x(50-2x)$ or $A=-2x^2+50x$ 7e. $x=12.5$ 7f. $y=25$ 7g. The pen has two sides of 12.5 ft and one side 25 ft. 7h. 312.5 ft^2

Do the Math 1. 2; $\left(-\dfrac{1}{2},0\right)$ and $(4,0)$ 2. 1; $(3,0)$ 3. 2; $(-0.28,0)$ and $(1.78,0)$ 4 – 9. See Graphing Answer Section 10. maximum; 11 11. minimum; −6 12. 25 and 25 13. −5 and 5 14a. ≈ 4.84 seconds 14b. ≈ 383.39 feet 14c. ≈ 9.74 seconds

Section 10.6

Five-Minute Warm-Up 1. $[-4,-2)$ 2. $(-3,1]$ 3. $(-2,\infty)$ 4. $(-\infty,5]$ 5. $\{x|x \geq 2\}$ 6. $\{1+\sqrt{6},1-\sqrt{6}\}$

Guided Practice 1a. above 1b. below 2a. graphical 2b. algebraic 3a. $f(x)=x^2+2x-3$ 3b. $(-3,0);(1,0)$ 3c. $(-1,-4)$ 3d. See Graphing Answer Section 3e. negative 3f. $[-3,1]$ 4a. positive 4b. negative 5a. $(x+7)(x-5)=0$ 5b. $x=-7; x=5$ 5c. $(-\infty,-7);(-7,5);(5,\infty)$ 5d. See Graphing Answer Section 5e. positive 5f. no 5g. $(-\infty,-7)\cup(5,\infty)$

Do the Math 1. $[-1,8]$ 2. $(-\infty,4)\cup(10,\infty)$ 3. $(-4,-1)$ 4. $\left(-\dfrac{7}{2},1\right)$ 5. $(-\infty,-2)\cup(3,\infty)$ 6. $\left(-\infty,\dfrac{3-\sqrt{29}}{2}\right]\cup\left[\dfrac{3+\sqrt{29}}{2},\infty\right)$ 7. $\left(-\infty,\dfrac{-3-\sqrt{69}}{6}\right)\cup\left(\dfrac{-3+\sqrt{69}}{6},\infty\right)$ 8. $(-\infty,\infty)$ 9. $\varnothing$ 10. $\{4\}$ 11. $(-\infty,-4)\cup(0,\infty)$ 12. $[-8,6]$ 13. $(-\infty,0]\cup[5,\infty)$ 14. $(-\infty,-9]\cup[7,\infty)$ 15. between \$50 and \$70 16. $\left[-\dfrac{4}{3},2\right]\cup[6,\infty)$ 17. $(-\infty,-2)\cup\left(-\dfrac{5}{3},2\right)$

Section 10.7

Five-Minute Warm-Up 1. $[-3,2)$ 2. $\{x|x<-1\}$ See Graphing Answer Section 3. $\left[-\dfrac{5}{3},\infty\right)$ 4. yes

Guided Practice 1a. 0; 0 1b. 0; 0 2. positive; negative (alternately: negative; positive) 3. positive; negative; negative; positive (alternately: negative; positive; positive; negative) 4. denominator to equal zero 5a. $x=-2$ 5b. $x=4$ 5c. −3 (answers may vary) 5d. negative 5e. negative 5f. positive 5g. 0 (answers may vary) 5h. positive 5i. negative 5j. negative 5k. 5 (answers may vary) 5l. positive 5m. positive 5n. positive 5o. negative 5p. $x=4$ 5q. See Graphing Answer Section 6. $\dfrac{-2x+12}{(x-1)(x+1)}>0$ 7. $\dfrac{x-11}{x+2}<0$ 8. See Graphing Answer Section 9. See Graphing Answer Section

Do the Math 1. $(-\infty,-8)\cup(-2,\infty)$ 2. $(-\infty,-12]\cup(2,\infty)$ 3. $(-5,10]$ 4. $(-\infty,-4)\cup\left(\dfrac{2}{5},5\right)$ 5. $\left(-1,\dfrac{2}{3}\right]\cup[6,\infty)$ 6. $(4,\infty)$ 7. $(-2,11]$ 8. $(-\infty,-6)\cup(-5,\infty)$ 9. $(-\infty,-3)\cup\left[-\dfrac{3}{2},0\right)$ 10. $\left(-\infty,-\dfrac{1}{2}\right)\cup(4,13]$ 11. $(-\infty,-3]\cup(8,\infty)$ 12. $\left(-\dfrac{2}{3},4\right)$ 13. $(-4,0)$ and $\left(\dfrac{5}{3},0\right)$ 14. $(-3,0)$ 15. 125 or more bicycles

Chapter 11 Answers
Section 11.1

Five-Minute Warm-Up 1. $\{x|x \neq 2, x \neq 1\}$ 2a. -47 2b. $-3a^2 + 12a - 11$
2c. $-3x^2 - 6xh - 3h^2 + 1$ 3a. not a function 3b. is a function

Guided Practice 1. $(f \circ g)(x)$ 2a. -3 2b. 25 2c. 25 3a. 39 3b. 38 3c. 38 4. A function is one-to-one if each unique element in the domain corresponds to a unique element in the range and each unique element in the range corresponds to a unique element in the domain. 5a. yes 5b. no 6. the horizontal line test 7. one-to-one 8. $f^{-1}(x); (b, a)$ 9a. $\{(15, -3), (-5, -1), (0, 0), (10, 2)\}$ 9b. $\{-3, -1, 0, 2\}$
9c. $\{15, -5, 0, 10\}$ 9d. $\{15, -5, 0, 10\}$ 9e. $\{-3, -1, 0, 2\}$ 10. $y = x$ 11a. $y = 4x - 8$ 11b. $x = 4y - 8$
11c. $x + 8 = 4y$ 11d. $y = \dfrac{x+8}{4}$ 11e. $f^{-1}(x) = \dfrac{x+8}{4}$

Do the Math 1. 253 2. 6 3. 1 4. -94 5. $(f \circ g)(x) = 4x - 3$ 6. $(g \circ f)(x) = \sqrt{x+2} - 2$
7. $(f \circ f)(x) = \dfrac{2(x-1)}{3-x}; x \neq 1, 3$ 8. one-to-one 9. not one-to-one 10. one-to-one
11. $\{(1, -10), (4, -5), (3, 0), (2, -5)\}$ 12. $(f \circ g)(x) = (g \circ f)(x) = x$ 13. $(f \circ g)(x) = (g \circ f)(x) = x$
14. $g^{-1}(x) = x - 6$ 15. $H^{-1}(x) = \dfrac{x-8}{3}$ 16. $f^{-1}(x) = \sqrt[3]{x+2}$ 17. $G^{-1}(x) = 3 - \dfrac{2}{x}$
18. $R^{-1}(x) = \dfrac{4x}{2-x}$ 19. $g^{-1}(x) = (x+3)^3 - 2$ 20. $V(t) = 36\pi t; 1080\pi \approx 3392.92 m^3$

Section 11.2

Five-Minute Warm-Up 1. 16 2. $\dfrac{1}{4}$ 3. 1 4. $\dfrac{1}{10}$ 5a. 6.0235 5b. 6.0234 6. $\dfrac{10}{x^2}$ 7. $\dfrac{2a^6}{5}$ 8. $64p^{15}$
9. $\left\{-\dfrac{1}{2}, \dfrac{5}{3}\right\}$

Guided Practice 1. a positive real number; 1 2. ≈ 2.63902 3. ≈ 2.66514 4. (b) 5. (a) 6. ≈ 2.718
7a. ≈ 20.09 7b. ≈ 0.14 8. the exponents are equal 9a. $3^{2x+1} = 3^3$ 9b. $2x + 1 = 3$ 9c. $2x = 2$
9d. $x = 1$ 9e. $\{1\}$ 10. ≈ 64.7 grams 11. $A = P\left(1 + \dfrac{r}{n}\right)^{nt}$; A = final amount; t = years that the interest accumulates; r = annual interest rate written in decimals; n = number of pay periods per year
12a. P=$100, r= .07, n=4, t=5 12b. $141.48 12c. P=$100, r= .07, n=360, t=5 12d. $141.90
Do the Math 1a. 9.518 1b. 9.673 1c. 9.735 1d. 9.738 1e. 9.739 2a. 20.086 2b. 4.482
3 – 5. See Graphing Answer Section 6. $\{-2\}$ 7. $\{4\}$ 8. $\{1\}$ 9. $\left\{-1, -\dfrac{1}{3}\right\}$ 10. $\{-2\}$ 11. $\left\{\dfrac{2}{3}\right\}$
12. $f(2) = 9; (2, 9)$ 13. $x = -4; \left(-4, \dfrac{1}{81}\right)$ 14a. ≈ 7287 million people 14b. ≈ 8259 million people

Section 11.3

Five-Minute Warm-Up 1. $\left\{x\big|x > -\dfrac{3}{2}\right\}$ 2. $\{3\}$ 3. $\{-4, -3\}$ 4. $\dfrac{1}{16}$ 5. 8 6. 4 7. $\dfrac{1}{4}$

Guided Practice 1. $x = a^y$ 2. $\log_2\left(\dfrac{1}{8}\right) = -3$ 3. $\log_a 3 = 4$ 4. $4^p = 30$ 5. $n^{-5} = 3$ 6a. $(0, \infty)$
6b. $(-\infty, \infty)$ 6c. $\{x|x < 5\}$ 7. $y = 5^x$; See Graphing Answer Section 8. $x = e^y$ 9. $x = 10^y$ 10. $\{-2\}$
11. $e^{-2} \approx 0.1353$ 12a. decibels (loudness) 12b. Richter Scale (earthquakes) 12c. pH (chemistry)

Do the Math 1. $\log_4 64 = 3$ 2. $\log_b 23 = 4$ 3. $\log z = -3$ 4. $3^4 = 81$ 5. $6^{-4} = x$ 6. $a^2 = 16$ 7. 2
8. 2 9. $(2, \infty)$ 10. $\left(-\infty, \dfrac{3}{5}\right)$ 11. -0.108 12. -0.693 13. $\{6\}$ 14. $\left\{\dfrac{27}{4}\right\}$ 15. $\{9\}$ 16. $\left\{\dfrac{7}{2}\right\}$ 17. $\{4\}$
18. $\{-2\sqrt{2}, 2\sqrt{2}\}$ 19. ≈ 9.2 on the Richter scale

Section 11.4

Five-Minute Warm-Up 1a. 1.140 1b. 1.139 2. $x^{\frac{1}{2}}$ 3. $a^{\frac{3}{4}}$ 4. 2^{-4} 5. $\left(\dfrac{3}{2}\right)^3$ 6. 1 7. 6 8. $y = 3^x$

Guided Practice 1a. $\sqrt{2}$ 1b. 0 1c. 1 1d. 7 2a. $\log_a M + \log_a N$ 2b. $\log_a M - \log_a N$
2c. $r \log_a M$ 3. $\log 9 + \log x$ 4. $\ln 2 - \ln x$ 5. $\log_4\left(\dfrac{1}{x-1}\right)$ 6. $\dfrac{\log_b M}{\log_b a}$ 7. ≈ 3.170 8. Answers may vary.

Do the Math 1a. -3 1b. $\sqrt{2}$ 1c. 10 2a. $2a$ 2b. $a + 2b$ 3. $\log_4 a - \log_4 b$ 4. $3\log_3 a + \log_3 b$
5. $3 + \log_2 z$ 6. $4 - \log_2 p$ 7. $5 + \dfrac{1}{4}\log_2 z$ 8. $\dfrac{1}{5}\ln x - 2\ln(x+2)$ 9. $\dfrac{1}{3}\left[\log_6(x-2) - \log_6(x+1)\right]$
10. $3\log_4 x + \log_4(x-3) - \dfrac{1}{3}\log_4(x+1)$ 11. 3 12. 4 13. $\log_2 z^8$ 14. $\log_2(a^4 b^2)$ 15. $\log_4\left[\sqrt[3]{z}(2z+1)^2\right]$
16. $\log_7(x^2)$ 17. $\ln\sqrt[3]{x^2 - 1}$ 18. $\log_5(x+1)$ 19. $\log_4\left(\dfrac{x^4}{16}\right)$ 20a. 0.827 20b. 1.631

Section 11.5

Five-Minute Warm-Up 1. $\{9\}$ 2. $\{-8, 3\}$ 3. $\left\{\dfrac{1}{4}, -2\right\}$ 4. $\{7, 0\}$ 5. $\{x | x < 3\}$

Guided Practice 1a. $\left\{\dfrac{5}{2}\right\}$ 1b. $\left\{\dfrac{1}{3}\right\}$ 1c. $\{64\}$ 2a. $x = \dfrac{\log 12}{\log 3} \approx 2.262$ 2b. $x = \dfrac{\ln 18}{3} \approx 0.963$
3a. $90 = 100\left(\dfrac{1}{2}\right)^{\frac{t}{19.255}}$ 3b. ≈ 2.927 sec. 4a. 3921 people 4b. $7500 = 2500e^{0.03t}$ 4c. 36.6 years or 1986
4d. 23.1 years or 1973

Do the Math 1. $\{13\}$ 2. $\{16\}$ 3. $\{8\}$ 4. $\left\{\dfrac{5}{8}\right\}$ 5. $\{5\}$ 6. $\dfrac{\log 8}{\log 3} \approx 1.893$ 7. $\dfrac{\log 20}{\log 4} \approx 2.161$
8. $\ln 3 \approx 1.099$ 9. $\log 0.2 \approx -0.699$ 10. $\dfrac{\log 5}{2\log 2} \approx 1.161$ 11. $\dfrac{\log 5}{\log 4} \approx 1.161$ 12. $\{4\}$ 13. $\{2\}$ 14. $\{2\}$
15. $\ln 4 \approx 1.386$ 16. $\{12\}$ 17. $\{\pm 2\sqrt{2} \approx -2.828, 2.828\}$ 18. $\{6\}$ 19. 2052 20. ≈ 2.989 years

Chapter 12 Answers
Section 12.1

Five-Minute Warm-Up 1. $2\sqrt{10}$ 2. $6\sqrt{3}$ 3. $4p^2\sqrt{2}$ 4. $|2x - 5|$ 5. -10 6. 25 7. 6 cm^2 8. 12

Guided Practice 1. $d = \sqrt{(x_2 - x_1)^2 + (y_2 - y_1)^2}$ 2a. a and c 2b. b and d 2c. (a) 2d. (b)
2e. add (c) and (d) 2f. take the square root of (e) 2g. simplify 3. $4\sqrt{5} \approx 8.94$ 4. two equal parts
5. $x_1 + x_2$; $y_1 + y_2$ 6. $(-4, -3)$

Do the Math 1. 5 2. 25 3. 2 4. $2\sqrt{13} \approx 7.21$ 5. $2\sqrt{78} \approx 17.66$ 6. $\sqrt{5.69} \approx 2.39$ 7. $(3, 5)$ 8. $(2, 2)$
9. $\left(1, \dfrac{11}{2}\right)$ 10. $(2, -2)$ 11. $(2\sqrt{6}, 4\sqrt{2})$ 12. $(-0.7, 1.95)$ 13a. See Graphing Answer Section

13b. $d(A, B) = 5$; $d(B,C) = 3\sqrt{5}$; $d(A,C) = \sqrt{58} \approx 7.62$ **14.** $(-16, -3)$ and $(8, -3)$

Section 12.2

Five-Minute Warm-Up **1.** $49; (x-7)^2$ **2.** $\frac{25}{4}; \left(y-\frac{5}{2}\right)^2$ **3.** $(y+8)^2$ **4.** $2(x-3)^2$

5a. $\frac{225\pi}{4} \approx 176.71$ in.² **5b.** $15\pi \approx 47.12$ in.

Guided Practice **1a.** center **1b.** radius **1c.** $d = 2r$ **2.** $(x-h)^2 + (y-k)^2 = r^2$
3. $(x+1)^2 + (y-3)^2 = 25$ **4a.** $(2, -3)$ **4b.** 2 **4c.** See Graphing Answer Section
5. $x^2 + y^2 + ax + by + c = 0$ **6a.** 9; 1; 9; 1 **6b.** 3; 1; 9 **6c.** $(-3, 1)$; 3 **6d.** See Graphing Answer Section
7. No; all circles fail the vertical line test. **8.** Yes; Center: $(2, 0)$; radius = 3

Do the Math **1–8.** See Graphing Answer Section **1.** $x^2 + y^2 = 25$ **2.** $(x-1)^2 + y^2 = 4$
3. $(x+4)^2 + (y-4)^2 = 16$ **4.** $(x-5)^2 + (y-2)^2 = 7$ **5.** $C(0, 0), r = 5$ **6.** $C(5, -2), r = 7$
7. $C(6, 0), r = 6$ **8.** $C(2, -2), r = 0.5$ **9.** $C(-1, 4), r = 3$ **10.** $C(-2, 6), r = 2$ **11.** $C(7, -5), r = 8$
12. $x^2 + (y-3)^2 = 25$ **13.** $(x-2)^2 + (y+3)^2 = 9$ **14.** $(x-1)^2 + \left(y+\frac{1}{2}\right)^2 = \frac{169}{4}$
15. $A = 49\pi$ square units; $C = 14\pi$ units

Section 12.3

Five-Minute Warm-Up **1.** $(2, 1)$ **2.** up **3.** $x = 2$ **4.** $25; (x+5)^2$ **5.** $-12; -3(x-2)^2$ **6.** $\{8, 0\}$
Guided Practice **1a.** focus **1b.** directrix **1c.** vertex **1d.** axis of symmetry **2a.** left or right **2b.** right
2c. left **2d.** up or down **2e.** up **2f.** down **3a.** down **3b.** right **4a.** left **4b.** $y^2 = -4ax$ **4c.** $y^2 = -\frac{1}{2}x$
5a. $x^2 - 8x = -4y - 20$ **5b.** $x^2 - 8x + 16 = -4y - 20 + 16$ **5c.** $x^2 - 8x + 16 = -4y - 4$
5d. $(x-4)^2 = -4(y+1)$ **5e.** down **5f.** $(4, -1)$ **5g.** See Graphing Answer Section

Do the Math **1.** $x^2 = 20y$ **2.** $y^2 = -32x$ **3.** $y^2 = 2x$ **4.** $y^2 = 16x$ **5.** $x^2 = -8y$
6–13. See Graphing Answer Section **6a.** $V(0, 0)$ **6b.** $F(0, 7)$ **6c.** $y = -7$ **7a.** $V(0, 0)$, **7b.** $F\left(\frac{5}{2}, 0\right)$
7c. $x = -\frac{5}{2}$ **8a.** $V(0, 0)$ **8b.** $F(0, -4)$ **8c.** $y = 4$ **9a.** $V(-4, 1)$ **9b.** $F(-4, 0)$ **9c.** $y = 2$ **10a.** $V(-5, 2)$
10b. $F(-2, 2)$ **10c.** $x = -8$ **11a.** $V(-1, 3)$ **11b.** $F(-1, 5)$ **11c.** $y = 1$ **12a.** $V(2, 4)$ **12b.** $F(-2, 4)$
12c. $x = 6$ **13a.** $V(2, 0)$ **13b.** $F\left(2, -\frac{5}{2}\right)$ **13c.** $y = \frac{5}{2}$ **14.** 20 feet

Section 12.4

Five-Minute Warm-Up **1.** $\frac{81}{4}; \left(y-\frac{9}{2}\right)^2$ **2.** $36; (x+6)^2$ **3.** $\frac{1}{4}; 4\left(x-\frac{1}{4}\right)^2$ **4.** $y = 9(x+3)^2 - 6$
5. $(-3, -6)$ **6.** up **7.** $x = -3$

Guided Practice **1a.** foci **1b.** major axis **1c.** minor axis **1d.** center **1e.** vertices **2a.** a^2 **2b.** a
2c. $2a$ **2d.** $2b$ **2e.** major **2f.** $a^2 - b^2$ **2g.** the origin **3a–3b.** See Graphing Answer Section

4. $\frac{x^2}{144} + \frac{y^2}{169} = 1$ **5a.** $\frac{(x+5)^2}{9} + \frac{(y-2)^2}{25} = 1$ **5b.** $C(-5, 2)$ **5c.** $V_1(-5, 7); V_2(-5, -3)$

5d. $F_1(-5, 6); F_2(-5, -2)$ **6a.** $\frac{(x-1)^2}{9} + \frac{(y-3)^2}{4} = 1$ **6b.** $C(1, 3)$ **6c.** $V_1(4, 3); V_2(-2, 3)$ **6d.** $F(1 \pm \sqrt{5}, 3)$

7a. $\dfrac{(x-4)^2}{36}+\dfrac{(y+3)^2}{64}=1$ **7b.** $C(4,-3)$ **7c.** $V_1(4,-11); V_2(4,5)$ **7d.** $F(4,-3\pm 2\sqrt{7})$

Do the Math **1 – 4.** See Graphing Answer Section **1a.** $V(\pm 5, 0)$ **1b.** $F(\pm\sqrt{21}, 0)$ **2a.** $V(0,\pm 6)$

2b. $F(0,\pm 2\sqrt{5})$ **3a.** $V(\pm 8, 0)$ **3b.** $F(\pm 3\sqrt{7}, 0)$ **4a.** $V(0,\pm 9)$ **4b.** $F(0,\pm 6\sqrt{2})$ **5.** $\dfrac{x^2}{25}+\dfrac{y^2}{21}=1$

Do the Math **6.** $\dfrac{x^2}{24}+\dfrac{y^2}{25}=1$ **7.** $\dfrac{x^2}{45}+\dfrac{y^2}{49}=1$ **8.** $\dfrac{x^2}{100}+\dfrac{y^2}{64}=1$ **9 – 10.** See Graphing Answer Section

11a. $\dfrac{(x+3)^2}{9}+\dfrac{(y-4)^2}{25}=1$ **11b.** $C(-3, 4)$ **11c.** $V(-3,-1)$ and $(-3,9)$ **11d.** $F(-3,0)$ and $(-3,8)$

12a. $\dfrac{x^2}{519.84}+\dfrac{y^2}{225}=1$ **12b.** yes

Section 12.5

Five-Minute Warm-Up **1.** $\{-4, 4\}$ **2.** $\{-1,-5\}$ **3.** $16; (x+4)^2$ **4.** $\dfrac{4}{9}; \left(y-\dfrac{2}{3}\right)^2$ **5.** See Graphing Answer Section

Guided Practice **1a.** foci **1b.** transverse axis **1c.** center **1d.** conjugate axis **1e.** vertices **2a.** a^2 **2b.** a **2c.** $2a$ **2d.** $2b$ **2e.** left and right **2f.** up and down **2g.** a^2+b^2 **2h.** the origin **3a.** center: $(0,0)$ **3b.** a=4, b=3 **3c.** c=5 **3d.** vertices: $(\pm 4, 0)$; foci: $(\pm 5, 0)$ **3e.** $(\pm 5, 9), (\pm 5, -9)$ **3f.** See Graphing Answer Section

4. A boundary line that the graph approaches but does not cross as x (or $y) \to \pm\infty$. **5a.** $\dfrac{b}{a}x$ **5b.** $\dfrac{a}{b}x$

Do the Math **1 – 4.** See Graphing Answer Section **1.** $V(\pm 3, 0)$ **1b.** $F(\pm 5, 0)$ **2a.** $V(0,\pm 9)$

2b. $F(0,\pm 3\sqrt{10})$ **3a.** $V(\pm 6, 0)$ **3b.** $F(\pm 2\sqrt{10}, 0)$ **4a.** $V(0,\pm 3)$ **4b.** $F(0,\pm\sqrt{13})$ **5.** $x^2 - \dfrac{y^2}{15}=1$

6. $\dfrac{y^2}{36}-\dfrac{x^2}{28}=1$ **7.** $\dfrac{y^2}{16}-\dfrac{x^2}{4}=1$ **8.** $\dfrac{x^2}{8.1}-\dfrac{y^2}{72.9}=1$ **9.** parabola **10.** hyperbola **11.** hyperbola **12.** circle

13. parabola **14.** ellipse **15a.** $(0, 5)$ and $(0, -5)$ **15b.** $(0\pm\sqrt{29})$ **16.** $y=\pm\dfrac{3}{2}x$

Section 12.6

Five-Minute Warm-Up **1.** $\left(2,\dfrac{3}{2}\right)$ **2.** $\{(x, y)|-x+3y=4\}$ **3.** $(-4,-5)$ **4.** $\varnothing$

Guided Practice **1a.** a line **1b.** a circle **1c.** See Graphing Answer Section **1d.** 2 **1e.** $x^2+y^2=16$ **1f.** $x^2+(x-4)^2=16$ **1g.** $2x^2-8x=0$ **1h.** $2x(x-4)=0$ **1i.** $x=0$ or $x=4$ **1j.** $y=x-4$ **1k.** $y=0-4$ **1l.** $y=-4$ **1m.** $y=x-4$ **1n.** $y=4-4$ **1o.** $y=0$ **1p.** $\{(0,-4),(4,0)\}$ **2a.** a circle **2b.** an ellipse **2c.** See Graphing Answer Section **2d.** 2 **2e.** -1 **2f.** $\begin{cases}-x^2-y^2=-4 & (1)\\ x^2+4y^2=16 & (2)\end{cases}$

2g. $3y^2=12$ **2h.** $y^2=4$ **2i.** $y=\pm 2$ **2j.** $x^2+y^2=4$ **2k.** $x^2+(2)^2=4$ **2l.** $x=0$ **2m.** $x^2+y^2=4$ **2n.** $x^2+(-2)^2=4$ **2o.** $x=0$ **2p.** $\{(0,2),(0,-2)\}$

Do the Math **1.** $(-1, 1), (0, 2)$, and $(1, 3)$ **2.** $(6, 8)$ and $(8, 6)$ **3.** $(0,-4), (\pm\sqrt{7}, 3)$ **4.** $(1, 1)$ **5.** $(-2, -2)$ and $(2, -2)$ **6.** $(-2, 0)$ and $(2, 0)$ **7.** $(-3, 0)$ and $(3, 0)$ **8.** $(-2, -6)$ and $(1, 15)$ **9.** $\varnothing$ **10.** $(-5, 0)$ and $(5, 0)$ **11.** $\varnothing$ **12.** $(-4, -7), (4, -7), (-1, 8), (1, 8)$ **13.** -4 and 12 **14.** 20 meters by 12 meters

Chapter 13 Answers
Section 13.1

Five-Minute Warm-Up 1a. −5 1b. −9 2a. 1 2b. 1 2c. −3 2d. 5 3a. $\dfrac{1}{9}$ 3b. $-\dfrac{1}{27}$ 3c. $\dfrac{1}{81}$ 4. $\dfrac{25}{24}$

Guided Practice 1a. terms 1b. ellipsis 1c. finite 2. a_n 3. $1, \dfrac{3}{2}, \dfrac{7}{3}, \dfrac{15}{4}, \dfrac{31}{5}$ 4a. $a_n = 3n$

4b. $a_n = (-1)^{n+1}\left(\dfrac{1}{6}\right)^n$ 5. 1; 2; 3; 4; 4; 7; 12; 19; 42 6. $-5 + (-3) + (-1) + 1 + 3 + 5 = 0$ 7. $\sum_{i=0}^{6}(2i)$

Do the Math 1. −3, −2, −1, 0, 1 2. $5, 3, \dfrac{7}{3}, 2, \dfrac{9}{5}$ 3. 2, 8, 26, 80, 242 4. $\dfrac{1}{2}, 2, \dfrac{9}{2}, 8, \dfrac{25}{2}$ 5. $a_n = 5n$

6. $a_n = \dfrac{n}{2}$ 7. $a_n = n^3 - 1$ 8. $a_n = \left(-\dfrac{1}{2}\right)^{n-1}$ 9. 55 10. 50 11. 120 12. 4 13. 16 14. 69

15. $\sum_{k=1}^{9}(2k-1)$ 16. $\sum_{k=1}^{16}\dfrac{1}{2^{k-1}}$ 17. $\sum_{i=1}^{15}(-1)^{i+1}\left(\dfrac{2}{3}\right)^i$ 18. $\sum_{k=1}^{12}\left[3\cdot\left(\dfrac{1}{2}\right)^{k-1}\right]$ 19a. $5033.33

19b. $5415.00 19c. $11,098.20

Section 13.2

Five-Minute Warm-Up 1. 4 2. $-\dfrac{9}{5}$ 3a. $(14, 4)$ 3b. $(1, -4)$ 4. −36

Guided Practice 1a. d 1b. a_1 2. yes; $a_1 = 5$; $d = 3$ 3. 3, 6, 11, 18, 27, 38; 3, 5, 7, 9, 11
4a. $a_n = 5n - 8$ 4b. $a_8 = 32$ 5a. $a_1 = 12$; $d = 3$ 5b. $a_n = 3d + 12$ 6a. $a_n = 4n + 8$ 6b. $a_{10} = 48$
6c. $S_{10} = 300$ 7a. $a_1 = 3$, $a_{10} = -87$ 7b. $S_{10} = -420$

Do the Math 1. $d = 10$; 11, 21, 31, 41 2. $d = \dfrac{1}{4}$; $1, \dfrac{5}{4}, \dfrac{3}{2}, \dfrac{7}{4}$ 3. $a_n = 3n + 5$; $a_5 = 20$

4. $a_n = -3n + 15$; $a_5 = 0$ 5. $a_n = \dfrac{1}{2}n - \dfrac{7}{2}$; $a_5 = -1$ 6. $a_n = 4n - 9$; $a_{20} = 71$ 7. $a_n = -6n + 26$; $a_{20} = -94$

8. $a_n = -\dfrac{1}{2}n + \dfrac{21}{2}$; $a_{20} = \dfrac{1}{2}$ 9. $a_n = 3n - 8$ 10. $a_n = 4n - 17$ 11. $a_n = -5n + 22$ 12. $a_n = \dfrac{1}{4}n + \dfrac{15}{4}$
13. 5500 14. 10,425 15. −9200 16. 5440 17. −2905 18. −413 19. $x = 1$ 20. 1600 seats

Section 13.3

Five-Minute Warm-Up 1a. $\dfrac{2}{3}$ 1b. $\dfrac{4}{9}$ 1c. $\dfrac{8}{27}$ 2a. 3 2b. 12 2c. 27 3a. $\dfrac{8x^3}{5}$ 3b. $9r^8$ 4. 2

Guided Practice 1a. r 1b. a_1 2. yes; $a_1 = 36$; $r = \dfrac{1}{2}$ 3. $1, \dfrac{2}{5}, \dfrac{4}{25}, \dfrac{8}{125}$; $r = \dfrac{2}{5}$ 4a. $\{a_n\} = \left\{2\cdot\left(-\dfrac{1}{3}\right)^{n-1}\right\}$

4b. $-\dfrac{2}{2187}$ 5a. $a_1 = 6, r = 4$ 5b. 1,048,575 6a. $a_1 = 6$; $r = -\dfrac{1}{3}$ 6b. $\dfrac{9}{2}$ 7. $99,658.27

Do the Math 1. $r = -2$; −2, 4, −8, 16 2. $r = 2$; $\dfrac{2}{3}, \dfrac{4}{3}, \dfrac{8}{3}, \dfrac{16}{3}$ 3. $r = \dfrac{1}{6}, \dfrac{1}{3}, \dfrac{1}{18}, \dfrac{1}{108}, \dfrac{1}{648}$ 4. arithmetic

5. neither 6. geometric 7. neither 8. $a_n = 30\cdot\left(\dfrac{1}{3}\right)^{n-1}$; $a_8 = \dfrac{10}{729}$ 9. $a_n = (-4)^{n-1}$; $a_8 = -16,384$

10. 177,147 11. −1280 12. 0.0000000004 13. 88,572 14. 40,950 15. $\dfrac{3}{2}$ 16. $\dfrac{48}{5}$ 17. 5 18. $\dfrac{1}{3}$

19. $\dfrac{5}{11}$ 20. $10,497.60

Section 13.4

Five-Minute Warm-Up 1. x^2+4x+4 2. y^2-6y+9 3. $16x^2-40xy+25y^2$
4. $9n^2+4n+\frac{4}{9}$ 5. 35

Guided Practice 1a. 151,200 1b. 15 2a. 10 2b. 1716 3. $p^4+8p^3+24p^2+32p+16$
Do the Math 1. 120 2. 360 3. 45 4. 1 5. 10 6. 21 7. 50 8. 1 9. $x^4-4x^3+6x^2-4x+1$
10. $x^5+25x^4+250x^3+1250x^2+3125x+3125$ 11. $16q^4+96q^3+216q^2+216q+81$
12. $81w^4-432w^3+864w^2-768w+256$ 13. $y^8-12y^6+54y^4-108y^2+81$
14. $243b^{10}+810b^8+1080b^6+720b^4+240b^2+32$ 15. $p^6-18p^5+135p^4-540p^3+1215p^2-1458p+729$
16. $81x^8+108x^6y^3+54x^4y^6+12x^2y^9+y^{12}$ 17. 1.00501

Appendix A Answers

Five-Minute Warm-Up 1. −1 2. 5 3. r^5 4. $\frac{7x^4}{3}$ 5. 0 6. $\frac{a^3}{2b^2}$ 7. $2x+5-\frac{15}{x+1}$

Guided Practice 1. standard; 0 2. $x-c$; $x+c$ 3a. −4 3b. 6
4a. $x^4+3x^3-9x^2+0x+18$; 1, 3, −9, 0, 18 4b. $3\overline{)1\ 3\ -9\ 0\ 18}$ 4c. $3\overline{)1\ 3\ -9\ 0\ 18}$

4d. $3\overline{)1\ 3\ -9\ 0\ 18}$ 4e. $3\overline{)1\ 3\ -9\ 0\ 18}$ 4f. $3\overline{)1\ 3\ -9\ 0\ 18}$
$\underline{3}$ $\underline{3}$ $\underline{3\ 18\ 27\ 81}$ 1
1 $1\ 6$ $1\ 6\ \ 9\ 27\ 99$

4g. quotient: $x^3+6x^2+9x+27$; remainder: 99 4h. $x^3+6x^2+9x+27+\frac{99}{x-3}$ 5. $f(c)$ 6. $f(c)=0$
7. remainder = −9 8a. no 8b. yes

Do the Math 1. $x-7$ 2. $x+6+\frac{7}{x-4}$ 3. x^2-x-20 4. $a^2-8a+15$ 5. $x^2-3x-4-\frac{5}{x+3}$
6. $a^3+8a^2-a-8-\frac{9}{a-8}$ 7. $3x^2+15x+18$ 8. 52 9. −68 10. $x-3$ is not a factor 11. $2x^2+13x+21$
12. x^3+3x^2+2x+1

Appendix B Answers

Five-Minute Warm-Up 1a. $A=s^2$; $P=4s$ 1b. $A=lw$; $P=2l+2w$ 1c. $A=\frac{1}{2}bh$; $P=a+b+c$
1d. $A=\frac{1}{2}h(B+b)$; $P=a+b+c+B$ 1e. $A=ah$; $P=2a+2b$ 1f. $A=\pi r^2$; $C=2\pi r$ or πd
2a. 15.961 2b. 15.961 3a. −0.10 3b. −0.09 4a. 10 4b. 9 5a. 100.7 5b. 100.7 6a. $5.77 6b. $5.76
Guided Practice 1a. A point has no size, only position, and is usually deginated by a capital letter.
1b. A line only has length, no height or depth, and extends infinitely far in both directions. 1c. A ray is a half-line with one endpoint and extending infinitely far in one direction. 1d. A line segment is a portion of a line with a beginning and an end. 1e. Figues that are exactly the same size and shape. 1f. An angle that measures exactly 90°. 1g. An angle whose measure is between 0° and 90°. 1h. An angle whose measure is between 90° and 180°. 1i. Two angles whose measures sum to 90°. 1j. Two angles whose measures sum to 180°. 1k. Two lines which never intersect. 1l. Two lines that intersect to form right angles.
2a. $\angle c$ and $\angle f$; $\angle d$ and $\angle e$ 2b. $\angle a$ and $\angle h$; $\angle b$ and $\angle g$ 2c. $\angle a$ and $\angle e$; $\angle b$ and $\angle f$; $\angle c$ and $\angle g$; $\angle d$ and $\angle h$ 3a. the sum of the measures of the angles of a triangle equals 180 degrees or $x°+y°+z°=180$
3b. $x-15$; $\frac{x}{2}+45$ 3c. $x-15+x+\frac{x}{2}+45=180$ 3d. $x=60$ 3e. yes 3f. 45°; 60°; 75° 4. Trinagles which are exactly the same size and shape. The corresponding sides have the same length and the corresponding angles have the same measures. 5. ASA, SSS, SAS 6. Similar triangles have the same shape

but are different sizes. Two triangles are similar if corresponding angles have the same measures and the corresponding sides are proportional. Two triangles are similar if two pairs of corresponding angles are equal.   **7.** square feet or ft^2 **8.** cubic meters or meters^3 (m^3)

Do the Math **1.** 71° **2.** 39° **3.** $(90 - p)°$ **4.** $(105 - n)°$ **5.** 95° **6.** 74° **7.** $(180 - r)°$ **8.** $(140 - s)°$
9. 83° **10.** 46° **11a.** 40 miles **11b.** 96 square mile **12a.** 52 yards **12b.** 169 square yards **13a.** 36 m
13b. 60 square m **14a.** 28 in. **14b.** 36 suqare in. **15a.** 3π miles ≈ 9.42 miles **15b.** 2.25π miles$^2 \approx 7.07$ miles^2
16a. 108 cubic meters **16b.** 168 square meters **17.** 60 cubic feet **18a.** 54π cubic inches ≈ 169.65 inches^3
18b. 54π square inches ≈ 169.65 inches^2 **19.** 52.36 cubic inches

Appendix C Answers
Section C.1

Five-Minute Warm-Up **1.** 17 **2.** Yes **3.** See Graphing Answer Section **4.** $y = x + 1$
5. slope: $\frac{1}{3}$; y-intercept: $(0, 3)$ **6.** -15 **7.** $\{1\}$

Guided Practice **1a.** inconsistent; the lines are parallel **1b.** consistent and dependent; the lines coincide
1c. consistent and independent; the lines intersect **2a.** $-y = -3x - 14$ **2b.** $y = 3x + 14$ **2c.** $5x + 2y = -5$
2d. $5x + 2(3x + 14) = -5$ **2e.** $5x + 6x + 28 = -5$ **2f.** $11x + 28 = -5$ **2g.** $11x = -33$ **2h.** $x = -3$ **2i.** 5
2j. $(-3, 5)$ **3a.** (1) $-10x - 5y = 20$; (2) $3x + 5y = 29$ **3b.** $-7x = 49$ **3c.** $x = -7$ **3d.** $2(-7) + y = -4$
3e. $y = 10$ **3f.** $(-7, 10)$ **4.** The variables will both be eliminated and a false statement will remain. **5.** The variables will both be eliminated and a true statement will remain **6.** $\{(x, y) | -x + 3y = 1\}$ or $\{(x, y) | 2x - 6y = -2\}$

Do the Math **1a.** no **1b.** yes **2a.** no **2b.** no **3.** $(2, 0)$ **4.** $(1, -4)$ **5.** $(4, 2)$ **6.** $\left(\frac{5}{3}, -\frac{5}{2}\right)$
7. $\emptyset$ **8.** $\left(\frac{2}{9}, -\frac{13}{9}\right)$ **9.** $(5, -11)$ **10.** $\{(x, y) | y = \frac{1}{2}x + 2\}$ or $\{(x, y) | x - 2y = -4\}$ **11.** $\emptyset$
12. no solution **13.** exactly one solution **14a.** $y = -\frac{1}{2}x + \frac{5}{2}$ **14b.** $y = 2x$ **14c.** $(1, 2)$

Section C.2

Five-Minute Warm-Up **1.** -10 **2.** $(-5, -15)$ **3.** $\emptyset$ **4.** $\{(x, y) | 4x + y = 3\}$ or $\{(x, y) | 8x + 2y = 6\}$
Guided Practice **1.** a plane **2a.** consistent; independent **2b.** inconsistent **2c.** consistent; dependent
3a. $-3z - 9y - 9z = -27$ **3b.** $3x + 5y + 4z = 8$ **3c.** $-4y - 5z = -19$ **3d.** $-5x - 15y - 15z = -45$
3e. $5x + 3y + 7z = 9$ **3f.** $-12y - 8z = -36$ **3g.** $12y + 15z = 57$ **3h.** $-12y - 8z = -36$ **3i.** $7z = 21$
3j. $z = 3$ **3k.** $y = 1$ **3l.** $x = -3$ **3m.** $(-3, 1, 3)$ **4.** inconsistent; $\emptyset$ **5.** consistent and dependent
6a. $x = 5z - 3$ **6b.** $y = 10z - 8$ **6c.** $x = 5z - 3, y = 10z - 8$ **7a.** $a + b + c = 600$
7b. $80a + 60b + 25c = 33{,}500$ **7c.** $80a + \frac{3}{5}b(60) + \frac{4}{5}c(25) = 24{,}640$

Do the Math **1a.** Yes **1b.** Yes **2.** $(-2, 6, 6)$ **3.** $\left(\frac{13}{2}, 0, -\frac{3}{2}\right)$ **4.** $(0, -3, 1)$ **5.** $\emptyset$
6. $\left(2, \frac{3}{2}, \frac{1}{2}\right)$ **7a.** $a + b + c = 2$; $4a + 2b + c = 9$ **7b.** $a = 3; b = -2; c = 1$ **7c.** $f(x) = 3x^2 - 2x + 1$
8. Potato: 1.5; Chicken: 1.5; Coke: 1

Section C.3

Five-Minute Warm-Up **1.** $-2, 1, -3$ **2.** 12 **3.** $y = \frac{7}{5}x - 2$ **4.** $x = -\frac{4}{9}y - \frac{4}{3}$ **5.** $-6x + 27y - 3z$

6. $-10x + 5y - 15z$ **7a.** -29 **7b.** 13

Guided Practice 1. 2×4 **2.** standard; 0   **3.** See Graphing Answer Section   **4.** $\begin{cases} x + y = 2 \\ -3x + y = 10 \end{cases}$

5a. Interchange any 2 rows.   **5b.** Replace a row by a non-zero multiple of that row.   **5c.** Replace a row by the sum of that row and a non-zero multiple of another row.   **6a – 6b.** See Graphing Answer Section   **7.** When there are ones on the main diagonal (when the row and column number are the same) and zeros below the ones. This is also called triangular form.   **8a – 8d.** See Graphing Answer Section   **8e.** $(1, 2, -2)$
9. $\{(x, y) | x + 4y = 2\}$ **10.** $\emptyset$

Do the Math   **1 – 2b.** See Graphing Answer Section   **3.** $\{(x, y) | 5x - 2y = 3\}$ **4.** $\left(\frac{4}{3}, -\frac{5}{3}\right)$
5. $\emptyset$ **6.** $(0, -5, 4)$ **7.** \$7000 in savings; \$3000 in Treasury bonds; \$2000 in the mutual fund

Section C.4

Five-Minute Warm-Up   1. -1 **2a.** undefined   **2b.** 0 **3.** $\frac{6}{5}$ **4.** $-\frac{31}{20}$ **5.** $\{-3\}$ **6.** $\{28\}$ **7.** $\emptyset$

Guided Practice   1a. 14 **1b.** -25 **2a.** coefficient   **2b.** the constants   **2c.** second; y **3.** $\begin{cases} 3x - 5y = 9 \\ x + 2y = 2 \end{cases}$

4a. $\begin{vmatrix} 3 & -5 \\ 1 & 2 \end{vmatrix}$ **4b.** $\begin{vmatrix} 9 & -5 \\ 2 & 2 \end{vmatrix}$ **4c.** $\begin{vmatrix} 3 & 9 \\ 1 & 2 \end{vmatrix}$ **5a.** $x = \frac{D_x}{D}$ **5b.** $y = \frac{D_y}{D}$ **6.** See Graphing Answer Section.

7. See the determinants in the Graphing Answer Section   **7b.** 22 **7c.** -66 **7d.** 11 **7e.** 22 **7f.** $\frac{-66}{22} = -3$
7g. $\frac{D_y}{D} = \frac{11}{22} = \frac{1}{2}$ **7h.** $\frac{D_z}{D} = \frac{22}{22} = 1$ **7i.** $\left(-3, \frac{1}{2}, 1\right)$ **8.** inconsistent; $\emptyset$ **9.** consistent; dependent; infinitely many

Do the Math   1. 14 **2.** -4 **3.** -5 **4.** $(5, -4)$ **5.** $\left(\frac{7}{3}, \frac{5}{6}\right)$ **6.** $(3, 2, -1)$ **7.** Cramer's Rule does not apply   **8.** $x = -3$ **9.** area is 12.5 sq units

Graphing Answer Section

Section 1.3 Guided Practice
4.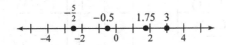

Section 1.3 Do the Math
7.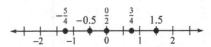

Section 2.7 Guided Practice
7b.

	Rate, mph	Time, hours =	Distance, miles
Slower boat	r	2.5	$2.5r$
Faster boat	$r + 4$	2.5	$2.5(r + 4)$
Total			50 miles

Section 2.7 Do the Math
5.

	Rate, mph	Time, hours =	Distance, miles
Beginning of trip	r	2	$2r$
Rest of the trip	$r - 10$	3	$3(r - 10)$
Total		5	580

Section 2.8 Warm-Up
7.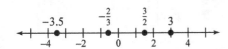

Section 2.8 Guided Practice
2. 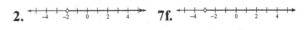 7f.

Section 2.8 Do the Math

1. $(5, \infty)$
2. $(-\infty, 6]$
3. $[-2, \infty)$
4. $(-\infty, -3)$
9. $\{x \mid x < 3\}; (-\infty, 3)$
10. $\{x \mid x > 3\}; (3, \infty)$
11. $\{x \mid x \leq -4\}; (-\infty, -4]$
12. $\{x \mid x > -2\}; (-2, \infty)$
13. $\{x \mid x \geq 5\}; [5, \infty)$
14. $\{x \mid x \geq -1\}; [-1, \infty)$
15. $\left\{x \mid x > \dfrac{11}{2}\right\}; \left(\dfrac{11}{2}, \infty\right)$
16. $\emptyset$ or $\{\ \}$
17. $\{p \mid p \text{ is any real number}\}; (-\infty, \infty)$

Section 3.1 Warm-Up
1.

Section 3.1 Guided Practice
1.

Section 3.1 Do the Math

1.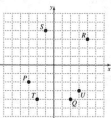
2.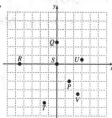

8.

x	y	(x, y)
−3	−17	(−3, −17)
1	−1	(1, −1)
2	3	(2, 3)

9.

x	y	(x, y)
−2	2	(−2, 2)
2	−1	(2, −1)
4	$-\dfrac{5}{2}$	$\left(4, -\dfrac{5}{2}\right)$

Section 3.2 Warm-Up

6.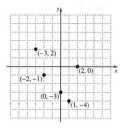

Section 3.2 Guided Practice

1.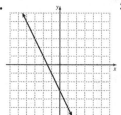
5c.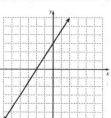
6.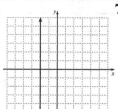
7.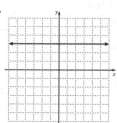

Section 3.2 Do the Math

3.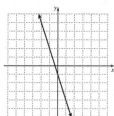
4.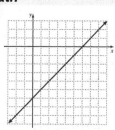
11.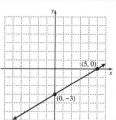
12.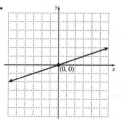

13.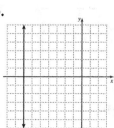
14.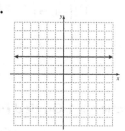

Section 3.3 Guided Practice

9.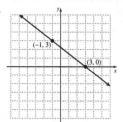

GA-2   Sullivan/Struve/Mazzarella, *Elementary & Intermediate Algebra*, 3e
Copyright © 2014 Pearson Education, Inc.

Section 3.3 Do the Math

3a – 3b.
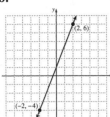

4a – 4b.
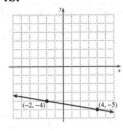

7.
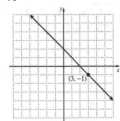

8.
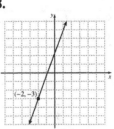

Section 3.4 Guided Practice

4c.
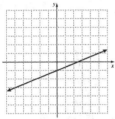

5d.
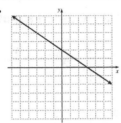

Section 3.4 Do the Math

5.
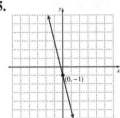

6.
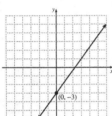

7.
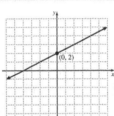

8.
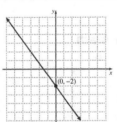

15e.
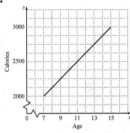

Section 3.5 Do the Math

1.
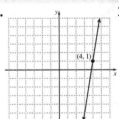

2.
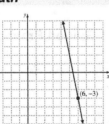

3.
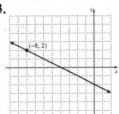

4.
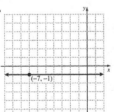

11b.
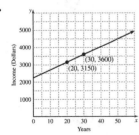

Section 3.6 Do the Math
1-3.

Slope of the Given Line	Slope of a Line Parallel to the Given Line	Slope of a Line Perpendicular to the Given Line
$m = 4$	4	$-\dfrac{1}{4}$
$m = -\dfrac{1}{8}$	$-\dfrac{1}{8}$	8
$m = \dfrac{5}{2}$	$\dfrac{5}{2}$	$-\dfrac{2}{5}$

Section 3.7 Warm-Up

4.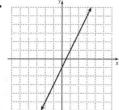

Section 3.7 Guided Practice

6h.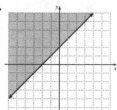

7c.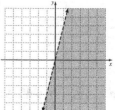

Section 3.7 Do the Math

5.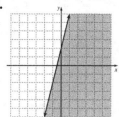
6.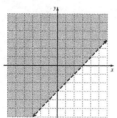
7.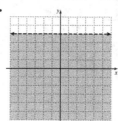
8.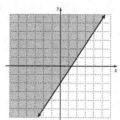

9.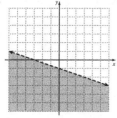
10.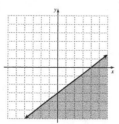
11.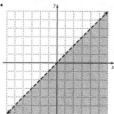
12.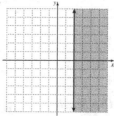

Section 4.1 Warm-Up

1a.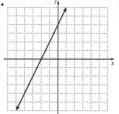
1b.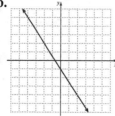

Section 4.1 Guided Practice

2g.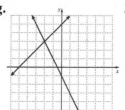
3.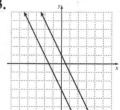

Section 4.1 Do the Math

3. 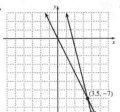   4. 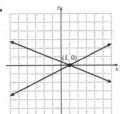   5. 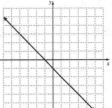   6.

11.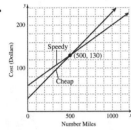

Section 4.4 Guided Practice
4c – 4d.

	Distance (miles)	Rate (mph)	Time (hours)
With the wind	2400	$a + w$	6
Against the wind	2400	$a - w$	8

Section 4.5 Guided Practice
1a–c.

	Number	Cost per Person in Dollars	Amount
Adults	a	7.50	$7.5a$
Students	s	4.00	$4s$
Total	40		202

2a–c.

	Number of coins	Value per Coin in Dollars	Total Value
Quarters	q	0.25	$0.25q$
Dimes	d	0.10	$0.1d$
Total	42		6.75

3a–c.

	Price $/Pound	Number of Pounds	Revenue
Almonds	6.50	a	$6.5a$
Peanuts	4.00	p	$4p$
Mix	6.00	50	300

4a–c.

	Number of ml	Concentration	Amount of Pure HCl
25% HCl solution	x	0.25	$0.25x$
40% HCl solution	y	0.40	$0.4y$
30% HCl solution	90	0.30	27

Section 4.5 Do the Math

1.

	Number	·	Cost per Item	=	Total Value
Bracelets	b		10		$10b$
Necklaces	n		15		$15n$
Total	69				895

2.

	Principal	·	Rate	=	Interest
Savings Account	s		0.0275		$0.0275s$
Certificate of Deposit	c		0.02		$0.02c$
Total			1700		37.75

3.

	Number of Pounds	Price per Pound	=	Total Value
Peanuts	p	5		$5p$
Trail Mix	t	2		$2t$
Total	40	3		$3(40)$

Section 4.6 Warm-Up

3a.

3b.

Section 4.6 Guided Practice

7.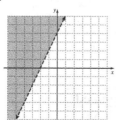

8d.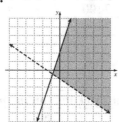

Section 4.6 Do the Math

3.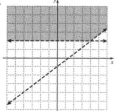

4.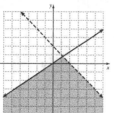

5.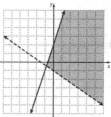

6.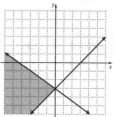

7a.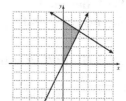

Section 6.2 Guided Practice

3a.

Factors whose product is −56	1, −56	2, −28	4, −14	7, −8	8, −7	14, −4	28, −2	56, −1
Sum of factors	−55	−26	−10	−1	1	10	26	55

4a.

Factors whose product is −24	1, −24	2, −12	3, −8	4, −6	6, −4	8, −3	12, −2	24, −1
Sum of factors	−23	−10	−5	−2	2	5	10	23

Section 6.3 Guided Practice

Factors of –24	1, –24	2, –12	3, –8	4, –6	6, –4	8, –3	12, –2	24, –1
Sum of 2	–23	–10	–5	–2	2	5	10	23

Section 8.1 Warm-Up
1.
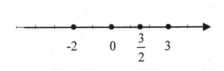

Section 8.1 Guided Practice
1.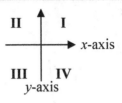

5.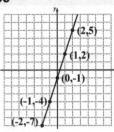

Section 8.1 Do the Math
2. 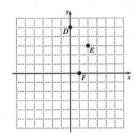   6. 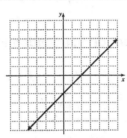   7. 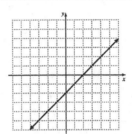   8.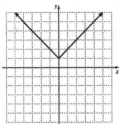

9.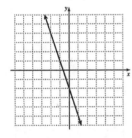

Section 8.2 Warm-Up
1. 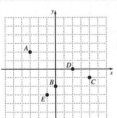   2. 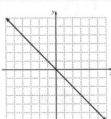   3. 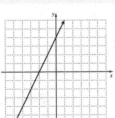   4.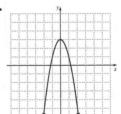

Section 8.2 Guided Practice
7.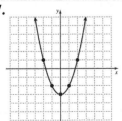

Sullivan/Struve/Mazzarella, *Elementary & Intermediate Algebra*, 3e

Section 8.2 Do the Math

3. 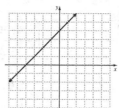 4. 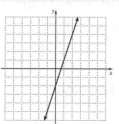 9. 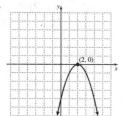 10.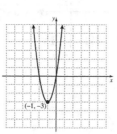

11. 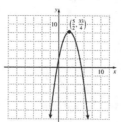 12. 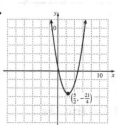 13. 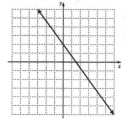 14.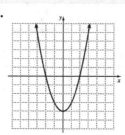

Section 8.4 Warm-Up

3. 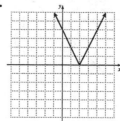 4.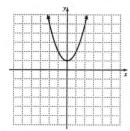

Section 8.4 Guided Practice

4.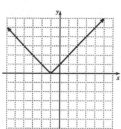

Section 8.4 Do the Math

6. 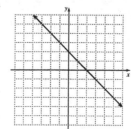 7.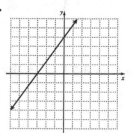

Section 8.5 Warm-Up

1. 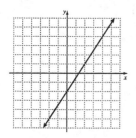 2.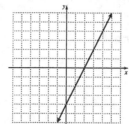

Section 8.5 Guided Practice

3c.

Section 8.5 Do the Math

1.

Section 8.5 Do the Math

2.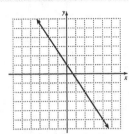
3.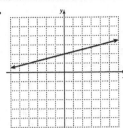
4.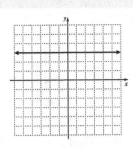
8.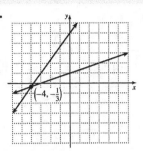

Section 8.6 Warm-Up
2.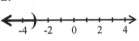

Section 8.6 Guided Practice
3a.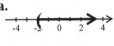
3b.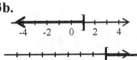
4d-f.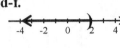

Section 8.6 Guided Practice
6d.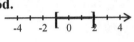
7e-g.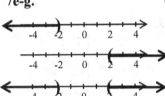

Section 8.7 Guided Practice
6d.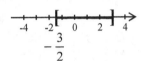

$$-\frac{3}{2}$$

8d.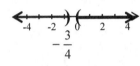

$$-\frac{3}{4}$$

Section 9.7 Warm-Up
5b.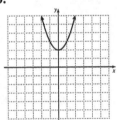

Section 9.7 Guided Practice
5b.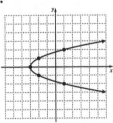
6b.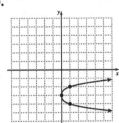

Section 9.7 Do the Math

9.
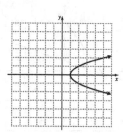

10.
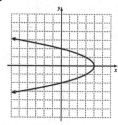

11.
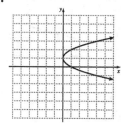

12.
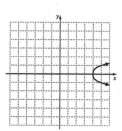

Section 10.4 Warm-Up

1.
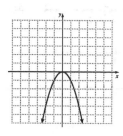

2.
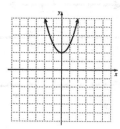

Section 10.4 Do the Math

1.
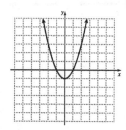

2.
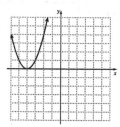

3.
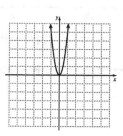

4.
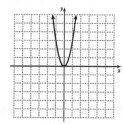

5.
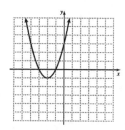

6.
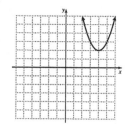

Section 10.5 Guided Practice
3k.

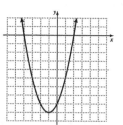

Section 10.5 Do the Math

4. 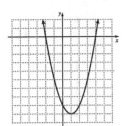   5. 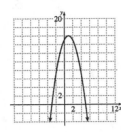   6. 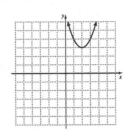   7.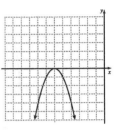

Section 10.5 Do the Math

8. 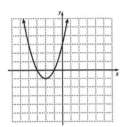   9.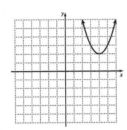

Section 10.6 Guided Practice

3d.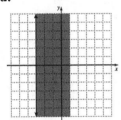

5d.

Interval	$(-\infty, -7)$	$x = -7$	$(-7, 5)$	$x = 5$	$(5, \infty)$
Test Point	-8	-7	0	5	6
Sign of $x + 7$	$-$	0	$+$	$+$	$+$
Sign of $x - 5$	$-$	$-$	$-$	0	$+$
Sign of product	$+$	0	$-$	0	$+$
Conclusion	Included	Not Included	Not Included	Not Included	Included

Section 10.7 Warm-Up

2.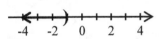

Section 10.7 Guided Practice

5q.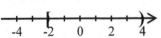

8.

Interval	$(-\infty, -4)$	-4	$(-4, 2)$	2	$(2, 3)$	3	$(3, \infty)$
Test Point	-5	-4	0	2	2.5	3	4
Sign of $x + 4$	$-$	0	$+$	$+$	$+$	$+$	$+$
Sign of $x - 3$	$-$	$-$	$-$	$-$	$-$	0	$+$
Sign of $x - 2$	$-$	$-$	$-$	0	$+$	$+$	$+$
Sign of quotient	$-$	0	$+$	Undef.	$-$	0	$+$
Conclusion	Not Included	Included	Included	Not Included	Not Included	Included	Included

9.

Section 11.2 Do the Math

3. 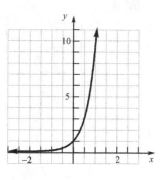 **4.** 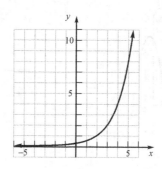 **5.**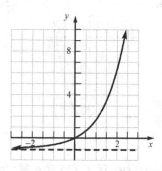

Section 11.3 Guided Practice

7.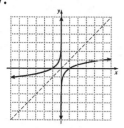

Section 12.1 Do the Math

13a.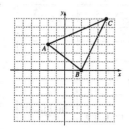

Section 12.2 Guided Practice

4c. 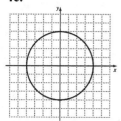 **6d.**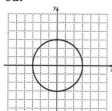

Section 12.2 Do the Math

1. 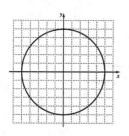 **2.** 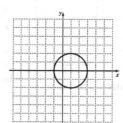 **3.** 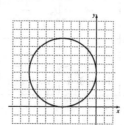 **4.**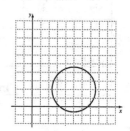

Section 12.2 Do the Math

5.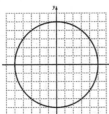
6.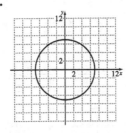
7.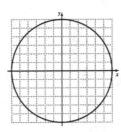
8.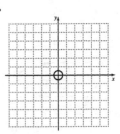

Section 12.3 Guided Practice

5g.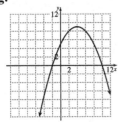

Section 12.3 Do the Math

6.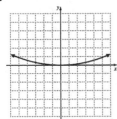
7.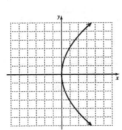
8.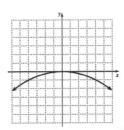
9.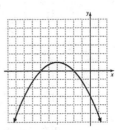

10.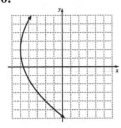
11.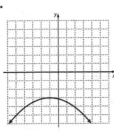
12.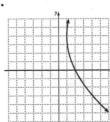
13.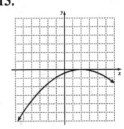

Section 12.4 Guided Practice

3a.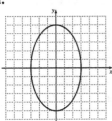
3b.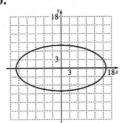

Section 12.4 Do the Math

1.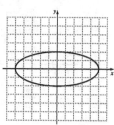
2.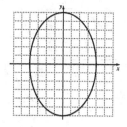
3.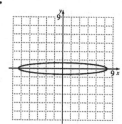
4.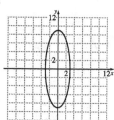

9.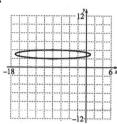
10.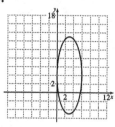

Section 12.5 Warm-Up

5.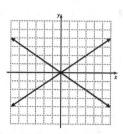

Section 12.5 Guided Practice

3.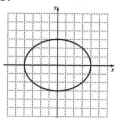

Section 12.5 Do the Math

1.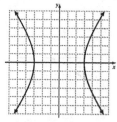
2.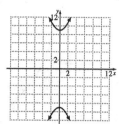
3.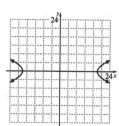
4.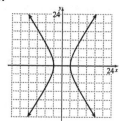

Section 12.6 Guided Practice

1c.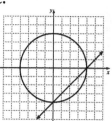

2c.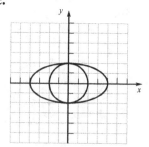

Section C.1 Warm-Up

3.

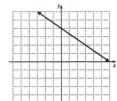

Section C.3 Guided Practice

3.
$$\begin{bmatrix} 2 & 1 & 1 & | & 3 \\ 0 & 4 & -7 & | & -1 \\ 1 & 3 & 0 & | & 0 \end{bmatrix}$$

6a.
$$\begin{bmatrix} 2 & -1 & | & 5 \\ -2 & 14 & | & -8 \end{bmatrix}$$

6b.
$$\begin{bmatrix} 3 & -8 & | & 9 \\ 1 & -7 & | & 4 \end{bmatrix}$$

Section C.3 Guided Practice

8a.
$$\begin{bmatrix} 1 & 1 & 1 & | & 1 \\ 2 & 2 & 0 & | & 6 \\ 3 & 4 & -1 & | & 13 \end{bmatrix}$$

8b.
$$\begin{bmatrix} 1 & 1 & 1 & | & 1 \\ 0 & 0 & -2 & | & 4 \\ 0 & 1 & -4 & | & 10 \end{bmatrix}$$

8c.
$$\begin{bmatrix} 1 & 1 & 1 & | & 1 \\ 0 & 1 & -4 & | & 10 \\ 0 & 0 & -2 & | & 4 \end{bmatrix}$$

8d.
$$\begin{bmatrix} 1 & 1 & 1 & | & 1 \\ 0 & 1 & -4 & | & 10 \\ 0 & 0 & 1 & | & -2 \end{bmatrix}$$

Section C.3 Do the Math

1.
$$\begin{bmatrix} 6 & 4 & | & -2 \\ -1 & -1 & | & -1 \end{bmatrix}$$

2a.
$$\begin{bmatrix} 1 & -1 & 1 & | & 6 \\ 0 & -1 & -1 & | & 15 \\ 3 & 2 & -2 & | & -5 \end{bmatrix}$$

2b.
$$\begin{bmatrix} 1 & -1 & 1 & | & 6 \\ 0 & -1 & -1 & | & 15 \\ 0 & 5 & -5 & | & -23 \end{bmatrix}$$

Section C.4 Guided Practice

6.
$$(1) \cdot \begin{vmatrix} -3 & 3 \\ 4 & -2 \end{vmatrix} - (-1) \cdot \begin{vmatrix} 5 & 3 \\ 1 & -2 \end{vmatrix} + (2) \cdot \begin{vmatrix} 5 & -3 \\ 1 & 4 \end{vmatrix}$$

Section C.4 Guided Practice

7a.
$$\begin{vmatrix} 1 & 2 & -1 \\ 2 & -4 & 1 \\ -2 & 2 & -3 \end{vmatrix}$$

7c.
$$\begin{vmatrix} -3 & 2 & -1 \\ -7 & -4 & 1 \\ 4 & 2 & -3 \end{vmatrix}$$

7d.
$$\begin{vmatrix} 1 & -3 & -1 \\ 2 & -7 & 1 \\ -2 & 4 & -3 \end{vmatrix}$$

7e.
$$\begin{vmatrix} 1 & 2 & -3 \\ 2 & -4 & -7 \\ -2 & 2 & 4 \end{vmatrix}$$